THE Cowboy Way

THE Cowboy Way

AN EXPLORATION
OF HISTORY AND CULTURE

Edited by Paul H. Carlson

TEXAS TECH UNIVERSITY PRESS

Copyright © 2006 Texas Tech University Press

This book is typeset in Berkeley oldstyle-ITC and Copperplate Gothic. The paper used in this book meets the minimum requirements of ANSI/NISO Z39.48–1992 (R1997). ∞

Designed by Tamara Kruciak

Library of Congress Cataloging-in-Publication Data
The cowboy way : an exploration of history and culture / edited by Paul H. Carlson
    p. cm
    Includes bibliographical references and index.
    ISBN 0-89672-425-5 (cloth : alk. paper)
    ISBN 0-89672-583-9 (pbk: alk. paper)
    1. Cowboys–West (U.S.)–History. 2. Cowboys–West (U.S.)–Social life and customs. 3. Ranch life–West (U.S.). 4. West (U.S.)–Social life and customs. 5. West (U.S.)–Ethnic relations.
    I. Carlson, Paul H.
    F596.C8773 2000
    978–dc21

                                                                99-40584
                                                                       CIP
ISBN-13 978-0-89672-425-9 (cloth)
ISBN-13 978-0-89672-583-6 (paperback)

Printed in the United States of America
06 07 08 09 10 11 12 13 14 / 9 8 7 6 5 4 3 2 1
TS

Texas Tech University Press
Box 41037
Lubbock, Texas 79409–1037 USA
800.832.4042
ttup@ttu.edu
www.ttup.ttu.edu

For

Diane and Jon
Steve and Sally
Kevin and Veronica

Contents

Preface

This book is about a national symbol, about a leading icon of popular culture. It is about myth, about common workingmen who became American folk heroes. It is about cowboys. It is about their history, their image, and their way of life.

The title comes from the country and western song "My Heroes Have Always Been Cowboys." In the opening line of the popular waltz Waylon Jennings sings: "I grew up a-dreamin' of bein' a cowboy and lovin' the cowboy way." Accordingly, the book treats music and dance, minorities and work, movies and myth, and other topics associated with the cowboy way.

The book is divided into sixteen essays, which can be read in any order, and there is a bibliography with comments. Four of the essays appeared previously in the *West Texas Historical Association Year Book* and are here revised and used with permission. The essays represent a combination of historical narrative, synthesis, and analysis. They touch on how cowboys of the late nineteenth and early twentieth centuries lived: their food, their clothing, their housing—some of the humbler, but still significant, aspects of cowboy life.

Although there is no dominant thesis, there are many themes. The themes celebrate hard work and bravery, they emphasize youth, and they reflect the excitement and danger as well as the simple living and mundane events that were all part of cowboying in the late nineteenth century. One significant theme is how an enterprise as ephemeral in nature as open range cattle raising continues to fascinate both scholars and the general reading public.

Thus, the book reexamines some older themes in light of modern scholarship, and it also presents some newer topics that reflect fresh interests and issues. The work is highly selective and in no sense all-inclusive. The aim was to offer a perspective on cowboys different from what is generally available and to provide short, readable essays that both entertain and inform. The idea was to reach a wide audience.

Cowboys, both mythic and real, have become part of an American epic that is commemorated from Denver to Dresden, from Montreal to Melbourne. Their image is burned deep into America's collective consciousness, and annually at cowboy poetry gatherings and at national cowboy symposiums Americans celebrate the life and culture of western ranch hands.

The abiding interest has a long history. It can be seen first in the attraction of dime novels and Buffalo Bill's Wild West Exhibition, then in the enormous popularity of Owen Wister's *The Virginian* (1902), and subsequently in the success of popular western novels of the type by Zane Grey and Max Brand, in western films (made in Italy, Germany, Hollywood, and elsewhere), in television programs, in public television documentaries, and in other formats, including the highly effective use of cowboys as advertising symbols. Serious scholars—including historians, sociologists, literary critics, and others—have studied cowboys and the symbols and myths that surround them.

In the popular view cowboys were men on horseback. In fact, most of the time they spent their days on foot working at such farm-related chores as repairing fences and cutting hay. Even in Wister's defining cowboy novel, for example, the hero of the story— the prototypal cowboy—herded neither cows nor cattle of any kind.

Nonetheless, in both his actual and his imagined life the cowboy has become a popular hallmark for defining what it means to be a "real" American male. Perceived as a tough, mobile, and independent outdoorsman, he has become a symbolic yardstick against which modern men might measure their own manhood. The most easily recognized of all western character types, the cowboy is something of a knight of the road who, with a large hat, tall boots, and a big gun, has ridden into legend and into the history books.

Acknowledgments

In putting this book together I enjoyed the support and assistance of many people. I owe much to the authors, of course, for without them there would have been no book. They willingly and thoughtfully contributed the essays, and all of them were patient with me and supportive of the work. I am thankful.

J. Boyd Trolinger, who also authored one of the essays, assisted in many ways, but especially in applying his significant computer skills to organizing and preparing the final manuscript. In addition, he offered ideas, opinions, and perspective. Christine Wyse Trolinger and Todd Walker provided related help. They made important contributions.

Staff members of the Southwest Collection at Texas Tech University provided pictures, information, and assistance. They have always aided researchers and writers enthusiastically, and indeed many of the pieces collected here were written from materials in the Collection's archives. On this project Janet Neugebauer, Freedonia Paschall, and Southwest Collection Director William Tydeman were especially encouraging.

Professor Milada Polisenska, a visiting scholar from the Czech Republic, brought with her to America some old cowboy movies made in Europe and starring European actors. With accompanying lectures, but no subtitles, she showed the films to American students, who easily identified the conventional plots, traditional themes, and familiar errors that often make their way into such productions. Her lectures and infectious enthusiasm were both helpful and delightful.

Joan Weldon, Peggy Arias, and Robert Hall, administrative staff members in the history department at Texas Tech, aided in numerous ways, both large and small, as did Andrew Young. Their assistance speeded completion of the work.

Once again my wife Ellen provided enormous amounts of encouragement, good cheer, and inspiration. I am grateful.

1 Myth and the Modern Cowboy

PAUL H. CARLSON

In July 1879, thirteen-year-old W. H. Childers and his nine-year-old brother mounted their horses near Sivells Bend in North Texas and headed southwest to the family's cattle range some twenty miles away. Their father had directed the boys to ride along the fence line, to inspect and repair it, to turn back drifting cattle if that was necessary, and to check the water holes. For the next several months the young cowhands, following their daddy's instructions, worked alone on the empty grasslands of western Cooke County, mending fences on the thirty-six-section spread, watching some three thousand longhorn cattle, and living in a tiny line shack. The Childers brothers were "cowboys" indeed, real nineteenth-century cowboys.[1]

Modern-day cowboys come in a variety of guises. In part because of low wages and long hours, two-thirds of them are recent immigrants, underpaid—like the Childers brothers—overworked, and often unemployed in the winter. A few are summertime ranch hands on break from school who, when working, wear t-shirts and tennis shoes more often than they wear the traditional cowboy garb of boots and large hats. A few more are permanent ranch hands, many of whom are caught up in myths about the Old West, and, partly as a result, dress in Wrangler jeans tucked inside tall boots, long-sleeved shirts with vests, and large Stetson-style hats.[2]

Then, too, there are urban, rhinestone cowboys. This group frequents honky-tonks like Billy Bob's in Fort Worth—places where they can be found dancing in pointy-toed boots and wearing tight-fitting jeans, western-cut shirts, fancy broad-brimmed hats with a feather in the band, and belt buckles the size of two-pound coffee can covers. Although probably few of them have ridden horseback seemingly endless miles to mend fences in cold and snow or to look for lost calves in wind and rain, the handsome young men there dress like the imagined cowboys of old, and by doing so make the cowboy myth a participation sport.

Clearly, it seems, there is something amiss about our modern-day perceptions of nineteenth-century cowboys. Prior to the twentieth century, cowboys—that is, real ranch hands—when they could afford it, wore dancing shoes and not boots to dances. At old-time western dances in the nineteenth century many cowboys tied bandanas around their arms to signal that they would dance the female part. The modern-day, costumed cowboys one sees at honky-tonks, at local restaurants, or in college classrooms hardly resemble real cowboys of the Old West, who removed their hats when they entered the parlor, when they sat at the dining room table, or when they went dancing. Western music singer-songwriter Mac Davis, in one of his sad pieces, complains about the modern, rhinestone cowboys and laments that he "would rather be a rodeo clown."

Who, then, are these pretenders, these modern-day cowboy wannabes who show up at western dance halls decked out in some Hollywoodesque uniform the likes of which never existed in the Old West? Even today, many working cowboys do not dress like the modern dance hall variety. Summertime cowboys working cattle on a large Nebraska ranch a few years ago, for example, wore t-shirts from the local high school athletic department, baseball caps turned backwards on their heads, and dirty white, high-top basketball shoes with the laces untied. Where were their boots and hats and long-sleeved shirts with vests?

So, what happened to the real cowboy? How is it that the modern rhinestone cowboy became the ideal? How is it, in other words, that the myth became the reality, that the imagined cowboy became the real cowboy? Pictures dating from as early as 1918, for example, already show cowboys dancing in fancy hats and tall boots and wearing huge belt buckles. Obviously, after the close of the cowboy's classical period the transformation came both early and quick.

The age of the classical cowboy—the western open-range cowhand—was brief, lasting just over two decades. Cowboys of the classical period, about 1865-1890 or so, were young, their average age about twenty-four. They remained cowboys on average about seven years. They seldom became prominent citizens or local pillars of society.[3] They wore ill-fitting and diverse garments—"California trousers," some were called. They wore no exaggerated belt buckles and no feather in their hats. Classical cowboys of the Old West bear little resemblance to today's honky-tonk version or, it seems, even modern-day working cowboys influenced by the mythic cowboy of our imaginations.[4]

Real cowboys were dirty, overworked laborers who, writes William Forbis, "fried their brains under a hot prairie sun."[5] Often unemployed in winter, many went to town, where they took odd jobs, like painting. John Clay, who lived among them, writes: cowboys were "a devil-may-care, roistering, gambling, immoral, revolver-heeled, brazen, light-fingered lot, who usually came to no-good end."[6]

Clay's version reflects a contemporary, nineteenth-century view of cowboys. J. Frank Dobie, among the first scholars to study early American cowboys, wrote that many of the first cowboys were young cattle thieves.[7] Until the mid-1880s the term *cowboy* in the West often meant "drunkard," "outlaw," "cattle thief," or something similar. Accordingly, in the mid-1880s President of the United States Chester A. Arthur classified cowboys with cattle thieves: "The cattle were stolen," he wrote, "by a party of outlaws, cowboys, and Indians." Few contemporaries wanted to be cowboys. Indeed, they were "horsemen" more than they were "cowboys."

If he worked cattle in the West, a person wanted to be known as a "waddy" or "cowhand" or simply "hand," or in some cases "herder." Some employees were "cowpunchers," or in South Texas "vaqueros." A few were called "buckaroos." Some, mostly the youngest, were "hoss stinks."[8]

Thus, during much of the classical period, respectable ranch hands were not cowboys. Consider Murdo Mackenzie. Recognized by the Food and Fiber National Institute of Achievement as making greater contributions to the western cattle industry than any other individual, Mackenzie managed the large Matador Land and Cattle Company for some thirty years.[9] As manager, he sent monthly reports to the company officers in Scotland. He also wrote letters, sent memos, and requested directions. During the first twenty years that he directed the great ranch, about 1890 to 1910, Mackenzie did

Mundo Mackenzie, Alexander Mackey, and N. Johnstone in 1892. They headed the Scottish-owned Matador Land and Cattle Company. Courtesy of Southwest Collection, Texas Tech University. SWPC File no. 64.

not call his men "cowboys"—not once. Not once in twenty years of correspondence, as far as his biographers can find, did he refer to his employees as cowboys. He wrote of his "hands," or of his "men," or less frequently of his "cowhands."

From Mackenzie's perspective cowboys were young, inexperienced, and often new to the cattle industry. They were boys—thus, "cowboys." Cowboys were poorly paid, itinerant workers who went on trail drives. They were often kids, sixteen to twenty years old on the average. Ellis Petty, for example, noted that he made his "first cattle drive when only twelve years old." H. P. Cook said: "I went 'up the trail' when I was only ten years old."[10] Here were the "cowboys." Many trail hands, that is, more than one in three, were

African American or Hispanic. Some scholars argue that forty percent of those who went "up the trail" to Kansas or elsewhere were black or Hispanic or Indian or Chinese.[11] The ranch hands were too valuable to send up the trail, or they refused to go. Besides, most of the cattle-trailing was done by contractors, not by ranchers. Few cowboys went "up the trail" more than once or twice.[12]

The term *cowboy* is an old one. In America in its earliest use, according to James Wagner, who has studied its etymology, it referred to cattle thieves, to Tories who fought with the British in the American Revolution, to other people held in low esteem. According to Wagner, the first reference to "cow-boy" (as it was then spelled) in an American dictionary defined the word simply as "a boy who tends cattle."[13]

Cowboy in its modern sense, its mythic sense, developed in the late 1880s. By 1887, the open-range cattle frontier was dead, killed by overextension, overgrazing, severe weather, and, when the bottom dropped out of the market, low prices. After 1887, cattlemen, as opposed to cowboys, took over the industry and changed what had been a wild adventure to a settled, stable business.

The concept of cowboys also changed. Buffalo Bill Cody's Wild West Exhibition, which had begun in 1883, was partly responsible. It included young men herding cattle, and Cody referred to them as cowboys. Beginning in 1884, he advertised one of the men, Buck Taylor, as "King of the Cowboys." Young, white, and virile, Taylor became a popular and featured attraction of the traveling exhibition and a hero to boys everywhere the show appeared.[14]

Then in 1887 Prentis Ingraham, a hack writer, produced a dime novel about Buffalo Bill's star. Without much originality, he entitled the story *Buck Taylor, King of the Cowboys.* The western yarn, published in Beadle's Half Dime Library, was very popular, and the cheap little book sold well and widely. Almost overnight the cowboy image was transformed.

What Buffalo Bill and Prentis Ingraham started, Charlie Russell's paintings, Owen Wister's famous novel *The Virginian,* Western films, pulp novels, and radio shows continued until the cowboy changed from a rogue to a hero. We have, it seems, sort of corrupted him in reverse. We have made him better than he was. Cowboys were not cattlemen; they were laborers, itinerant workers, seasonal employees. They stole cattle from their employers, and some of them took off at the first sign of trouble. The real cowboy was a common, nineteenth-century working stiff who was often illiterate, often unemployed, and often on the lowest rung of the community's socioeconomic hierarchy.[15]

Nonetheless, there is something about the cowboy (that is the mythic, imagined cowboy) that attracts us to him and his lifestyle—however distorted and stereotyped they have become. The mountain man hero, although there are modern rendezvous we can visit, does not claim the attention that the cowboy does. The western soldier, although there are forts to tour and Fourth Cavalry musters to attend, does not warrant such imitation. Pioneer farmers, French fur traders, Catholic missionaries, founding fathers, and even Davy Crockett-type scouts have, when compared to cowboys, little appeal to our collective sense. World War II heroes, as reflected in the G.I. Joe myth, were popular only until the Vietnam War.

But the mythic cowboy continues to appeal to us. Even people who never lived on a ranch; never rode a horse to rope a cow; never subsisted months at a time by choice on cornbread, beans, and bacon; never slept in a two-room, half dugout house—even these people make a hero out of the cowboy, the lowly cowpuncher, underpaid, poorly fed, and overworked. In part, the answer, as Lawrence Clayton has written, lies in our subconscious need for myths, including our imagined version of the cowboy.[16]

Also, there is a mystic and elusive vision of the cowboy as a medieval knight-errant. In the Middle Ages only the wealthy or the nobility rode horses. Miguel Cervantes's fictional Don Quixote, with his small, round companion Sancho Panza, dreamed of dressing in armor, being mounted on horseback, and riding in combat to right the world's wrongs. There may be a lot of Don Quixote in modern cowboys. Moreover, if you please, the horse is something of a subconscious sexual symbol—an extension of one's body, a power between one's legs. Mounted on horseback, the relatively small rider controls a large, beautiful, and majestic steed.

For another thing, we tend to confuse the cowboy and the cattleman of the Old West. The cowboy was a worker, a laborer. And if

A typical chuckwagon scene, ca. 1915. Photo by Frank Reeves, courtesy of Southwest Collection, Texas Tech University. SWCPC File no. FRO56.

cattle raising was an industry, he was an industrial worker. In fact, he formed associations and went on strikes.[17] (Here is a modern-day irony that apparently bothers few of us. Today most cowboy wannabes are opposed to labor unions.) How do we reconcile the cowboy of old who was a union boss with the modern-day cowboy who writes poetry and supports right-to-work laws? The cowboy was a landless wage earner. If he had the chance, he would have voted a Socialist ticket.

The cattleman was the rancher. He usually lived in a big house on the ranch, but he was often gone on business or pleasure to Fort Worth, Denver, or Chicago, or even to London or Paris. He had, if not money, at least position and power. In most parts of the West, when he voted, he voted a conservative ticket.

Texas Panhandle author John Erickson sees the differences between ranchers and modern-day cowboys. He has, in fact, been a working cowboy. He writes:

Ranchers are often prominent leaders in the community; cowboys are not. Ranchers often sit on governing boards of businesses, churches, and schools; cowboys do not. Ranchers are frequently the subject of articles in livestock journals, while cowboys are seldom mentioned. The rancher and his wife may belong to the country club, but the cowboy and his wife don't. The rancher has his circle of friends, the cowboy has his, and they do not overlap . . . [T]he rancher can take the day off . . . whenever he wishes, but the cowboy can't. . . . The rancher and the cowboy may dress alike, talk alike, and even think alike, but at six o'clock in the evening, one goes to the . . . barn while the other attends a meeting in town.[18]

Modern-day wannabes have so corrupted the cowboy that the nineteenth-century version comes out like a late twentieth-century cattleman: well-dressed, well-heeled, well-positioned. Such a cowboy never existed.

Looking back over the more than one hundred years since the close of the cowboy's classical era, one must conclude that cowboys have in fact entered the realm of mythology and semi-truth. They have done it in much the same way as the assassination of Julius Caesar, the tales of a youthful George Washington, and the battle of the Alamo. The classical cowboy is no longer seen as a dirty, malnourished bore in need of a hot bath. Rather, he is seen as neatly dressed in expensive boots, a distinctively cut shirt, designer jeans, and a fancy hat that no longer serves a useful purpose.

We have made him into something he never was. We have transformed the real cowboy—that is, a thirteen-year-old Childers kid who "fried his brains under a hot prairie sun"—into a rhinestone caricature of our symbolic knight. We celebrate what was essentially a pimple-faced, hormone-crazed adolescent—or, in Winifred Kupper's words, "the Peter Pan of the Range."[19]

Clearly, we have invented the modern cowboy. He is an imagined character, one created by misconception, myth, and falsehood. He is a symbol of freedom, independence, strength, and action, and our image-building has made the myth useful to advertising executives: the continuing life of "the Marlboro Man," for instance, suggests that the cowboy myth sells products and suits our contemporary lifeways. Today, as a result, many of us want to be cowboys. At least we dress like we want to be cowboys.

Our modern version, however, is not a thirteen-year-old kid and his younger brother alone on a Cooke County cattle range working their father's herds without compensation. In our twentieth-century fabrication we have changed and improved waddies and hoss stinks of the late nineteenth century, turned them into modern, carefully costumed cowboys, and made them a central part of a popular national myth about the American West. In today's complex world the cowboy—at least the mythic, idealized version of him that we have conjured up in our collective memory—appeals to our urban souls. We thrill in this enduring myth of the Old West.

Notes

1. Jim Lanning and Judy Lanning, eds., *Texas Cowboys: Memories of the Early Days* (College Station: Texas A & M University Press, 1984), 14-22.

2. *Lubbock Avalanche-Journal,* March 21, 1994.

3. William Forbis, *The Cowboys* (New York: Time-Life Books, Inc., 1973), 17.

4. See, for example, pictures in ibid., 16, 28, 93, 170, 205; picture no. 2341, Crosby County Historical Museum, Crosbyton, Texas.

5. Forbis, *The Cowboys,* 7.

6. John Clay, *My Life on the Range* (Chicago: privately printed, 1924), 56. See also Edward Everett Dale, *Cow Country* (Norman: University of Oklahoma Press, 1942, 1965), 116, 122.

7. J. Frank Dobie, *The Longhorns* (New York: Bramhall House, 1942), 27-28; J. Frank Dobie, *A Vaquero of the Brush Country* (Austin: The University of Texas Press, 1981), 43-68.

8. See, for example, Lanning and Lanning, eds., *Texas Cowboys,* 7, 28, 134, 142.

9. Dale, *Cow Country,* pp. 99-100; Paul H. Carlson, "From Farm Worker to Cattle Tycoon," *Ranch Magazine* 70 (October 1988): 42.

10. Lanning and Lanning, eds., *Texas Cowboys,* 87, 108.

11. See Philip Durham and Everett L. Jones, *The Negro Cowboys* (New York: Dodd, Mead, 1965), 44-45; Dale, *Cow Country,* 35, 45, 47-48, 116.

12. Jimmy Skaggs, *The Cattle-Trailing Industry: Between Supply and Demand, 1866-1890* (Lawrence: University Press of Kansas, 1973), 1-3, 51-52, 67; Lanning and Lanning, eds., *Texas Cowboys,* 82, 87, 101, 108.

13. James Wagner, "Cowboy—Origin and Early Use of the Term," *West Texas Historical Association Year Book* 63 (1987): 91-98.

14. Richard Hine, *The American West: An Interpretive History,* 2d ed. (Boston: Little, Brown and Company, 1984), 144-46, 292-94.

15. Ibid. See also William S. Savage Jr., *The Cowboy Hero: His Image in American History and Culture* (Norman: University of Oklahoma Press, 1979), 109-112, 111, 165-66.

16. Lawrence Clayton, "Today's Cowboy: Coping with a Myth," *West Texas Historical Association Year Book* 60 (1984): 183.

17. Hine, *The American West,* 148; Robert E. Zeigler, "The Cowboy Strike of 1883: Its Causes and Meaning," *West Texas Historical Association Year Book* 47 (1971): 32-46.

18. John R. Erickson, *The Modern Cowboy* (Lincoln: University of Nebraska Press, 1981), 5-6. See also Dale, *Cow Country,* 74-76.

19. Winifred Kupper, *The Golden Hoof* (New York: Alfred A. Knopf, 1945), 72.

2 *Cowboy*: Origin and Early Use of the Term

JAMES R. WAGNER

People today often picture a cowboy as a hard-working but somewhat undisciplined fellow who is so fiercely independent that he will ride off into the sunset any time he becomes bored with his surroundings. It is a romantic image. Thanks first to dime novels and later to movies and television, the term *cowboy* has also been romanticized, at least in the United States. A study of the origin and early use of the term reveals quite another meaning, one that suggests that the first cowboys had little to do with raising or marketing of livestock.

The earliest use of the term dates to the Revolutionary War in the 1770s in New York and New England. During the Revolutionary War, when a person spoke of a cowboy, he was speaking of someone who was hated and feared not only by his enemies but also on occasion by his friends.

The first group of men called cowboys were Tories, or Loyalists, from Westchester County, New York, who fought with the British during the Revolutionary War.[1] These green-coated soldiers were, perhaps, more interested in their own personal enrichment than they were in supporting Great Britain. They ravaged Westchester County, stealing and killing indiscriminately.[2] James Fenimore Cooper seems to have been historically accurate when he intimated in his first novel, *The Spy*, that Revolutionary "cow-boys" were called "cow-boys" only because they had a penchant for stealing other people's cattle.[3]

Use of the word *cowboy*, a derisive name for Loyalist military units, may have begun in Westchester County, but as the struggle

Previously published in *West Texas Historical Association Year Book*, 63 (1987): 91–100.

between Loyalists and Patriots became increasingly bitter, the term was applied to all Loyalist units in the state and perhaps in other states as well.

The cowboy has become such an idolized figure in modern America that a number of regions in the United States have claimed that the first cowboys were residents there. Scholars from several different sections of the country have written monographs claiming that the word *cowboy* itself was a product of the livestock industry of their locality. Several parts of the country enjoyed conditions that may or may not have led to the use of the term *cowboy* to describe those persons actively involved in tending livestock as a profession. The fact is, however, that lack of documentary proof of the use of the term leaves the question unanswered.

Even colonial New England, at first glance, seems to have had conditions existing that would have been very conducive to coinage of the term *cowboy*. It is known that, as early as the 1680s in the Massachusetts Bay Colony, young boys or disabled persons were assigned the task of tending their community's cattle kept in one common herd.[4] Theoretically, conditions were ripe for coining the word to describe the youngsters who tended the herds while able-bodied men were out farming. Probably the reason the word *cowboy* did not appear in common usage at the time related to the fact that tending cattle was considered, at this time and place, to be a lowly job given to those who were unskilled or handicapped.

In Massachusetts of the 1680s, the cow itself was not a very highly respected animal. The word "cow" was then, and is sometimes today, used to describe a fat, sloppy, loathsome woman or a prostitute. It is therefore not hard to imagine that boys occupied in tending cattle would not have wanted to be called cowboys. Being called cowboys in such a way would not only have identified them with an occupation they did not want to be identified with, but also it would have indicated that they were only boys and, as such, basically unskilled and/or inexperienced.

In their book titled *Cattle and Men*, Charles Towne and Edward Wentworth claim that the word *cowboy* was in use a number of years before the coming of the Revolutionary War.[5] They stated that stock raisers in both North and South Carolina developed a system of managing their herds by alternately penning them in and letting them loose to roam the countryside. This system placed a great dependence on mounted stockmen to round up the cattle and take them to holding pens at the appropriate times. It is the mounted men whom Towne and Wentworth say were called cowboys as early as the 1760s. Conditions seem to have been even more favorable for the establishment of the word *cowboy* to describe the men who handled cattle on horseback than was the case with those cow-tending boys of New England. But as good as their claim is, Towne and Wentworth seem to be unable to offer any primary source material that actually shows that people were using the word *cowboy* before the Revolution broke out. The fact that men tended cattle in the South in a manner that reminds one of the way western ranchers handled their cattle does not give us the proof necessary to validate such an assertion. Towne and Wentworth further weaken their claim when they include a contradictory footnote in a chapter subsequent to the one containing the material about cattle raising in the South. The footnote reads: "The origin of the 'cowboy' is a historical mystery. During Revolutionary days, the British guerrillas were sometimes called 'cowboys' and the Americans 'skinners.'"[6]

According to J. Frank Dobie, the first Texas cowboys were nearly all young men who were undisciplined and self-willed. The cowboys, he writes, not only rounded up cattle left behind by Mexican ranchers when they fled Texas during the revolt against Mexico in 1836, but also attacked Mexican ranchers who had remained loyal allies of Texas, stealing their cattle and sometimes killing the Mexicans as well.[7] Such raids were carried out with some frequency by cowboys from as early as 1836 to at least 1845.[8] Dobie states that the raids by cowboys were long remembered for their violence. He writes, "Many of these first cowboys thought no more of killing a Mexican than 'upping' an Indian or using the double of a rope on a rattlesnake."[9]

Terry G. Jordan also writes that the first cowboy made a living by rounding up abandoned Mexican cattle or by stealing the cattle from active Mexican ranches. He claims, however, that "by 1860 the meaning [of the word *cowboy*] had been altered to mean any Anglo cowhand."[10] He supports this contention with this quotation: "'The young men that follow this "Cow-Boy" life,' wrote a stock-raiser in that year, 'notwithstanding its hardships and exposures, generally become attached to it.'"[11] The quotation seems to be more a proof of the fact that young men or boys were called cowboys but does not indicate that every Anglo who tended cattle quickly came to be called a cowboy. Dobie writes that the negative reputation brought about by the first cowboys has never quite disappeared. If both men are correct, then it should follow, logically, that it is too difficult to transform the word *cowboy* into an acceptable description for every Anglo that tended cattle.

There is another theory that says that the cowboys the stock-raisers were talking about were boys and young men called "maverickers." These youngsters were, at times, occupied in rounding up and branding strays that wandered loose over a vast majority of South Texas.[12] Young men made up the majority of the people who went out in search of unclaimed cattle. The occupation was a way to earn a fair wage, but competition between such independent cowboys was ruthless. At times these cowboys engaged in activities that, although not illegal, could not be classed as morally correct.[13] If such an explanation is acceptable, little transformation in the reputation of cowboys in Texas occurred since use of the term began in 1836.

With the negative connotations that were attached to the word *cowboy* in both New York and Texas, the word seems to have hovered just outside the realm of acceptable, polite dictionary words. *Cowboy* was, for a number of years, a pejorative word that did not make its way into polite vocabulary. It was not included in the first edition of Noah Webster's Dictionary in 1828. Neither does it appear in the 1846 or 1852 editions of the *American Dictionary of English*. In fact, the earliest dictionaries that were available in which the word *cowboy* is defined as someone who tends cattle are the 1888 edition of *A New English Dictionary on Historical Principles* and the 1901 edition of *The American Dictionary of English*.

If it had been a technical word that described some specific technique of livestock raising, the word *cowboy* would have gone from pejorative to polite usage and would have been accepted into the dictionary quickly. Such was not the case. It was not a word that described a technical aspect of some science, profession, or industry. It remained a slang word often used in a derogatory manner. Moreover, as long as young cowboys were thought to shift from tending cows to rustling them when they were unemployed, the word *cowboy* kept its derogatory meaning.[14]

Existing evidence suggests that as late as the early 1870s the word *cowboy* was not just a term describing a man or boy who raised livestock. To some officers in the Army, a cowboy was considered a bum, a ruffian, and a nuisance. Such was certainly the case with an officer, Colonel Wesley Merritt, who wrote a report to his commanding officer on May 4, 1874, saying, "Although there are honorable exceptions, the majority of ranchers and cowboys are idle, shiftless, and lazy." Expressing a similar assessment of the morals of cowboys, Lieutenant S. H. Lincoln wrote in a letter from Fort Concho on September 16, 1875, that, "My camp was attacked last night by Indians or Cow-Boys."

One must not think that all men occupied in raising and marketing cattle were nasty, dishonest thieves. Such was by no means the case. The truth of the matter is that many honorable people were involved in all phases of the livestock raising industry from its beginning, but the evidence indicates that they most certainly did not call themselves cowboys.[15] There was a sizable group of words used by westerners, and Texans in particular, that described the amount of expertise and competence a man had achieved in the ranching business.

The words are well-known. They include cowman, cowhand, hand, vaquero, cowpuncher, and yes, cowboy. The cowman, everyone agrees, was the one who owned the cows. Cowhands and hands were veteran stockmen who knew the ins and outs of raising livestock. *Vaquero,* which is the Spanish word for cowhand, was used by many Anglos in South Texas. Since a number of Mexicans had attained a high degree of efficiency in riding, roping, and other skills that were necessary to get a longhorn of the right age out of the brush during roundup, many cowhands thought of vaqueros as the best cowhands.[16] Anglos in South Texas especially were proud to be called vaqueros. J. Frank Dobie recorded many of the experiences of ex-cowhand John Young, who preferred to be called a vaquero

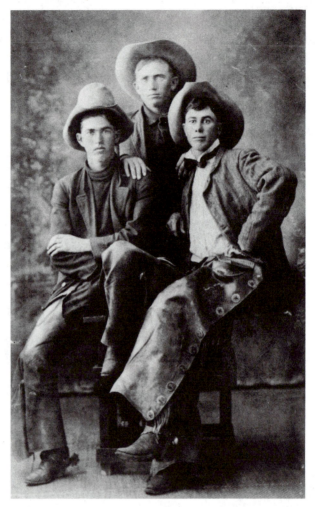

As this ca. 1916 photo clearly shows, cowboys were and are often young. Courtesy of Southwest Collection, Texas Tech University. SWCPC File no. 64 (B), Envelope 1.

rather than cowhand, although they seemed to mean roughly the same thing. Young preferred to be likened to his Mexican comrades because they were good at what they did.[17] Young did not want to be called a cowboy.

A cowpuncher was, according to Young, not just a cowboy. Although modern people have used the word interchangeably with the word "cowboy," Young writes that a cowpuncher was a person who herded cattle either on or off a ship by using a cattle prod. Other authors claim the cowpuncher loaded cattle on trains.[18] The cowpuncher, though not as highly skilled or as respected as a cowhand, gained some recognition that he had reached a level of efficiency higher than a mere cowboy, who was a novice in the cattle business.

As a group, the cowboys stood on the lowest rung of the livestock raising ladder. They did so because they were literally boys who helped tend cows. Also, it must be said that many African Americans and Mexicans came to work as inexperienced youngsters and were not necessarily given a chance to advance rapidly to more important chores. Thus, minority cowboys in Texas remained, perhaps, mixed up with the young cowboys when they might well have been efficient enough to gain cowhand status. Cowhands seem to have stayed home, for the most part, during the long trail drives, because trail driving was most often handled by young cowboys, some just barely in their teens. Many were about sixteen or seventeen years old, and only a few were in their mid-twenties.[19] A few cowhands went along on some trail drives to make sure that the drive stayed headed in the right direction, but most trailers were the young boys who could stand the unceasing pace of the trail and the monotony of doing the same dusty, dirty chore day after day.

Many young cowboys joined trail-driving companies. They made a living going back and forth from the ranches to the trail towns. Such a life was a hard one, and the cowboys did not, after a few years, continue going on trail drives.

At the end of a trail drive, upon arrival at their destination, cowboys tended to celebrate boisterously, and generally they caused consternation among citizens of the local cowtown. Townspeople wanted the money that young cowboys brought into town, but, for the most part, hated the cowboys' behavior and way of life. Towns in Texas and other places in the West also had difficulty with an oversupply of cowboys, who were unemployed part of the year because ranches needed fewer laborers in winter months. Jobs on the ranches at these times of the year went to the cowhands with the most experience. Some cowboys worked at odd jobs around the towns. They painted houses and took other jobs.[20] Others stole enough to get by. The attitudes of the townsfolk toward the cowboys—whom, sadly, they perceived to be anyone making a living on a cattle ranch—were graphically shown in Tombstone, Arizona. During the famous trouble between the Clantons and the Earps, Tombstone citizens labeled the feud the cowboys against the Earps.[21]

When reliable and thoughtful cowhands ventured into the towns, the townsfolk, who disliked and distrusted cowboys, became confused. They saw a different type of person from what they assumed cowboys would be. They did not see the difference in age, maturity, and reliability that separated cowhands from the cowboys.

Likewise, when the authors of the "dime novel," or "penny dreadful" as it was called, came west to look for subjects to use in their western thrillers, the writers found the seasonally employed young cowboy. They made him into an instant hero in the East, where inhabitants got western adventure vicariously through the inexpensive thrillers.[22] The thrillers were about young men going west for adventure or to become cowboys.[23] They depicted cowboys as heroes, who were completely different than the real item. In the later stages of the dime novel era, such thieves and killers as Jesse James and Billy the Kid became cowboy heroes.[24] The dime novel helped change the image of the cowboy. The word *cowboy* itself made a drastic change in reputation. The negative connotations that had followed it for so long dropped away quickly. The cowboy became a romantic figure of mythical proportions. Television and Hollywood movies completed the transformation of the cowboy from a young thief to a character of heroic proportion.

The early use of the term *cowboy* can be traced from its beginning in the Revolutionary War period through its use in dime novels. In tracing the term one must separate the actual cowboy from the cowman who owned the ranches that supplied the means of gathering the large herds of cattle together and from the cowhand and vaquero who pulled, pushed, and drove the longhorns out of the thickets into the herds and then sent them on their way to feed a beef-hungry northern and central United States. Extant records, personal accounts, ranch reports, diaries, dictionaries, and other traditional sources suggest that the term cowboy was used in a derogatory sense by nearly everyone connected with the cattle industry until the early twentieth century. Those who worked cattle on western ranges were cowpunchers, cowhands, vaqueros, or hands. They were not, until more recent times, really cowboys.

Notes

1. Richard Lederer, Jr., *Colonial American English* (Connecticut: A Verbatim Book, 1985), 60; *A New English Dictionary on Historical Principles*, 1888 ed. (New York: Clarendon Press), s.v. "cowboy"; Ernest Weekly, *An Etymological Dictionary of Modern English* (London: John Murry, 1921), s.v. "cowboy"; *Brewer's Dictionary of Phrase and Fable*, centenary edition, revised edition 1981, Ivor H. Evans, ed. (New York: Harper & Row, 1817), s.v. "cowboy." Over the years the term has been spelled a number of different ways. Some of these variations are listed as follows: cowboy (all one word), the variation used today (from J. Neal, *Brother Jonathon III*, 1825); cowboys (from *Thatcher Military Journal,* 1823); Cow Boys (from a quote from Washington Irving that reads, "A beautiful region . . . [Westchester County] now almost desolate by the scourging of the Skinners and the Cow Boys," 1857); Cow-Boys (from James Fenimore Cooper's *The Spy*, 1821); cowboys (from September 10, 1887 issue of *The Spectator*). Most of the variations occurred before the word *cowboy* became the accepted spelling. As late as 1902, however, the word "Cow-Boy" appeared in Owen Wister's *The Virginian* in its hyphenated and double capitalized version. In this chapter, the modern form will be used unless the word appears in a different form in a direct quote.

2. James Fenimore Cooper, *The Spy* (London: George Rutledge and Sons Limited, 1849), 14; *Brewer's Dictionary of Phrase and Fable*, s.v. "cowboy"; Earnest Weekly, *An Etymological Dictionary of Modern English*, op. cit.

3. Ibid.

4. Charles Wayland Towne and Edward Wentworth, *Cattle and Men* (Norman: University of Oklahoma Press, 1955), 131; Abiel Holmes, "The History of Cambridge," *Massachusetts Historical Society Collections* VII (1801), 26.

5. Towne and Wentworth, *Cattle and Men*, op. cit., 143; *The Turner Thesis*, edited and introduction by George Rogers Taylor, 3rd ed. (Lexington: D. C. Heath and Company, 1972), 12-13.

6. Towne and Wentworth, *Cattle and Men,* 168.

7. J. Frank Dobie, *The Longhorns* (New York: Bramhall House, 1942), 27-28; Terry G. Jordan, *Trails to Texas* (Lincoln: University of Nebraska Press, 1981), 74-75; Towne and Wentworth, *Cattle and Men,* op. cit., 168.

8. Jordan, *Trails to Texas,* 74-75; Dobie, *The Longhorns,* 27, and *A Vaquero of the Brush Country,* preface by Lawrence Clark Powell (Austin: University of Texas Press, 1981), 43-68.

9. Dobie, *The Longhorns,* 27-28, and *A Vaquero of the Brush Country,* 43.

10. Jordan, *Trails to Texas,* op. cit., 74.

11. Dobie, *The Longhorns,* op. cit., 43-69, and *A Vaquero of the Brush Country,* 276-77; Don Worcester, *The Chisholm Trail* (Lincoln: University of Nebraska Press, 1980), 8-9.

12. Ibid., 74-75.

13. Ibid.

14. Peter Watts, *A Dictionary of the Old West* (New York: Random House, 1977), s.v. "cowboy"; Dobie, *A Vaquero of the Brush Country,* op. cit., 276-277.

15. Watts, *A Dictionary of the Old West,* op. cit.; Charles Zurhorst, *The First Cowboy and Those Who Followed* (New York: Abelard-Schuman, 1973), 13.

16. Zurhorst, *The First Cowboy and Those Who Followed,* op. cit., 17; Dobie, *A Vaquero of the Brush Country,* op. cit., 193.

17. Towne and Wentworth, *Cattle and Men,* op. cit., 168; *Brewer's Dictionary of Phrase and Fable,* op. cit., s.v. "cowboy."

18. Worcester, *The Chisholm Trail,* op. cit., 17; Dobie, *A Vaquero of the Brush Country,* op. cit., 193; Cornelius C. Smith, *A Southwestern Vocabulary: The Words They Used* (Glendale, CA: Clark and Company, 1984), s.v. "vaquero."

19. *The Trail Drivers of Texas,* Marvin Hunter, ed. Introduction by Byron Price (Austin: University of Texas Press, 1985), xi; Worcester, *The Chisholm Trail,* op. cit., 81.

20. William H. Forbis, *The Cowboy* (New York: Time-Life Books, 1973), 7, 17; Worcester, *The Chisholm Trail,* 22-137.

21. Watts, *A Dictionary of the Old West,* op. cit., s.v. "cowboy."

22. Denis R. Rogers, *Munro's Ten Cent Novels* (Fall Rivers, Massachusetts: LeBlanc, 1958), 153-54; Kenneth L. Donelson and Alleen Pace Nilson, *Literature for Today's Young Adults* (Glenview, Illinois: Scott Foresman and Company, 1980 ed.), 234-38.

23. Rogers, *Munro's Ten Cent Novels;* Donelson and Nilson, *Literature for Today's Young Adults,* 153-54.

24. Ibid., 156 (Rogers) and 234-38 (Donelson and Nilson).

3 Vaqueros in the Western Cattle Industry

JORGE IBER

In his 1981 work, *Clio's Cowboys: Studies in the Historiography of the Cattle Trade,* Don D. Walker called upon historians to reexamine their interpretation of the cowboy and his daily life. This should be done, Walker believed, in order to present a more realistic portrayal to a public whose image of these mounted laborers has been wildly disfigured by novels, movies, and television. Walker's goal was to write about the cowboy in a way that got us "back to the way things were, not the way we prefer them to have been."[1] In the nearly two decades since, many historians have worked to present a more thorough picture of the varied aspects of life in the American West.

As a result, the West is now a more complicated (and interesting) place than it used to be.[2] The historical landscape of the region is now populated, not only by Anglos and Native Americans, but by Hispanics, Asians, and African Americans, by railroad workers, miners, and merchants, and by men and women. While it may upset some, this more "complex" reality goes a long way toward providing a more realistic telling of the history of the West. The region's story now encompasses "a darker set of images, a less exultant song."[3]

While strides have been made in fleshing out various portions of "the truth" of the West, the cowboy's representation in the minds of many Americans remains little changed. Often viewed as a romantic figure in U.S. literature and history, this personification of the "rugged individualist" who "tamed a forbidding land" continues shrouded in myth and lore. In a history that, by the late 1990s, had become more inclusive of peoples of color, the cowboy workforce is still often envisioned as composed primarily of men "of the purest Anglo-Saxon type, as in Owen Wister's *The Virginian*."[4]

21

Many Americans do not realize that perhaps one-third of all cowboys involved in the trail drives after the Civil War were either African American, Hispanic (on Texas ranches, Mexican American cowboys constituted around fifty percent of the workforce), or Native Americans.[5] A large segment of our nation's populace continues to regard the cowboy as a strictly "American" phenomenon, a fact borne out by television and print ads. In a recent spot for the Chevrolet Malibu, General Motors' advertising agency claimed that the United States was the country that "invented the cowboy." In his 1990 work, *Cowboys of the Americas*, Richard H. Slatta noted that in the United States it is common for print, radio, and television ads for rodeos to advertise the event as the "all American sport."[6] In reality, many of the tools, techniques, terms, and much of the culture of the cowboy in the United States, Canada, and even Hawaii can be traced directly back to the Spanish, "who first introduced cattle to the New World and who first developed cattle ranching in this hemisphere . . . ."[7] While Mexican or Mexican American vaqueros show up in cowboy movies and novels, they are often presented as evil and corrupt, in stark contrast with white mounted men.[8]

The earliest mounted men herding cattle in the Western Hemisphere bear little resemblance to the romantic, dime store novel cowboy. By the middle decades of the sixteenth century, the proliferation of livestock in the New World had caused an overabundance of cattle that simply "outdistanced the demand for meat."[9] Such large numbers "gave rise to customary rights for killing the animals" in many parts of the Spanish empire and helped create the gaucho (Argentina), vaquero (Mexico), llanero (Venezuela), and huaso (Chile) equestrian cultures.[10] While such horsemen performed both legal and illegal activities, their status in Spanish society held constant. Vaqueros working on ranchos or missions were often individuals of mixed races (mestizos or mulattos) and, "in the eyes of most . . . Spaniards, nothing more than a poor laborer on horseback."[11] The gaucho, in particular, earned an unsavory reputation. He often hunted wild cattle without governmental sanction and gradually depleted the supply of livestock. The gaucho, in the eyes of the region's elites, was an impediment to progress and profits.[12]

Vaquero roping cattle at Atontonlico Ranch, Hacienda de San Juan de Michis, Durango, Mexico, ca. 1937. Courtesy of Southwest Collection, Texas Tech University. SWCPC File no. FR012.

In an effort to bring an unwieldy element of its population under control, various governmental entities throughout the Spanish empire issued pronouncements, proclamations, and laws. In 1529 the Mexico City council established the *mesta* (stockmen's organization) in part to "punish frauds and crimes which are committed with much frequency."[13] During the seventeenth century the Buenos Aires council issued licenses to limit hunting and labeled "the landless rural population vagrant and criminal" in order "to control the gaucho."[14] This broadly based, multi-century effort brought change (and some structure) to the existence of vaqueros, gauchos, llaneros, and huasos. As elite groups in different parts of the empire "began to extend their domination over the resources of the plains . . . the rise of ranches and organized livestock industry changed . . . life and work."[15]

The vaquero's transplantation to the northern reaches of New Spain did little to alter his social standing. The movement of Spanish settlers into what we now know as the American West commenced in the 1590s and continued for the next two centuries. During this extended period the Spaniards brought their culture, laws, religion, and cattle, first to New Mexico, then to Arizona, Texas, and California. The vaquero of Mexico became "an integral part of spreading cattle-related culture," and followed other *Españoles* north.[16] He remained mired in poverty, with debt often securing his ties to ranch owners. "Most vaqueros were soon in debt, and as the years passed it was not unusual for a vaquero to be born into debt, because the children inherited their parents' obligations."[17] The ethnic background of these mounted men further solidified their lowly status. The Spanish population on the northern frontier proved insufficient to handle the growing stock on the prairies. During the 1700s, in an effort to increase and supplement their labor force, Franciscans trained mission Indians in herding tactics and other chores. The final secularization of missionary facilities after the Mexican Revolution helped provide rancheros (both Mexican and American) with a reservoir of talent and experience they put to use during the post Mexican-American War era.[18]

The arrival of Anglos in Texas (and points farther west) after the 1820s produced a "fusion" of two cattle herding cultures.[19] The union proved both fruitful and tempestuous. The Mexican-Anglo ranching marriage provided much of the know-how used during cattle drives of the 1860s and 1870s, but the relationship destroyed the economic status and lifestyle of many rancheros in Texas and California.[20] The replacement of one group by another at the pinnacle of the cattle industry fostered the myth of the "purely" American cowboy and cattleman. In his 1986 work, *Los Mesteños: Spanish Ranching in Texas, 1721-1821,* Jack Jackson asserts that the notion of the Anglo origin of Texas ranching "gained credence . . . until it achieved mythic proportions, sweeping from the vast plains not only the herds, but also the herdsmen themselves."[21] Rancheros in California suffered a similar fate. Many Californio elites lost control of their lands. Excessive taxation, plus the legal fees required to defend their "imperfect" land titles, bankrupted the Spanish-speaking aristocracy.[22] With their properties sold or confiscated, and their prestige gone, many of their sons chose the path of banditry to fight against both perceived and real injustices.[23] Some took their cowboying skills and joined other Spanish-speakers in the service of new cattle barons.

As a result, during the last half of the nineteenth century vaqueros used their knowledge and training to perform a wide range of services for American employers. They herded stock, broke horses, branded calves, rounded up strays, mended fences, and worked most positions on cattle drives. On at least one occasion, vaqueros trained a "tenderfoot" employer in cattle raising and handling techniques.[24] In return for performing such difficult labors, white cowboys could expect an average monthly pay of twenty-five to thirty dollars during the 1870s. By the 1880s and 1890s cowhand wages increased to about thirty-five to forty dollars per month. Vaqueros rendered similar services but earned, on average, one half to two thirds of "white man's pay."[25] Spanish-speakers did not lack the ability to do the jobs, but racist ideology reduced both their wages and their opportunities for promotion. Many Americans depicted the vaqueros as "greasers" who were "lazy, thieving, untrustworthy, and incompetent. . . . Despite their reputation as superior ropers and horsebreakers, vaqueros seldom rose above the common cowboy ranks to become foremen or trailbosses."[26]

Over the past sixty years various authors have provided glimpses of the vaquero at work in locations throughout the West. Two of the most important are J. Frank Dobie and Dane Coolidge. More recently, David Montejano, Jane Clements Monday, Betty Bailey Colley, James Beckstead, and Connie Brooks have examined cowboy and vaquero life. Their studies illustrate the dichotomous feelings many *Americanos* held regarding vaqueros as employees and coworkers.[27] To many whites, Spanish-surnamed cowhands were valuable cogs in the machinery of the ranch and trail; for others, the cultural baggage of Manifest Destiny and the Protestant work ethic led to "a situation where ethnic or national prejudice provided a basis for separation and control."[28]

In *A Vaquero in the Brush Country*, Dobie examines the significant role of Mexicans and Mexican Americans on South Texas ranches. To the ranchers who came to dominate the cattle industry after the Civil War, the vaqueros were "generally speaking . . .

reliable and loyal and as trustworthy" as any other cowboys.[29] At the King and Kenedy ranches the owners and their families developed particularly close ties with their workers. Richard King learned the cattle business from the Mexicans in his employ, and he respected his vaqueros for their skill, bravery, and dedication.[30] In addition to their more routine chores, King often used his men as escort when carrying the ranch's payroll, "which often could amount to as much as $50,000."[31]

During the final decades of the nineteenth century vaqueros on the King and Kenedy ranches enjoyed working conditions that greatly exceeded those of their colleagues. King paid his Kineños the same rate as whites and, more importantly, did not force his men to face seasonal (winter) unemployment, a factor that permitted the men an opportunity to marry and start families while retaining their positions. Management encouraged marriage and family formation by providing vaqueros with jacales (straw and wood huts) first and later with wood frame houses and schools for their children.[32] The vaqueros' work was hard and physically demanding, but the Mexican and Mexican American horsemen contrasted their fate with the plight of other Spanish-speakers in South Texas and felt "secure in the knowledge that at least their families would be housed and fed."[33]

Other vaqueros encountered much more difficult conditions. In his 1987 work, Anglos and Mexicans in the Making of Texas, 1836–1986, David Montejano notes that the "paternalism of Anglo patrones (patrons) . . . did not obscure the anti-Mexican . . . sentiments and divisions of the ranch world."[34] Ranchers often provided separate facilities for their Mexican and white cowboys. In the managerial hierarchy of the ranch the "Anglo always stood over the Mexican."[35] Even at the King and Kenedy ranches, which were famous for their paternalism, only one Spanish-surnamed individual earned a promotion to foreman before 1926.[36]

Limited promotional opportunities and segregated facilities were not the only forms of discrimination. In keeping with the ideas of racial superiority, white cowboys often questioned vaqueros' dedication to their jobs, as well as criticizing their "backward" cow handling skills. In his 1939 work, California Cowboys, Dane Coolidge captured the attitudes of Arizona and Texas hands regarding their

Spanish-speaking comrades. During a 1914 roundup and drive near Nogales, Arizona, Coolidge listened to strong negative comments about Mexican deficiencies in cattle handling, driving, cutting, and roping. In all areas, the vaqueros, claimed Coolidge's informants, lacked the necessary skills and knowledge. The "knowledgeable" sources believed that "there wasn't a Mexican in the outfit that knew the first thing about cattle."[37] No matter how many times whites showed the vaqueros how to rope (the proper, "Texas" way), they did not seem to grasp the fundamentals of the "superior" technique. In utter frustration, many of the Texans gave up, because "there is no use trying to make them over . . . they see no reason they should change their ways to suit some Tejano whom they despise."[38]

Such tales served to solidify the notion that Americans were the true cattlemen of the West. The *Españoles* and their *Mexicano* descendants, according to this view, knew little, if anything, about handling stock. J. W. Fourr, a cattle inspector in Douglas, Arizona, summed up the white man's view of the cattle industry's history by stating that

> the round-up system is all American. Branding up calves out of a round-up is Texan. And until today a Mexican cannot drive a big herd of cattle. . . . They do not know how to string out a herd. I worked with them all of my life and never saw one, old or young, that could handle big bunches. . . . Our ways of handling horses and cattle are superior.[39]

Unfortunately, inspector Fourr did not have the opportunity to see vaqueros on Captain King's Santa Gertrudis ranch.

The almost magical and mythical romance and freedom (as well as the hard work, limited opportunity, and low wages) of the open range lasted about fifty years after the Civil War. In 1874, in DeKalb, Illinois, Joseph F. Glidden began producing barbed wire, and with the invention, "in a few short years the face of the West began to change."[40] Some writers have suggested two principal reasons for the end of the open range: 1) a desire by cattlemen effectively to control and breed their stocks, and 2) the arrival of large numbers of settlers and farmers on the frontier.[41] The economic impact of these developments sounded the death knell for the vaquero and cowboy life throughout much of the West. For a brief period (through the 1890s), vaqueros benefited from such changes, for they had previously performed the same work as whites for less pay. Still, the expansion of railroads and new opportunities in commercial agriculture reduced the number of employees necessary at most ranches.[42]

The closing of the range presented mounted laborers with a crucial question: what could such men do to support themselves? Ironically, many turned to the railroads, "the instrument of the final displacement," in order to survive.[43] Others moved from more densely populated areas to locations where open range operations managed to survive for a few more years. In her 1993 work, *The Last Cowboys: Closing the Open Range in Southeastern New Mexico, 1890s–1920s*, Connie Brooks describes the plight of a dying breed of men. Her research provides insight into the calculations, considerations, and decisions of cowboys trying to survive in a radically altered economic reality. Brooks collected information on thirty-two cowboys, mostly ex-Texans, who migrated to the western edge of the Llano Estacado in New Mexico during the late 1890s and early 1900s. A few of the men hoped to continue cowboying, but most had to forgo their aspirations. Eventually, they all took up homesteads or gravitated into other professions. All of the individuals under study were white, and their race, in part, permitted them to enter a wide range of professions, such as law enforcement, politics, banking, entrepreneurship, and skilled occupations. The end of the open range may have been psychologically traumatic to some white cowboys, but it appears that many made a fairly smooth transition to new lives and pursuits.[44]

Displaced vaqueros and rancheros faced greater challenges in adapting to new realities. In his 1966 work, *The Decline of the Californios: A Social History of the Spanish-Speaking Californians, 1846-1890*, Leonard Pitt delineates the circumstances confronting the Spanish-surnamed cattlemen of the Golden State. The need to survive drove the men to seasonal work such as sheepshearing, which "generally made the vaquero a supernumerary" in a changing economy. For the sons of former landowners, the reduction in status proved an almost unbearable burden. Scions of Californio gentry "put away finery and reappeared in brown overalls and red bandanas ready for action."[45] The development of newer industries, such as oil, did not provide significant opportunities for the desperate men, and during the 1870s and 1880s the Southern Pacific Railroad did not hire them in large numbers. Mexicans and Mexican Americans worked only "on the last leg of construction . . . but the hundred railroad hands permanently employed at the downtown switching yard . . . came mainly from the eastern United States, not Mexico or California."[46]

The change from open range to a more diversified economy proved disastrous for most rancheros and vaqueros, but not all of them faced similar circumstances. The *Kineños* and *Kenedeños* thrived because of their close relationship to the King and Kenedy families. By the 1990s some of these *familias* had worked on these ranches for five or six generations. Their numbers have declined; at present only sixty vaqueros are required to handle the more than sixty thousand head of cattle on the King Ranch. Over the years the ranches have diversified into other businesses, including cotton, grain, sugar, citrus, and most recently, tourism. While many of the great-grandsons of the original vaquero families no longer ride the range, they work in other aspects of ranch operations. Many of the descendants, both male and female, have moved into managerial and supervisory positions either in the field or in an office environment. The educational facilities (and college scholarships) provided by the two properties' managers assure the continuation of training, in agricultural management and other fields, for future generations of *Kineños* and *Kenedeños*.[47]

The vaquero has traveled a long and difficult road from the era of the *mesta* in Mexico City to the tourist tram of the King Ranch. The mounted laborers and their families have endured low social status, minimal pay, government sanctions, racism, and economic displacement. Through it all, their knowledge and skills, handed down from generation to generation, have made them a vital part of the cattle industry. The vaquero of the American West has survived attempts by historians, fiction writers, and others to dispel their efforts and contribution. At the dawn of the twenty-first century, the descendants of the *Españoles* who introduced cattle to the Western

Hemisphere, continue to ply their trade (although in reduced numbers) with honor, dignity, and pride, contrary to the notions of advertising agencies and rodeo promoters.

Notes

1. Don D. Walker, *Clio's Cowboys: Studies in the Historiography of the Cattle Trade* (Lincoln: University of Nebraska Press, 1981), 91.

2. Three of the best examples of this "new" Western history are Richard White, *"It's Your Misfortune and None of My Own": A New History of the American West* (Norman: University of Oklahoma Press, 1991), Patricia Nelson Limerick, *The Legacy of Conquest: The Unbroken Past of the American West* (New York: W.W. Norton & Company, 1987), and Patricia Nelson Limerick, ed., *Trails: Toward a New Western History* (Lawrence: University Press of Kansas, 1991). Of particular interest in *Trails* is Elliott West's "A Longer, Grimier, But More Interesting Story," 103-11.

3. Walker, *Clio's Cowboys,* 56.

4. Kenneth R. Porter, "The Labor of Negro Cowboys," in Clyde A. Milner II, ed., *Major Problems in the History of the American West* (Lexington, Massachusetts: D.C. Heath and Company, 1989), 343-58, quote on page 344. Another study detailing the role of African Americans in the cattle industry is Philip Durham and Everett L. Jones, *The Negro Cowboys* (Lincoln: University of Nebraska Press, 1965). See also Connie Brooks, *The Last Cowboys: Closing the Open Range in Southeastern New Mexico, 1890s–1920s* (Albuquerque: University of New Mexico Press, 1993), 36-40.

5. Porter, "Negro Cowboys," 344. See also Jane Clements Monday and Betty Bailey Colley, *Voices from the Wild Horse Desert: The Vaquero Families of the King and Kenedy Ranches* (Austin: University of Texas Press, 1997), xxiv.

6. Richard H. Slatta, *Cowboys of the Americas* (New Haven, Connecticut: Yale University Press, 1990), 127.

7. David Dary, *Cowboy Culture: A Saga of Five Centuries* (Lawrence: University Press of Kansas, 1989), 4.

8. Slatta, *Cowboys of the Americas,* 165-67 and 204-9.

9. Dary, *Cowboy Culture,* 16.

10. Slatta, *Cowboys of the Americas,* 2-11.

11. Ibid., 22. See also Dary, *Cowboy Culture,* 13, 38, and 48.

12. Slatta, *Cowboys of the Americas,* 15-16 and 33-34.

13. Dary, *Cowboy Culture,* 11.

14. Slatta, *Cowboys of the Americas,* 15.

15. Ibid., 11. See also Dary, *Cowboy Culture,* 15-16.

16. Dary, *Cowboy Culture,* 26.

17. Ibid., 29.

18. Ibid., 48 and 68. See also Jerald Underwood, "The Vaquero in South Texas with An Interpretation by John Houghton Allen," in *West Texas Historical Association Yearbook,* Volume 68, 1992, 93-99.

19. Slatta, *Cowboys of the Americas,* 19.

20. Don Worcester, *The Texas Longhorn: Relic of the Past, Asset for the Future* (College Station: Texas A & M University Press, 1987), 3. David Montejano, *Anglos and Mexicans in the Making of Texas, 1836-1986* (Austin: University of Texas Press, 1987), 52-56. Jack Jackson, *Los Mesteños: Spanish Ranching in Texas, 1721-1821* (College Station: Texas A & M University Press, 1986), 586.

21. Jackson, *Los Mesteños,* 593.

22. Leonard Pitt, *The Decline of the Californios: A Social History of the Spanish-Speaking Californians, 1846-1890* (Berkeley and Los Angeles, California: University of California Press, 1966), 83-113. Pitt argues that the arrival of the American economic and legal system was not solely to blame for the downfall of the Californios. Many of these rancheros also spent a great deal of money on luxuries, bad investments, and gambling. These circumstances also contributed to their downfall.

23. Robert J. Rosenbaum, *Mexicano Resistance in the Southwest: "The Sacred Right of Self Preservation"* (Austin: University of Texas Press, 1981).

24. Monday and Colley, *Voices from the Wild Horse Desert,* xx. See also Jackson, *Los Mesteños,* 595.

25. Monday and Colley, *Voices from the Wild Horse Desert,* xxix.

26. Slatta, *Cowboys of the Americas,* 165-66.

27. See, for example, J. Frank Dobie, *A Vaquero of the Brush Country* (Austin: University of Texas Press, 1981), Dane Coolidge, *California Cowboys* (Tucson: The University of Arizona Press, 1985), and James Beckstead, *Cowboying: A Tough Job in A Hard Land* (Salt Lake City: University of Utah Press, 1994), Montejano, *Anglos and Mexicans in the Making of Texas,* and Connie Brooks, *The Last Cowboys.*

28. Montejano, *Anglos and Mexicans in the Making of Texas,* 82.

29. Dobie, *A Vaquero of the Brush Country,* 60.

30. Monday and Colley, *Voices from the Wild Horse Desert,* xx-xxi.

31. Ibid., 9.

32. Ibid., 112-17 and 126-28.

33. Ibid., 12.

34. Montejano, *Anglos and Mexicans in the Making of Texas,* 82.

35. Ibid., 35.

36. Monday and Colley, *Voices from the Wild Horse Desert,* xxx.

37. Coolidge, *California Cowboys,* 67.

38. Ibid., 87.

39. Ibid., 149-50.

40. Dary, *Cowboy Culture,* 311.

41. Ibid., 309-10. See also Robert R. Dykstra, *The Cattle Towns* (Lincoln: University of Nebraska Press, 1968).

42. Montejano, *Anglos and Mexicans in the Making of Texas,* 90-91 and 106-7.

43. Ibid., 91.

44. Brooks, *The Last Cowboys,* 9, 30, and 43-46.

45. Pitt, *The Decline of the Californios,* 254-55.

46. Ibid., 256. For information on the conditions vaqueros faced in Texas, see Montejano, *Anglos and Mexicans in the Making of Texas,* 114.

47. Monday and Colley, *Voices from the Wild Horse Desert,* 161-97.

4 Black Cowboy: Daniel Webster "80 John" Wallace

DOUGLAS HALES

On a cold March morning in 1876, his thoughts dominated by cattle and cowboys, Daniel Webster Wallace stole away from home. He joined a cattle drive and the event forever changed his life.

Born into slavery on September 15, 1860, Wallace grew up on a small farm near the Texas Gulf Coast in Victoria County. Following the emancipation of slaves at the close of the Civil War in 1865, he moved with his former owners, the O'Daniels, some eighty miles north to a larger farm in Fayette County. The O'Daniels, records suggest, were "good people," and the Wallace family trusted them. In fact, Wallace's mother remained with Mary O'Daniel for the rest of her life, and in turn Mrs. O'Daniel raised Wallace's younger brothers after their mother's death.

On the Fayette County property, young Wallace, like many other African Americans in the rural South of the post-Civil War era, seemed destined for a life of farm work in the region's cotton and corn fields. But such pursuits held little interest for him. Besides, Wallace knew that black cowboys were working in his part of Texas. No doubt he had seen them for himself as cattle and cowboys passed the O'Daniel place on a regular basis. The brief glimpses of cowboy life that he could observe from the farm must have appealed to his spirit.

So, on that cold March morning in 1876, after completing his predawn chores, fifteen-year-old Daniel Wallace left to pursue his dreams. His life's journey led him to the cowboy way and into the western cattle industry, where he became one of the few successful and enduring African American ranchers in the nineteenth-century West.[1]

His initiation as a cowboy came quickly. When Wallace approached the cattle drovers, the trail boss asked him to show that he could handle a horse. Wallace found the tough little cow pony he was assigned much harder to handle than the old plow horses back on the farm. He climbed in the saddle, but not knowing what to do next he waited for the horse to move. When it moved, it bucked— vigorously. Refusing to be thrown, however, Wallace dug his feet into the stirrups, held on to the saddle horn, and somehow stayed aboard the half-wild mount. Impressed with Wallace's determina- tion and tenacity, the trail boss gave him a job.

The cattle drive went from Victoria County through Wallace's Fayette County to land still controlled by American Indians. It ended near Buffalo Gap at the corner of Runnels and Taylor Coun- ties in west-central Texas near modern Abilene. Now with cattle trailing experience for a reference and $1.50 in his pocket, Wallace went looking for a permanent job as a cowboy.[2]

Traveling southeast to Lampasas County, Wallace found a posi- tion with Sam Gholson, a small rancher with a reputation as an excellent "Indian fighter." After Wallace had been in Gholson's employment a short time, John Nunn from South Texas came through Lampasas with a herd of cattle. He was looking for unclaimed grassland on which to start a ranch. As Nunn had found elsewhere, however, local ranchers in the Lampasas area told him to keep his herd moving. In dire need of additional cowhands to con- tinue his quest, Nunn, on Gholson's suggestion, hired Wallace, who immediately left with Nunn and the cattle in search of good grass.

Wallace became one of Nunn's most valuable workers. Nunn, known as a "Christian gentleman" who allowed no profanity in his presence, assigned Wallace the important job of "wrangler" or "hoss stink," a position that included keeping track of horses, taming wild ones, and being the first to awaken each morning so that he might rouse the camp. For Wallace the work was fine. But Nunn, after again being rejected from occupied grasslands, moved to Sand Rock Springs, twelve miles from the present town of Snyder in Scurry County. Here he settled in and here Wallace worked with him for an additional seventeen months.[3]

Because of the often harsh and dangerous conditions on cattle drives and on the open range, cowhands understood that it was essential to cooperate with each other. Therefore, black cowboys were often exempted from the severe anti-black attitudes exhibited through much of American society in the post–Civil War era.

As a black man, however, and often the only African American in the outfit, Wallace was on occasion subjected to racial attacks. When they occurred, he confronted his attacker head on. One such event took place while Wallace worked for Nunn. A large man known as a "bully from a neighboring herd" came to a Nunn cow camp not far from Sand Rock Springs, and in front of Wallace he "bragged of the many things he had done to cowboys of color." Wallace, who stood six feet, three inches tall but was only seventeen years old and a bit thin, immediately challenged the big man to a fight. Then, in short order Wallace defeated the unpopular bully and, according to one source, won the respect of both his adversary and his fellow cowboys. In fact, he became something of a legend on the southern plains.[4]

Wallace proved himself in other ways. He was an excellent roper, for example, and several times he accepted friendly challenges to rope a buffalo while riding his horse. One time when he tried the dangerous trick, he came close to losing his life. Most of the time he completed the daring act with success.[5]

In 1878, Wallace received a letter from Mary O'Daniel notifying him that his mother was seriously ill and wanted to see him. Accordingly, he left Nunn's employ to make the long journey back to Fayette County. Unfortunately, the letter, which was several months old, had arrived late, and his mother had passed away before he could make it back. After taking care of his mother's affairs, Wallace took a position with Clay Mann, a well-known rancher in the Colorado City area. Seventeen years old and already a seasoned cowboy, Wallace traveled some four hundred miles to begin work at the new job.

Over the next eleven years Mann and Wallace developed a close personal and working relationship. As his reputation grew, Wallace came to be called "80 John," a name taken from Mann's popular "80" cattle brand. He began work on Mann's ranch near the present-day town of Colorado City. The Silver Creek operation was a ranch in the making. At first no housing was available, and ranch hands had to make do with abandoned dugouts left by buffalo hunters, or "more often," according to Wallace, "we used our wagons and the ground with our blankets for a bed and a saddle for a pillow. . . . It was common to find a snake rolled up in your bedding or to be awakened early in the morning by the howl of the wolf or the holler of the panther."

Indian raids were still quite common in this part of Texas, and Wallace always slept with a gun under his head. He remembered Indians stealing "all of our horses in '78 and most of them in '79, but we stayed there all the year of '80." Perhaps about the same time, but probably a few years earlier, Comanche warriors stole one of his favorite horses. He tracked the band for several days before giving up the chase.[6]

One of his most dangerous encounters with Native Americans took place in Mexico. In the early summer of 1883, Mann acquired a ranch in the Mexican state of Chihuahua. He sent his brother John Mann, Wallace, and several other cowboys with four thousand head of cattle to establish his presence on the property. Unfortunately, Indians controlled the same area and took a dim view of intruders, and Mexican citizens along the way warned the Americans of the danger. According to Wallace, "we knew not what to do but go on where Mann told us to go. . . ." After crossing the Chihuahuan desert the Americans arrived at the ranch site in late August, built a small headquarters, and kept a close vigil for Indian people. They saw no Indians until October 18, when a number of warriors appeared and "raided our camp, burned our wagons and pens, killed our milk cows, took all the horses we weren't riding and waylaid us in the hills." In an attempt to break out, John Mann and another man were killed, while, according to Wallace, "the rest of us ran as fast as we could to the open prairie, away from the mountains."

After the Indians had left, Wallace managed to return to bury the scalped and mutilated bodies of his fallen comrades. Mann's cowboys rounded up what cattle they could find and moved to a neighboring ranch before heading back to Texas a year later. In 1886, Mann sold the ranch to Randolph Hearst.[7]

Daniel Webster Wallace, known as "80 John" Wallace, highly respected cattleman and rancher of Mitchell County, Texas. Courtesy of Southwest Collection, Texas Tech University. SWCPC File no. 39(A), Envelope 7.

Wallace and his favorite horse "Peck," who "was no bigger than a half a bushel," rode range lines all across Mann's holdings, land from eastern Mitchell County to the Double Mountain Fork of the Brazos River in Kent County. Eventually Wallace became Mann's right-hand man and his companion on cattle drives and expeditions for more land.

Mann trusted Wallace, whose straightforward response to assignments was, "I will do my best." Mann also entrusted Wallace with the protection of his family. In addition, he trusted Wallace with the buying and selling of cattle. Wallace recalled numerous times he was sent alone with large sums of Mann's money. To buy new stock, once he went to Midland, Texas, with thirty thousand dollars in cash in his pocket. When Mann died in 1889, his wife thought so highly of Wallace that she asked him to take over the training of her two young sons.[8]

Wallace credited Mann with teaching him the fundamentals of cattle raising and good ranch management. In fact, Wallace began his own cattle herd through a "gentleman's agreement" with Mann. For a two-year period he earned $720 but drew only $120, applying the remainder to purchase cattle. Mann allowed Wallace's stock with the "Wallase" brand to be run along with his own cattle. As the herd grew, Mann encouraged his young cowhand to purchase land, and in 1885 Wallace bought two sections of railroad land southwest of present-day Loraine in Mitchell County. Later he homesteaded the property.[9]

In the winter months during this time Wallace attended a school for blacks in Navarro County. There he met and fell in love with Laura Dee Owen, a recent graduate of the school. On April 8, 1888, the couple married. Not long afterward Mann died, and Wallace and Laura took steps to become full-time ranchers on their own spread. Now in business for himself, "80 John" Wallace changed his brand from "Wallase" to a "D-triangle.' He cut native wood, probably mesquite, for posts to fence his land, and he worked part-time for wages at cattle pens and for other ranchers, including the Slaughter and Spade ranches. And, when a local bank loaned him money to buy Durham and Hereford heifers, he moved to improve his herd.

Daniel Webster "80 John" Wallace became a connected and well-respected resident of Mitchell County. In 1905 he purchased two additional sections of land on Buck and Silver creeks, tributaries of the Colorado River, and this property became the Wallace Silver Creek Ranch. The increased acreage and stored feed helped Wallace through some tough droughts.[10]

Wallace also joined the Texas and Southwestern Cattle Raiser's Association. For the next thirty years he attended meetings of the organization and earned the respect of its mainly white members. Indeed, an event that occurred while he was on a train en route to one such meeting illustrated the respect. Railroad cars in Texas at the time were segregated, with separate cars for blacks and whites. But several white ranchers came into the black car where Wallace was riding and greeted the black cattleman. When told they would have to return to the whites only car, one man refused, stating, "I have known 80 John for thirty years. We ate and slept on the ground together. I see no reason that makes it impossible for me to set here now."[11]

Like other cattlemen, Wallace had his share of troubles. The most difficult time came with a drought in 1894. He first moved his cattle to better pasturage in Oklahoma, but the grass there did not last, and he had to sell his herd. Because of depressed prices, he sold at a loss and faced possible ruin. As in other critical times in his life, once again he received help from some friends. Winfield Scott, an old acquaintance and retired rancher living in Fort Worth, furnished him with a ten-thousand-dollar line of credit to restock his ranch, and soon Wallace prospered again. To deal with future droughts, Wallace allocated twelve hundred acres of prime land for feed crops. Another drought in 1917 posed no real setback for Wallace, even when, for better grazing, he had to send a portion of his herd to New Mexico.

Wallace prospered. He hired cowboys, both black and white, bought more land, and brought in tenants to farm some of the more productive bottomlands. In 1915 he resurveyed some of his land, particularly the original railroad land. Upon completing the new survey, he found that one hundred acres of the original purchase had not been paid for. Fearing it might be taken from him at a later date, he notified the state and purchased the land. It was a wise move, for in 1953, several years after his death, wildcatters found oil on those same one hundred acres. His descendents benefited from his prescience.[12]

Meanwhile, as the droughts illustrate, a successful ranching operation needs good water. Always in short supply in West Texas, water was a precious commodity. Wallace, however, had a knack for finding it. He developed a reputation in the use of a "divining rod," and he was often being sought by other ranchers who wanted him to find water for them. He also used his "skill" to locate sites for windmills on his own property.[13]

The Great Depression of the 1930s brought hard times to the American cattle industry. In his seventies during the Depression, Wallace was nonetheless determined to ride out the troubles. Fortunately, through frugal and efficient management he ended the period debt-free.

After Franklin D. Roosevelt's election to the presidency in 1932, the federal government moved to help cattlemen through the New Deal programs. Wallace refused most aid, but he did accept help through the Emergency Buying Program of 1934, a law that allowed the President to establish the Surplus Relief Corporation. Through this program, the government hoped to raise cattle prices by buying surplus cattle. In turn, Wallace saw no reason not to sell to the government at thirteen dollars per head some seventy-six of his older and in some cases sick cattle.

Through such aid, Wallace, unlike many (if not most) cattlemen during the period, remained profitable. In fact, during the Depression, he bought and paid for two additional sections of land that adjoined his property near Loraine in Mitchell County. As a result, with the coming of World War II, the Wallace ranching operation was in a position to benefit from the high demand for beef during the booming war-time economy.[14]

Daniel Webster "80 John" Wallace died on March 28, 1939. At the time he owned sixteen sections of land, including farmland operated by ten different tenant families. Today the ranch is operated by his descendants.

Wallace learned his ranching lessons well. He came to understand the importance of making improvements for both the protection of his livestock and the comfort of his family. For example, he was a good carpenter, and through the years he built a large ranch house, numerous barns, and cattle pens, and made other improvements on his property.

His life was an amazing and successful one. At the time of his death, Wallace had four children and four grandchildren. Even though he himself had very little formal education, he sent each of his children through school, including college, and he assisted several other young people with their education. In Colorado City, Wallace donated the land on which an African American school was built. In Loraine, he financed the building of the First Baptist Church, which also served as a school for black children.[15] He was a young black cowboy who became a highly respected cattleman.

The information available on African American ranchers like "80 John" Wallace is limited. Compared to white cattlemen, black ranchers never existed in large numbers. For Wallace there was little time to worry about his historical significance—being one of the few black cowboys who built a large cattle ranching operation. Moreover, few historians before the 1960s studied or wrote about African American contributions to the West. During Wallace's lifetime little was mentioned about black ranchers or cowboys, at least in Texas, except for the contribution of *The Cattleman,* a magazine for ranchers. In 1936, John M. Hendrix wrote a bland but fair article entitled "Tribute Paid to Negro Cowmen." It is a brief account of Wallace's life, but the short piece also treats such legendary black cowboys as Bill Pickett, William Coleman, and Tige Avery.

Beginning in the 1950s, interest in black cowmen began to increase. The *West Texas Historical Association Year Book* in 1952 published a sketch on Wallace, one he dictated to a friend shortly before his own death in 1939. Hertha Webb's M.A. thesis, a largely forgotten piece on Wallace from Prairie View A & M University, was completed in 1957. Mainly, it represents interviews with Wallace's daughter Mary Wallace Fowler, but the brief work adds some information to Wallace's life story. Hettye Wallace Branch, another Wallace daughter, published a small book about her father in 1960;

it is a romanticized view of cowboy life and adds little new information. William S. Savage published "The Negro Cowboy on the Texas Plains" in *The Negro History Bulletin* in 1961, but he relied mainly on previously published material.

Each of these works is to be commended for bringing a black cowboy turned cattleman to the attention of scholars and the public at large. At the same time the shortage of information on black cowboys shows the price society pays when an important segment of the country's population is excluded from serious study.

Daniel Webster "80 John" Wallace, an African American, moved from slave to farm laborer and then from cowboy to cattleman. In some ways his story was typical of American cowboys: he started young and on a trail drive before working for others; he collected his own small herd while employed as a cowboy; and he eventually acquired his own cattle operation, one he built into a successful ranching enterprise. In other ways his story is atypical: here was a black cowboy who became a respected cattleman and a major contributor to his community and to African American schools. He was one of the few western ranchers who survived the shakeout of the mid-1880s and the drought and national financial depression of the 1890s. His ranch some 110 years later remains in his family's possession. Clearly, "80 John" Wallace was one of the most remarkable black cowboys of the American West, and his story deserves retelling.

Notes

1. R. C. Crane, ed., "D. W. Wallace ('80 John'): A Negro Cattleman on the Texas Frontier," *West Texas Historical Association Year Book* 28 (October 1952): 113-14; Hettye Wallace Branch, *The Story of "80 John"* (New York: Greenwich Book Publishers, 1960), 13-14; Alwyn Barr, *Black Texans: A History of African Americans in Texas, 1528-1995*, 2nd ed. (Norman: University of Oklahoma Press, 1996), 91, 149. The Crane article is actually a reprint of a sketch dictated by Wallace shortly before his death. In this article Wallace identifies his former owner as Mary Cross, though most sources identify her as Mary O'Daniel.

2. Branch, *The Story of "80 John,"* 13-14; John M. Hendrix, "Tribute Paid to Negro Cowmen," *The Cattleman* 22 (February 1936): 25.

3. William S. Savage, "The Negro Cowboy on the Texas Plains," *The Negro History Bulletin* 24 (April 1961): 157; Crane, "D. W. Wallace ('80

John')" 114-15; Branch, *The Story of "80 John,"* 15. Gholson's reputation as an Indian fighter seemed to be based purely on word of mouth.

4. Branch, *The Story of "80 John,"* 15-18.

5. Ibid.

6. Ibid., 18-22; Crane, "D. W. Wallace ('80 John')," 116-18; Ira B. Jones and Rupert Richardson, "Colorado City, the Cattleman's Capital," *West Texas Historical Association Year Book* 19 (October 1943): 38.

7. Crane, "D. W. Wallace ('80 John')," 117-18; Branch, *The Story of "80 John,"* 22, 29-30. The accounts of Wallace and Branch differ in their descriptions of the Mexico episode. Where differences occur this author used the first-hand account of Wallace.

8. Branch, *The Story of "80 John,"* 28, 36; Hertha Auburn Webb, "D. W. '80 John' Wallace—Black Cattleman, 1875-1939" (M.A. thesis, Prairie View A & M University, 1957), 22-25.

9. Branch, *The Story of "80 John,"* 30, 36; Webb, "D. W. '80 John' Wallace—Black Cattleman," 22. Wallace purposely spelled his brand using an "s" instead of "c."

10. Webb, "D. W. '80 John' Wallace—Black Cattleman," 22, 30-32; Branch, *The Story of "80 John,"* 30, 36, 37-39.

11. Branch, *The Story of "80 John,"* 37-38.

12. Webb, "D. W. '80 John' Wallace—Black Cattleman," 36-37.

13. Ibid.; Branch, *The Story of "80 John,"* 30-32, 41.

14. Webb, "D. W. '80 John' Wallace—Black Cattleman," 33-36.

15. Ibid., 32-33, 38-39; Branch, *The Story of "80 John,"* 48-49.

Cattle of the Matador Ranch being herded across the South Saskatchewan River in Canada, ca. 1920. Courtesy of Southwest Collection, Texas Tech University. SWCPC File no. 64(f), Envelope 1.

5 Indian Cowboys of the Northern Plains

THOMAS A. BRITTEN

To adherents of the dime novel/Hollywood depiction of the American West, the image of an *Indian cowboy* may appear to be an oxymoron. The term does not seem to fit with much of the traditional interpretation of the American West, especially the popular ideas about cowboys—cowboys found in the novels of Zane Grey and Louis L'Amour or in old Westerns featuring John Wayne or Ben Johnson. Nowhere, it seems, is the need for a conceptual "reconditioning" more apparent than in our understanding of the great cattle drive and cowboy era. As Peter Iverson has noted, "cattle ranching is a story usually told and nearly always understood in one color, and that color is white."[1] Consequently, the unique contributions and experiences of non-white cowboys and ranchers is either ignored or relegated to the periphery, thus distorting the truth and undermining one's acquisition of a more broadly based and accurate appreciation of western history.

American Indians, in particular, have been overlooked in studies of the cowboy and cattle drive era—an odd omission considering the long history of Native American stock-raising activities in the West. By practically all accounts, the sixteenth-century expeditions of Spaniards Francisco Coronado and Juan de Oñate were the first to introduce cattle (and reintroduce horses) into what is now the United States. By the early seventeenth century, Indian peoples of the American Southwest had managed to acquire horses, which quickly became "objects to be stolen, traded for, fought over, dreamed about."[2] In a century's time, Native Americans residing on the northern Plains had also gained possession of the horse. Impressed with the animal's strength and speed, Indian peoples acquired large horse herds, unaware of the profound changes they would bring to virtually every aspect of their societies and cultures.

Obtaining horses was a momentous event in the histories of Plains tribes, and the circumstances surrounding their acquisition became fixed in oral traditions. Joseph Medicine Crow, a respected elder and historian of the Crow tribe, writes that his people acquired horses sometime around 1725 from Indians near Great Salt Lake. The Crows named the horse *Ichilay*, which means "to search with," perhaps because they intended to use the animals to search for game or for enemies.[3]

Blackfoot tradition regarding the acquisition of horses resembles, in a general way, that of the Crows. A party of Blackfeet warriors traveled south to "Always Summer Land" and came across enemies riding on what they believed were "elk dogs." They succeeded in stealing some of the horses and returned home. The Lakotas, meanwhile, acquired horses through trade with neighboring Cheyennes. Although they were initially afraid of the horse, the Lakotas believed it to be a holy creature possessing great spiritual power. Consequently, they gave it the name *sunka wakan,* or "holy dog."[4]

Horses revolutionized Plains Indian cultures, and nowhere was the change more evident than in their subsistence hunting activities. While bison hunting had been performed on foot for generations, the acquisition of horses allowed hunters to follow herds over far greater distances, to transport larger quantities of meat, and to kill bison year-round. The new advantages allowed Indian communities the time (which otherwise would have been spent hunting) to develop larger, more stable social groups and increasingly sophisticated forms of cultural expression. For many Plains peoples, therefore, the century and a half that followed their acquisition of horses was both an economic and cultural "Golden Age," or traditional period, as it is now called.[5]

The traditional era proved short-lived. During the closing decades of the nineteenth century, northern Plains tribes suffered a series of cataclysms that altered their lives, undermined their futures, and led many to look to the past as a means of dealing with the dramatic changes swirling around them. The cataclysms included military defeat at the hands of the United States Army, confinement on reservations, and the near extermination of the bison. If these were not enough, the federal government in the late nineteenth century stepped up its efforts to assimilate Native Americans into the majority society. The only alternative to assimilation, some government officials warned, was extermination.

The government employed several strategies to expedite assimilation. Reservation agents, for instance, encouraged Native Americans to take up agricultural pursuits and become economically self-sufficient, Indian children were sent to off-reservation boarding schools, and missionaries labored to convert the government's "wards" to Christianity. But the assimilationist agenda was flawed in several important respects. Most Native Americans were not interested in assimilation, choosing instead to hold on to the rich cultural heritages passed down from their elders. Graduates of off-reservation boarding schools returned home as foreigners to their own people, caught in a "shadow world" of cultural ambiguity.[6] The climate and terrain of the northern Plains, meanwhile, proved resistant to agricultural pursuits, as the failed white homesteaders of the 1860s gave ample testimony.

In response to the difficulties, policy-makers applied a new economic and social strategy aimed at encouraging Native American self-sufficiency and economic independence while providing men a means of achieving status and recognition among their people. The new strategy was cattle ranching. The 1868 Fort Laramie treaty included a stipulation that the federal government would furnish Lakota families a cow and a pair of oxen. A decade later, more cattle were issued, and by 1886 there were over forty-six hundred head of cattle on the Pine Ridge reservation alone. Most of the cattle issued were Texas longhorns, a breed famous for speed, stamina, and unpredictable behavior. According to Captain D. C. Poole, a reservation agent in the Dakotas in 1881, the cattle had "wide, branching horns, long legs and lank bodies . . . seldom weighing over a thousand pounds gross weight. They were as fleet as an elk, and as easily frightened."[7]

Ranching on the Blackfeet reservation in Montana, meanwhile, began in 1888 when the government purchased cattle for them.

Unloading cattle at LeBeau, South Dakota, ca. 1909. Courtesy of Southwest Collection, Texas Tech University. SWCPC File no. 64.

Within four years, renowned Indian expert George Bird Grinnell later testified, there was "hardly an individual in the tribe who [was] not the possessor of one or more cows." In 1890, John Murphy, the agent at the Fort Berthold reservation in North Dakota, requested that the government pay the Mandans, Arikaras, and Hidatsas (commonly referred to as the Three Affiliated Tribes) in stock rather than in cash for land cessions made four years earlier. The federal government approved the plan, and a year later Murphy purchased four hundred cows and forty bulls. Before long, each family owned at least one cow or steer.[8]

The opinions of government policy-makers and reservation agents were by no means unanimous concerning the merits of introducing cattle on northern Plains reservations. Critics argued that ranching reinforced nomadism and the horse culture, which they believed were detrimental to assimilation. In their minds, taking up the plow and growing cash crops was the best solution to the so-called "Indian problem." Proponents of Indian cattle ranching, on the other hand, maintained that stock-raising appealed to many Native American men, and thus provided a useful transition from nomadism to a sedentary and "civilized" existence. Captain D. C.

Poole summed up such reasoning, as well as the popular belief in the progressive nature of cultures: "Why endeavor to make [Indians] unwilling agriculturalists, in place of leading them to a pastoral life, for which they show considerable inclination, and which has always come in the first state of advancement from barbarism to civilization?"[9]

Open range cattle ranching did indeed hold out a relatively strong appeal for peoples who had only recently been forced to abandon their dependence on the bison economy and relocate to reservations. Like the bison hunts of old, cattle ranching was a communal activity that required young men to work together and to depend on each other during cattle drives and annual roundups. Women, meanwhile, often accompanied Indian cowboys at roundup time, pitching tents nearby, where they butchered cattle, cut the meat into strips, and then hung it to dry as they had once done with bison. Hidatsa ranching families utilized virtually every part of a cow. Stomachs were used as canteens, heartskins as lunchbags, and split cow horns were used in fashioning headdresses. Cowboying was masculine and equestrian work that required mobility, self-reliance, stamina, and courage. It provided young men a means of achieving wealth and status, not unlike being a successful hunter or warrior. As Gordon MacGregor observed in his study of the Oglalas on the Pine Ridge reservation in South Dakota, "of all the whites the Indian knew, the cowboy, who was their equal as a horseman and a marksman, who lived in the out-of-doors, who moved independently and without fear of the Indian, made the greatest appeal...."[10]

Indian cowboys, in some respects, resembled their white, black, and Hispanic counterparts. Some worked with other family members on small individually owned ranches or with larger herds owned communally by the entire tribe. Several gained employment with the much larger non-Indian owned ranches, which leased vast expanses of the northern Plains and, at times, entire reservations. After the American Civil War, Texas Confederates had drifted back home, disillusioned about the recent past and unsure of the future. For many of them, the cattle business provided economic livelihood and an opportunity to adjust to life in the era of Reconstruction.

Indian men, meanwhile, also faced difficult adjustments to reservation life and uncertain futures. Felix LeBeau, a Lakota from the Cheyenne River reservation in South Dakota, learned to ride broncs in hopes of getting a job as a cowboy and to escape attending a boarding school. He secured employment with the Texas-based Matador Land and Cattle Company and stayed with the outfit for four years. He wore "traditional" cowboy garb, consisting of a shirt, pants, boots, a hat, and pigskin chaps. Felix recalled that on trail drives only heifers were butchered for meat, which, when mixed with tomatoes, corn, bacon, and sowbelly, made an appetizing meal. Reiterating a criticism common among virtually all cowboys, he complained that Texas longhorns were wild and "skittish." Their horns were such a menace, he recalled, that when they were sent through a chute at roundup time, a cowboy with an ax would chop off their sharp points. LeBeau made good money. He received approximately seventy-five dollars a month for his services, and during winter months he returned home to tend to his own small herd.[11]

Other Cheyenne River cowboys worked for large cattle companies. Hoe Yellowhead spent six years at Richard Pratt's Carlisle Indian Industrial School in Pennsylvania before returning home to work for Dan Zimmerman's DZ outfit. He later found more lucrative employment as a bronc rider in a Wild West show. Guy Buffalo worked for several large ranching companies, including the HO, the Diamond A, and the Big Stone Cattle Company. In addition to familiar complaints about longhorn skittishness, Buffalo cited prairie fires as a major problem confronting cowboys on the northern Plains. During one such episode, Guy and his co-workers killed and cut open a large steer and, after tying it securely to their horses, dragged it across the ground to put out the fire. After spending nearly twenty years as a cowboy, Buffalo went to school at the Santee Normal Training School in Nebraska, and he became the first Native American ordained minister in South Dakota.[12]

Lakota cowboys Amos Eagle Boy, Leo Shooter, and Leo Little Eagle resided on the Standing Rock reservation in North Dakota and obtained work on large cattle ranches while they were boys. Eagle Boy began working cattle for the Matador Land and Cattle

Company when he was only fourteen years old. Shooter gained employment with Ed Lemmon's L-7 ranch when he was sixteen. Little Eagle came from a ranching family. In addition to running cattle, his father raised pigs and chickens. As a young man, Little Eagle, along with his cousin Straight Pine, worked for the Montana-based King Cattle Company. His daily routine was typical of other northern Plains cowboys. During the summers he rode fence-lines and made needed repairs. During dry spells, he hauled water to thirsty cattle. In the autumn he helped mow hay that, in harsh winter months, would be fed to cattle and horses.[13]

The story of Standing Rock cowboy Joe Wicks provides additional insight into the life of Indian cowboys during the first decades of the twentieth century. Joe, of Scotch-Irish ancestry, had been orphaned as a child and placed in an orphanage near St. Paul, Minnesota. Around 1903, he was adopted by a Russian immigrant family, who brought him to North Dakota. His adoptive parents apparently were abusive to him, and in 1914 the courts placed him in the home of a rancher named Steve Wicks. He soon met and befriended some Indian cowboys who were the "first people who ever treated him as a human being and a man, and treated him as a brother." In 1918 he accompanied friends George Defender (who would later become a world-class bronc rider) and Henry Fast Horse to live on the Standing Rock reservation, where he took a job as a cowboy for the Styles Cattle Company. From then on he "made up his mind that he was an Indian" and his Lakota friends taught him "how to be a good bronc rider." Hoe later married a Lakota woman, and in 1925 she gave birth to a son they named Chaske, which means "first-born."[14]

After long trail drives where cattle were delivered to railheads for shipment east, Native American cowboys, like their non-Indian counterparts, enjoyed relaxing in a local hotel where they could bathe and get a good meal. At night they would join other cowboys at saloons to discuss their recent hardships, boast of heroic deeds, and, at times, gamble their earnings away. Wolf Chief, a Hidatsa rancher, appreciated the presence of women after spending long weeks on the trail. "Girl waiters waited on us," he recollected, "and I thought they had sweet faces and nice clothes. I wished they were

[at] home, and I would get after one as my wife or sweetheart. But I knew their customs were different. . . . But at home if I were young, I would."[15]

Native American ranchers and cowboys also shared many of the hardships that plagued other northern Plains stock-raisers. When reservation agents distributed government cattle to Indian stock-raisers, they were branded with an ID (Indian Department), USIS (United States Indian Service), or TID (Trust Indian Department) brand. At the Fort Berthold reservation in North Dakota, cattle belonging to the Three Affiliated Tribes were branded a second time. Each family was assigned a brand, commonly made up of its initials. Wolf Chief's brand, for example, was WC. To more easily distinguish what tribe the owner came from, however, each group affixed its brand to a different part of the cow. The Mandans branded on the right shoulder, the Hidatsas on the right thigh, and the Arikaras on the left thigh.[16]

But branding, unfortunately, did not always deter rustlers. Belknap "Ballie" Buck, a *métis* (a person of mixed Indian and European blood) of Assiniboine and Gros Ventres ancestry from Montana, worked as a cowboy for the Sand and Taylor (ST) outfit. In 1893 he accompanied a posse to recover several horses that had been stolen. After trailing their quarry for several days, Buck and his cohorts cornered the rustlers, and after an exchange of gunfire, they recovered the animals. Indian cowboys were not immune from such extralegal activities. In her study of cattle ranching on the Fort Berthold reservation, Irene Castle McLaughlin states that some Indian cowboys resisted the encroachment of white ranchers by stealing horses, killing cattle, or running their own herds on lands being leased by white stock-raisers.[17]

Harsh weather conditions were a constant threat to the economic security of cattlemen on the northern Plains. The winter of 1903-1904 killed an estimated 1,720 head of cattle at the Fort Berthold reservation, nearly twenty-five percent of the entire herd. Ballie Buck referred to the winter of 1906-1907 as the "Great Slaughter" due to the deaths of so many cattle. Unable to stamp through the snow to reach grass, cattle reportedly ate the bark from trees and died with their "bellies full of cottonwood." Sharp ice "scraped the hair off cattle's legs and off their faces," Buck recalled, and "blood would run down their faces and legs and the sores would cut open continuously." Driven by high winds and minus-forty-degree temperatures, cattle wandered into nearby towns, becoming a safety hazard for worried citizens. A late snow in the spring of 1914, meanwhile, resulted in the deaths of hundreds of South Dakota cattle when they drifted into a big lake bed and trampled one another to death.[18]

Although rustlers and periodic blizzards plagued—and occasionally bankrupted—northern Plains stockmen, these were random problems that most ranchers accepted as part of the gamble associated with stock-raising. Indian cattle ranchers, however, faced an additional obstacle that undermined their stock-raising efforts in a systematic and, many Native Americans would argue, equally ruthless manner. That obstacle was the federal government's Indian policy. On the one hand, policy-makers demanded that Native Americans adopt cattle ranching as a means of expediting assimilation and self-sufficiency. Yet they simultaneously carved up reservations into individual allotments that were too small for Indian families to raise cattle effectively, and forced tribes to lease what remained of their lands to non-Indian cattlemen.

The infamous Dawes Severalty Act of 1887 proved disastrous to Indian cattle ranchers on the northern Plains. Prior to allotment, most tribes held their cattle herds in common and grazed their livestock on reservation ranges. Following allotment, however, reservation lands were "individualized" into small tracts of 160 to 320 acres for each family, hardly sufficient land on which to raise adequate-sized herds. Indian ranchers, moreover, rarely possessed the necessary capital to fence their holdings and secure feed when their pastures were depleted of grass. The case of the Blackfeet reservation in Montana is illustrative of the fate of many Indian ranchers following allotment. The federal government had praised Blackfeet stock-raisers in the mid-1890s for their successful adoption of cattle ranching. After the reservation was allotted in 1907, however, Blackfeet cattle-raising ventures plummeted, and ranchers were forced to kill their cattle for food, or offer them to storekeepers in payment of their bills.[19]

The leasing of reservation lands to non-Indian ranchers also contributed to the demise of Indian stock-raising practices. Grazing leases were made between cattlemen on the southern Plains and the Kiowas, Comanches, Southern Cheyennes, and Southern Arapahos in the 1880s, but Congress did not get around to passing legislation legalizing the practice until 1891. Designed to permit young, old, or disabled allottees to derive some income from their allotments, the 1891 leasing act was soon revised to include the authorization of leases "by reasons of age, disability, or *inability*." Leases were to be short in duration—a maximum of five years—and tribal lands were to be leased only with tribal consent. Non-Indian ranchers could purchase permits that required lessees to pay one dollar a head to graze for an entire season or a fixed rent lease that allowed them to rent an entire area and graze as many head on the tract as they wished. Ironically, Indian stock-raisers who desired to run cattle on tribally owned lands also had to purchase permits. The restriction, according to historian Richmond Clow, discouraged and penalized legitimate Indian cattle operators from building their own herds.[20]

By the turn of the century, non-Indian ranchers were running hundreds of thousands of cattle on reservation lands. In 1905 the Three Affiliated Tribes leased a two-hundred-thousand-acre section of unallotted land (known as the "Big Pasture" or "Big Lease") to railroad magnate James Phelan. Phelan ran eight to ten thousand cattle on his lease, along with a thousand horses. Just three years earlier, Ed Lemmon acquired a lease to 865,000 acres of land on the Standing Rock reservation, and the Matador Land and Cattle Company leased lands on the Cheyenne River reservation. In 1907, Matador herds were reportedly grazing on virtually the whole Blackfeet reservation at Fort Belknap.[21]

Although the leasing of reservation lands to large non-Indian-owned cattle ranching operations provided needed income to impoverished tribes and employment for dozens of Native American cowboys, the long-term effect of the practice proved detrimental to the government's goal of achieving Indian economic self-sufficiency. Indian stock-raisers who struggled to eke out a living running small herds on their allotments must have questioned the utility of such an enterprise when thousands of non-Indian cattle were grazing on reservation lands with profits sent to cattle barons headquartered in Texas or back East. As Leo Little Eagle, a Standing Rock cowboy, observed, before farmers moved in and erected fences and whites began leasing reservation lands, "we had everything good. Don't have to talk about ranges . . . free range, everything's free. Had a lotta cattle, everybody had cattle, horses."[22]

Indian cowboys and ranchers in the modern era face many of the same challenges and problems that their parents and grandparents encountered when cattle were first introduced onto northern Plains reservations. Indian cattle-raising ventures continue to be modest family-run operations. At times, families form partnerships and run their cattle rent-free on the unused lands of non-ranching relatives and friends. Herds are generally small in size. Sixty percent of Blackfeet ranchers ran less than one hundred head each according to a government investigation conducted in 1954. In a similar study organized in 1960, over half the Indian ranchers from seven northern Plains reservations (Pine Ridge, Blackfeet, Fort Belknap, Fort Peck, Rosebud, Lower Brule, and Crow Creek) possessed herds numbering fewer than one hundred head of cattle. Many of these small ranchers, consequently, had to find additional employment to make a living. In order "to achieve an appreciable degree of economic independence," government agricultural experts advised, ranches of 150 to 200 head were needed—a tall order considering many Native Americans' inability to acquire additional pasture lands or credit from local banking establishments.[23]

Financial restraints also help account for the fact that most modern Indian stock-raisers on the northern Plains attempt to graze their herds on pastures year-round. Approximately seventy percent of the Blackfeet ranchers questioned in 1957 used the same ranges for winter and summer pasturage. Crow ranchers—regardless of the size of their herds—did likewise.[24] Although not the most efficient means of fattening cattle, grazing is the simplest, cheapest, and least labor-intensive. While some Indian ranchers supplement grazing by planting hay (particularly to feed cattle during winter months), most do not. Raising hay requires equipment to sow and harvest the crop, and most Indian ranchers lack the financial resources to purchase the necessary implements. The results of inadequate feeding practices have been disastrous to Indian cattle ranchers. Their cattle are less resilient to disease, are more likely to eat poisonous plants, and produce weaker calves. Indian ranchers who run cattle on pasture grasses year-round, consequently, often suffer lower cow-calf yields and higher cattle mortality rates.[25]

An even greater obstacle to Indian cattle ranching (and hayraising for that matter) remains the issue of leasing. Non-Indian lessees, carrying on a tradition dating back to the 1891 leasing act, continue to run cattle on the majority of reservation lands on the northern Plains. On the Pine Ridge reservation, for example, nearly two-thirds of the reservation in 1963 was devoted to ranching, but only one-third was controlled by Indian ranchers. Leasing continued to be an even greater problem for the Crows. According to a study conducted in 1964, non-Indians were utilizing eighty-seven percent of the tribe's grazing lands, ninety-five percent of irrigated land, and ninety-four percent of the dry farmland. Oddly, government agricultural consultants, while aware that non-Indians continue to dominate reservation resources, choose to ignore the pernicious influence such practices exert on Indian stock-raising activities. Instead, they argue blithely that if Native Americans would simply "adjust ranch organization, adopt approved ranch practices, and develop adequate [herds and ranches]" then they could "become completely self-supporting."[26]

Indian cowboys share a passion for rodeos with their non-Indian counterparts. Participation in rodeo events, Elizabeth Atwood Lawrence has written, is not unlike the old Plains Indian custom of counting coup. Counting coup was a daring feat that required bravery and high risk. Those warriors who successfully "counted coup," consequently, were assured of glory and status in the eyes of their people. Rodeo events, likewise, can be quite hazardous. At the annual Crow Fair rodeo, in fact, a praise song is sung just prior to the festivities as a petition to God that no participant be injured or killed that day. Belt buckles awarded to rodeo champions are worn ostentatiously as Crow men of rank once wore otter skins or feathers as signs of their achievements.[27]

Rodeos hold additional cultural significance to northern Plains tribes. At a Rosebud rodeo held at the turn of the century, traditional dances and games were held prior to rodeo events. After dinner, a sham battle of the Custer fight entertained spectators along with games of tug-of-war, a potato race, and chasing a greased pig. Chaske Wicks (the son of Joe Wicks mentioned earlier) remembered that during the 1930s Indian participants would set up tipis and "live just like the old Indians in the days gone by." In addition, they "performed cultural dances and told of their cultures, and . . . were able to sell their beadwork and so forth at these rodeos throughout North Dakota and Minnesota."[28]

An additional lure of rodeos may have been the lucrative purses offered to champions. Tom Three Persons, a Blood, was a world champion bronc rider in 1912. Described as being rather modest and quiet, Tom nonetheless wore a pair of red angora chaps, a blue shirt with a white kerchief tie, and a black hat while participating in the famous Calgary Stampede. After successfully riding a bronc named Cyclone (which had reportedly pitched its rider the previous 130 times), Three Persons walked off with $1,000 in gold, a trophy saddle, and a belt with a silver buckle.[29]

While most of the rodeos held on northern Plains reservations did not offer the big purses, such as those offered at the Calgary Stampede, they nonetheless remained a powerful incentive to cowboys who would otherwise expect to earn between fifty and a hundred dollars a month. David Blue Thunder raced his string of ponies at the Frontier Days and won "first money" at a rodeo held on the Rosebud reservation. Grover Red Fox, from the Cheyenne River reservation, won first money in bronc riding and third money in steer roping at a rodeo held in 1911. Cheyenne River cowboy Felix LeBeau, meanwhile, refused to participate in rodeos. "I don't care to ride in the rodeos," he stated pragmatically in an interview conducted in 1971. "You got to pay ten dollars to enter and if you win nothing, why then you got nothing. You're just out ten dollars."[30]

While rodeos, to a certain extent, have allowed Indian cowboys an opportunity to gain status among their people, stock-raising ventures on the northern Plains have become increasingly controversial and divisive on several reservations, forcing Indian cowboys to make difficult choices. In recent years, tribal land use practices have placed Indian cowboys and ranchers at odds with traditionalists, who view cattle raising as harmful to the environment and inconsistent with certain tribal traditions. In Yellowstone National Park, for example, the Montana Department of Livestock has killed over one thousand head of bison due to concerns that they may be carrying brucellosis, a contagious disease of bacterial origin. When transmitted to cattle, brucellosis can cause spontaneous abortion of calves. Unconvinced that the bison pose a serious threat, many Native Americans have harshly criticized the slaughter, characterizing the killings as "sacrilegious" and an "American tragedy." But many Indian cowboys and ranchers on the Crow and Fort Belknap reservations in Montana and the Wind River reservation in Wyoming support the Montana Department of Livestock's policy. From their perspective, the bison pose a serious threat to their own cattle and bison herds.[31]

On the Rosebud reservation in South Dakota, meanwhile, prairie dogs have become embroiled in a controversy pitting traditionalists against Lakota cowboys. One of the most ancient rites of the Lakotas is the *yuwipi* ceremony, which is held when someone has a problem and needs help. One of the obligations of the petitioner is to provide sacred food for the ceremony—usually a young dog—which is cut into pieces and then boiled before being eaten by those in attendance. On the Rosebud reservation, however, some traditionalists prefer using prairie dogs for the sacred meal. But prairie dogs have long been viewed as a menace to stock-raisers. Prairie dog towns can contain hundreds of burrows, which can be hazardous should a cow or horse step in them. Prairie dogs also destroy grass. Consequently, Rosebud cowboys have sometimes sought to reduce prairie dog populations by shooting or poisoning the rodents.[32]

Thus, cattle ranching on the northern Plains continues to exert tremendous influence and pressure on native peoples, who in turn have responded in diverse ways. On the one hand, as Peter Iverson has written, Indian involvement in ranching "helped bridge the transition from treaties and agreements" and the early reservation days to the final years of an assimilationist era. "[Indian cowboys] could be like the white men and not be white men. They were finding new ways to remain Indians." But while it has provided Indian cowboys with economic livelihood and a means of achieving status, stock-raising has also forced them to place "immediate needs and desires over older cultural regulatory patterns, shattering both traditional standards of behavior and static white stereotypes of Indians as the 'original conservationists.'"[33] As native stock-raisers on the northern Plains face the future, therefore, they will continue to face difficulties choosing whether they can continue to be both cowboys and Indians.

Notes

1. Peter Iverson, *When Indians Became Cowboys: Native Peoples and Cattle Ranching in the American West* (Norman: University of Oklahoma Press, 1994), 25.

2. Ibid., 9.

3. Joseph Medicine Crow, *From the Heart of Crow Country: The Crow Indians' Own Stories* (New York: Orion Books, 1992), 23, 100-1.

4. James W. Schultz, *Blackfeet and Buffalo: Memories of Life Among the Indians* (Norman: University of Oklahoma Press, 1962), 312; Raymond J. DeMallie, ed., *The Sixth Grandfather: Black Elk's Teachings Given to John G. Neihardt* (Lincoln: University of Nebraska Press, 1984), 315.

5. Russel L. Barsh, "The Substitution of Cattle for Bison on the Great Plains," in Paul A. Olson, ed., *The Struggle for Land: Indigenous Insight and Industrial Empire in the Semiarid World* (Lincoln: University of Nebraska Press, 1990), 104-5.

6. Robert Utley employed the phrase "shadow world" in a PBS documentary on the Carlisle Indian Industrial School entitled *In the White Man's Image*.

7. *Cattle Ranching by Indians on Pine Ridge Reservation, South Dakota*, Report Number 178 (Billings, Montana: United States Department of the Interior, Bureau of Indian Affairs, Missouri Basin Investigations Project, 1964), 3-4; D. C. Poole, *Among the Sioux of Dakota* (New York: D. Van Nostrand, 1881), 95.

8. Sidney J. Tietema, et al., *Indians in Agriculture, Cattle Ranching on the Blackfeet Reservation* (Bozeman, Montana: U.S. Department of the Interior, Bureau of Indian Affairs, Missouri Basin Investigations Project, 1957), 36; Irene Castle McLaughlin, "Colonialism, Cattle, and Class: A Century of Ranching on the Fort Berthold Indian Reservation" (Ph.D. dissertation, Columbia University, 1993), 179-80.

9. Iverson, *When Indians Became Cowboys*, 58-59; Poole, *Among the Sioux of Dakota*, 100.

10. McLaughlin, "Colonialism, Cattle and Class," 192-194; Carolyn Gilman and Mary Jane Schneider, *The Way to Independence: Memories of a Hidatsa Family, 1840-1920* (St. Paul: Minnesota State Historical Society Press, 1987), 17, 41, 91, 243; Gordon MacGregor, *Warriors Without Weapons: A Study of the Society and Personality Development of the Pine Ridge Sioux* (Chicago: University of Chicago Press, 1946), 108.

11. Felix LeBeau, Interview by Steve Plummer, June 2, 1971, interview 0689, transcript, Doris Duke Oral History Collection, Institute of American Indian Studies, University of South Dakota, Vermillion.

12. Joe Yellowhead, Interview by Steve Plummer, June 9, 1971, interview 0695, transcript; Guy Buffalo, Interview by Steve Plummer, June 28, 1971, interview 0704, transcript, Doris Duke Oral History Collection, Institute of American Indian Studies, University of South Dakota, Vermillion.

13. Amos Eagle Boy, Interview by Steve Plummer, August 3, 1971, interview 0761, transcript; Leo Shooter, Interview by Steve Plummer, June 20, 1972, interview 0831, transcript; Leo Little Eagle, Interview by Steve Plummer, June 21, 1972, interview 0832, transcript, Doris Duke Oral History Collection, Institute of American Indian Studies, University of South Dakota, Vermillion.

14. Chaske F. Wicks, Interview by M. Garner, February 16, 1969, interview 0357, transcript, Doris Duke Oral History Collection, Institute of American Indian Studies, University of South Dakota, Vermillion.

15. Gilman and Schneider, *The Way to Independence,* 246.

16. Ibid., 249; McLaughlin, "Colonialism, Cattle, and Class," 182.

17. Helen Parsons Neilson, *What the Cow said to the Calf: Stories and Sketches by Ballie Buck, a Legendary Indian Cowboy* (Gig Harbor, Washington: Red Apple Publishing, 1993), 13, 52-54; McLaughlin, "Colonialism, Cattle, and Class," 168-169.

18. McLaughlin, "Colonialism, Cattle, and Class," 185; Neilson, *What the Cow Said to the Calf,* 74; Bess Adrian, "Life Story of Mr. and Mrs. Sam White Horse," in Paul and Winifred Reutter, eds., *Early Dakota Days: Stories and Pictures of Pioneers, Cowboys, and Indians* (White River, South Dakota: 1962), 233.

19. Tietema, et al., *Indians in Agriculture: Cattle Ranching on the Blackfeet Reservation,* 36. The federal government blamed "poor management, drought, and severe winters" for the decline.

20. Richmond L. Clow, "Cattlemen and Tribal Rights: The Standing Rock Leasing Conflict of 1902," *North Dakota History* 54 (1987): 23-24;

Donald L. Parman, *Indians and the American West in the Twentieth Century* (Bloomington: University of Indiana Press, 1994), 4-5.

21. McLaughlin, "Colonialism, Cattle, and Class," 158; Iverson, *When Indians Became Cowboys*, 35-39, 41-47; Clow, "Cattlemen and Tribal Rights: The Standing Rock Leasing Conflict of 1902," 23-30.

22. Leo Little Eagle, Interview by Steve Plummer, June 21, 1972, interview 0832, transcript, Doris Duke Oral History Collection, Institute of American Indian Studies, University of South Dakota, Vermillion.

23. Tietema, et al., *Indians in Agriculture: Cattle Ranching on the Blackfeet Reservation*, 11; *Cattle Ranching on Pine Ridge Reservation*, 5, 16; *Crow Cattle Ranching Operations, Crow Indian Reservation, Montana* (Bozeman, Montana: Missouri River Basin Investigation Project, Report Number 187), 3.

24. Tietema, et al., *Indians in Agriculture: Cattle Ranching on the Blackfeet Reservation*, 12; *Crow Cattle Ranching Operations, Crow Indian Reservation, Montana*, 1, 29.

25. Sidney J. Tietema, *Indians in Agriculture: Blackfeet and Crow Indian Reservations* (Bozeman, Montana: Missouri River Basin Investigation Project, 1961), 6-7; Ralph E. Ward, et al., *Cattle Ranching on the Crow Reservation* (Bozeman, Montana: Missouri River Basin Investigation Project, 1956, Number 522), 37.

26. *The Dakota Indian Economy* (Brookings, South Dakota: Missouri River Basin Investigations Project, 1963, Bulletin 509), 7; *Crow Cattle Ranching Operations, Crow Indian Reservation, Montana*, 31; *Cattle Ranching on the Blackfeet Reservation*, 40-41.

27. Elizabeth Atwood Lawrence, *Hoofbeats and Society: Studies of Human-Horse Interactions* (Bloomington: Indiana University Press, 1985), 33.

28. Blanche Kaufman, "Reuben Quick Bear," in *Early Dakota Days*, 167-168; Chaske Wicks interview.

29. Neilson, *What the Cow Said to the Calf*, 127-30.

30. Bess Adrian, "David Blue Thunder Story," in *Early Dakota Days*, 19-20; Grover Red Fox interview; Felix LeBeau interview.

31. Peter Ames Carlin, et al., "Buffalo Soldier," *People Weekly* 47 (April 21, 1997): 172-75; Alexander Cockburn, "The Bison-killers," *The Nation* 264 (March 10, 1997); Lorrie Brown, Montana Department of Livestock, phone interview by author, January 29, 1998; Scott McMillion, *Bozeman Chronicle*, phone interview by author, February 2, 1998.

32. John Lame Deer and Richard Erdoes, *Lame Deer: Seeker of Visions* (New York: Washington Square Press, 1972), 172-77; William K. Powers, *Yuwipi: Vision and Experience in Oglala Ritual* (Lincoln: University of

Nebraska Press, 1982), 70; Calvin Iron Shell, Chairman of the Sioux City American Indian Center, interview by author, February 24, 1998.

33. Iverson, *When Indians Became Cowboys,* 84; David Rich Lewis, "Native Americans and the Environment: A Survey of Twentieth Century Issues," *American Indian Quarterly* 19 (Summer 1995).

6 English Cowboy: The Earl of Aylesford in the American West

JIM FENTON

An impeccably dressed Englishman, the Seventh Earl of Aylesford, on a warm summer day in 1883 got off a train at the cattle boomtown of Big Spring on the West Texas frontier. From then to early 1885, a period of some twenty months, the Earl—despite his polish and culture, despite his failed, even laughable efforts as a cattleman—won over the hearts of the raw, mischievous cowboys there, and in the process he became a favorite and popular local legend.

It is difficult to learn the full truth about the Earl's activities at Big Spring, located in Howard County about a hundred miles west of Abilene. But it quickly became clear that he was more than just another odd frontier character to hit town. He probably was assisted in his journey west from New York by none other than the powerful American railroad financier, Jay Gould, who had money invested in the Texas and Pacific Railroad, which had reached Big Spring on April 28, 1881.[1] If Gould did help him, that was to be expected, for the Earl was used to running with important people. And, like Texans, he did things in a big way.

The Earl was a "remittance man," that is, the black sheep of a titled European family who was banished because he was no longer welcome in polite society back home, but provided with a comfortable allowance on which to live. A remittance man was often looked down on, wherever he was exiled.[2]

There were vague rumors that the Earl had left home under pressure. But it was the custom on the frontier, with its assortment of difficult people, not to pry into a man's past. Such curiosity could be dangerous.

The Earl arrived at Big Spring, largely a tent community located at the lower end of the fabled Staked Plains (Llano Estacado), in late

Joseph Heneage Finch, Seventh Earl of Aylesford, ca. 1883, about the time he left England to live briefly in Big Spring, Texas. Courtesy of the UT Institute of Texan Cultures at San Antonio.

August 1883. John Birdwell, a saloon owner, greeted him. Birdwell, a former Texas Ranger and future local sheriff, apparently was asked by Gould's land commissioner, a Dr. Ennis, to help the Earl get settled. Both the Earl and Birdwell were avid hunters, thus providing another link between them.

Aylesford formally introduced himself to Birdwell as the Seventh Earl of Aylesford. Birdwell quickly let him know that such a name in cattle country could get someone hurt. Thereafter, the aristocrat was known as "Judge," a time-honored, unofficial title given to an important local person.

In the Judge's entourage were several close friends. One was the "rollicking, fun-loving" Lord Harry Gordon, as a reporter described him.[3] Another was the Episcopalian clergyman, Bishop Arthur Chichester Burnard. The bishop was the Earl's private secretary, spiritual confidante, and financial advisor. He served also to offset any mischief Gordon might think up for the Earl. Rounding out the group were five servants, including a Von Paussen, the Earl's German cook, and William Benham, his favorite butler.

The Earl also brought with him a railroad car filled with twenty to thirty purebred horses and dozens of valuable hunting dogs, plus "hunting paraphernalia in bewildering numbers."[4] Several weeks later his younger brothers, Daniel Harry and Clement Edward, arrived with additional baggage.

No arrangements had been made to take the Earl and his people to his new home, the Rush Ranch. Thus he sought lodging at one of the town's few wooden buildings and the one public place to get a room, the Cosmopolitan Hotel. But it was full. The situation called for decisive action, and the Earl was up to the task. Truth and legend indicate he bought the building on the spot from Mrs. Emma F. Dugan, without seeing the inside of a single room.[5] At one thousand dollars he paid twice its fair market value. Compounding these acts, the story goes, he returned the hotel to its former owner the next day with but one requirement: she must keep a set of rooms available for him and his people at all times, for those frequent occasions when they visited town.

Not all of this charming tale was accurate. While the basic narration was correct, the facts somehow got blurred and exaggeration crept in. Joe Pickle, a current Big Spring resident and retired editor of the *Big Spring Herald,* dug deeply into the official records of the city of Big Spring and Howard County to learn more fully the specifics of the Englishman's business deals. Pickle was able to find out that the Earl paid four thousand dollars for the Cosmopolitan Hotel, with one thousand dollars cash down. Thus he did not, as legend has it, pay just a thousand dollars cash for the building. And did he give the hotel back to Mrs. Dugan the next day? "Not so," claimed Pickle,

> English thrift was too much ingrained in him, despite extravagances. While he sometimes lost property because he could not pay his debts, he was not guilty of recklessly giving it away.[6]

Moreover, the Earl's financial advisor, Bishop Burnard, likely would have argued against such outrageous generosity. The Earl was indeed a big spender and a poor businessman, but not to the extent of giving away a valuable property. Regardless, it was misstatements and embellishments in the collective memory of Big Spring's earliest settlers that largely accounted for the Earl's legendary status.

Even less is known about the Earl's purchase of the Rush Ranch. It seems it was bought for him by a third party before he arrived at Big Spring. The ranch, located on Wildhorse Creek about twelve miles north of Big Spring, was on choice land. With it came a sizable herd of cattle. The cattle industry on the Great Plains was then the craze. It offered wealthy Easterners in the United States, and aristocrats and businessmen in England and in Scotland, quick profits.

The Rush Ranch consisted of about twenty-five hundred acres of stirrup-high grass and the cattle, whose numbers are unknown. Although the Earl busied himself mainly with "sport and . . . glass,"[7] he considered the ranch operation to be his "cattle farming experiment."[8] He later had R. F. Kennedy, a Scotsman, place a down payment of ten thousand dollars for him on a much larger ranch. This spread, consisting of thirty-seven thousand acres, was located at Monument Spring in the southeastern corner of New Mexico Territory, about a hundred miles west.[9]

Yet, despite living on the Rush Ranch with its herd of cattle and despite having a substantial interest in another ranch, there is no direct evidence that he actually worked cattle. He was, for all purposes, an absentee landowner. Thus he made little impact on the West Texas cattle industry.

The Earl's living accommodations at the Rush Ranch consisted of a large, newly built but plain, unpainted wooden building a story and a half high. The house had eight or nine simple rooms on either side of a central hallway. The walls along the aisle were decorated with so many guns and other hunting gear that the dwelling "bristled like an arsenal."[10] The gun collection was very valuable. One servant's sole job was to look after it. Despite being ordinary, the large house was tastefully decorated with sturdy, classic English furniture and other accessories. Behind the ranch house was a huge barn and lesser buildings, and there was a kennel located on two acres of land enclosed by a barbed wire fence that secured the dogs for hunting.

The Earl's passion for hunting dogs allowed him to make a minor contribution to his adopted country. In 1883, according to canine authority Mercedes Braun, he imported to the United States a "brace" (or pair) of blooded basset hounds.[11] The bassets were the finest available, having been sired by an English dog (named Jupiter) with the model features of the breed. It is commonly believed in kennel circles that the Marquis de Lafayette, a friend of George Washington, first introduced the basset hound to the United States after the Revolutionary War. But the breed was not registered by the American Kennel Club until 1885, so it seems it was not an important species in this country when the Earl brought in his pair. Bassets, short dogs with long bodies, hunt by scent. They track well in rough terrain and have great stamina. The Earl used his to hunt the cottontail rabbits abundant in the buffalo and grama shortgrass fields of the Rush Ranch.

But of all the Earl's acquisitions, it was his purchase of the sizable herd of cattle on the Rush Ranch that must be looked at most closely. This was no doubt the worst of his business deals. The evidence shows he paid forty thousand dollars for the cattle—"range delivery"—or sight unseen.[12] They may well have been a part of his purchase of the Rush Ranch, acquired through a go-between, even before his arrival at Big Spring. Thus he likely was given a short count. But he was not the first or last outside investor to be bested in a cow deal with an unschooled but range-savvy West Texas cattleman. And the fact that only twenty months later the herd was sold at auction for just $750.00 lends credence to the possibility that he got the short end of this deal. Even with declining prices then, this was a precipitous drop in value.

Another of the Earl's acquisitions requires scrutiny. Once he tried to get Birdwell to go hunting with him. The Texan begged off, citing pressing business at his saloon, but the Earl would not be denied. Legend has it he bought the place,[13] which sold mainly "'red liquor,' and a few cigars,"[14] for six thousand dollars and, well into the next day, helped give away the liquor. The quotable Texas historian J. Evetts Haley called this act of generosity "an early civic philanthropy of which the populace approved."[15] Lore had it, too, that when the hunting trip was over, the Earl gave the saloon back to Birdwell. But Pickle's research again clarified the point: the Earl did buy the saloon, but he did not give it back to the original owner.[16]

The Earl and Birdwell became good friends. The Englishman put the Texan on four acres of land in northwest Big Spring, where the latter broke wild horses to supplement the Rush Ranch stable. Moreover, Birdwell felt comfortable enough with his wealthy, royal host to introduce him occasionally to friends as the "Lord God of Aylesford."[17] The Earl seemed to enjoy the kidding, recalling that his drinking mates back in England had similarly mocked him.

The story about how the Earl came to own a meat market is a quaint one—and true. As an Englishman, the Earl's palate desired mutton. But this was a meat little valued, and largely unavailable, in Texas beef country. So the walled-up tent at Big Spring that served as the butcher shop could not fully accommodate the Earl's needs. He accordingly built a shop of his own at 121 Main Street. It was Big Spring's first permanent building and stands to this day. The Earl's German cook, Von Paussen, who best knew his Lord's taste in meat, apparently supervised the business.[18]

The Earl was an active man, regularly traveling to Big Spring and to the hunting fields nearby. On the trips, he dressed in a style alien to the local citizenry. For those rare occasions when he attended a social function in town, he wore formal black. On hunting trips, he wore gray corduroys and expensive cravats or scarves, the apparel of the English nobility on fox hunts. But it was likely his small, hornless English saddle, once compared to a "huge postage stamp," that drew most of the cowboys' laughter.[19]

Birdwell and another prominent local citizen, D. C. Earnest, a neighbor and friend, periodically went on hunting trips with the Earl; they served as guides. Once, after the Earl frightened off several pronghorns by firing prematurely, Earnest taught him how to hunt Texas style and success followed. During their next hunt, in a snowstorm, the two hunters crept to within two hundred feet downwind of twenty pronghorns and fired away. The Earl killed three; Earnest one. The Texan had never "seen such an enthusiastic hunter."[20] The Earl had learned by then that hunting the game native to the southern Great Plains at Big Spring was quite different from hunting fox in England and big game in Africa and Asia. The Earl was not accustomed to the degree of stealth and patience required by local hunting.

On one hunt, the Earl followed a quaint custom that made Earnest cringe. After his servant broke the ice on a nearby stream, the nobleman took his morning bath "with as much nonchalance and apparent pleasure as if he had been in a warm bathroom."[21]

The Earl loved liquor, mostly expensive scotch and bourbon, as much as he loved hunting. He reputedly drank a half gallon of whiskey each day, occasionally following it with a fifth or two of gin. A reporter for the *Chicago Inter-Ocean* recalled seeing a "pile of empty bottles" outside the Rush Ranch house "as big as a haystack."[22] But this claim assumed too much, for others on the ranch, in an age of hard drinking, surely drank their share of his alcohol.

The Earl took his whiskey straight and held it well, seldom being seen fully drunk. But Jeff Davis Milton, a young Texas Ranger destined to become a renowned Southwestern lawman, recalled that the Earl and Birdwell occasionally "got drunk together." He once saw them "in a gentlemanly sort of way sitting in the shallow Colorado River, sobering up—with a bottle between them."[23]

In time, the Earl complained that the cowboys would not let him be one of the boys. Apparently they played pranks on him and his people, as cowboys on the town sometimes enjoyed such mischief. He found a sympathetic listener in D. C. Earnest. Earnest told him to wait until the upcoming spring roundup to gain the cowboys' acceptance. But in the meantime, the Earl should get rid of his fancy riding clothes and his tack and

Go down to Colorado [City] and let Pete Snyder sell you a
Stetson hat, spend $5 for a pair of Petmecky spurs, and
don't forget to let that bowlegged Dutchman, Fred Myers,
make you a pair of Texas boots. Hide that 'muley' thing you
call a saddle and spend $50 for a Texas saddle with a Cali-
fornia tree, and be sure the saddle has Beef Buyers saddle
pockets.[24]

When the Earl asked why the saddle pockets were called Beef
Buyers, Earnest told him that the previous year a cattle buyer working for a large stockyard had ridden out to appraise the local steer
crop, seated on a saddle with big side pockets covered with black
hair. Thereafter, all the cowboys had to have Beef Buyers pockets on
their saddles. "You know," Earnest concluded, "us punchers are hell
on style."[25]

After completing the purchases, the Earl, then suitably attired,
was in April 1884 formally inducted into the local cattle fraternity.
About one hundred cattle bosses and their cow punchers, men on
the ranches within the watershed of the Clear Fork of the Red River
and the Concho River, met on Morgan Creek in Mitchell County for
the annual spring roundup when all the newborn calves would be
branded. There another of the Earl's new friends, Columbus "Lump"
Mooney, wagon boss of C. C. Slaughter's famous Long S Ranch, officially introduced the assembled men to the Earl of Aylesford from
London, England. Mooney told them, "The Earl bought the Rush
Ranch. Treat him right; he is a friend of mine."[26] Besides, he had a lot
of tobacco.

But though the Earl's status greatly improved, the roundup initially failed. Before the work could be completed, the grateful nobleman rolled out a ten-gallon keg of whiskey and the hands got so
drunk that their bosses had to take them home to sober up.

Regardless, the Earl was now a fully accepted member of the
West Texas cowboy subculture. This was no small feat. For cowboys
tended to hold in deep disdain, and sometimes treat badly, a person
who seemed odd or pretentious. However, with his new title,
"Judge," with his stylish cowboy clothing and gear, with his favorable introduction at roundup by a respected cattle boss, and with

his generosity with his liquor, he at last turned the cowboys in his favor. The townspeople followed suit.

Each of these factors helped the Earl win the cowboys' friendship. Yet it was his willingness to share his liquor with them that was the clincher. For, as Shine Philips noted, "anybody who gives away enough can be popular."[27]

Thereafter, the *Times* of London reported, he was "exceedingly popular with the cowboys,"[28] whom he occasionally drank with at the bars in town. He once bought a group of them seventeen rounds of drinks. His rustic friends responded with the loyalty typical of the American cowboy, supposedly even getting into several shoot-outs at Big Spring over some fancied slight of him. Indeed, their support for him was so complete that, a newsman for the *Chicago Inter-Ocean* wrote, they "would spill their blood in his behalf as readily as he opens his bottle for them."[29] But there was some West Texas "wind" in this claim, too, because Big Spring, although a raw setting, saw very little gunplay.

The Earl had other endearing qualities. Beneath a patina of oddness, he was a cultured, polite person. A witty, charming man, he was good company. Among the cowboys, a people who placed great emphasis on friendship, these were valuable traits. Standing just over six feet tall and being handsome and powerful, his classic looks posed no problems for him. Nor did his scant interest in the local women, in whose presence he was always polite, detract from his masculinity. He was a man's man.

But he was also a troubled man, for, unknown locally, he had serious financial and medical problems. The first indication of these trends came in late 1884, when the guardian of his estate back in England, upon hearing of his wasteful spending at Big Spring, substantially reduced his fifty-thousand-dollar yearly allowance. Thus the aristocrat returned to England that winter and sold his country estate near Maidenhead, outside London, and borrowed at least thirty thousand dollars from Daniel Harry, the older of his two younger brothers. Moreover, while returning from watching the Derby races at Epsom, he got into a scuffle at a railroad station and broke a leg.[30] He had not quite recovered from this injury when he returned to Big Spring.

His health, however, had begun to fail even before he left Texas. He was being treated by his physician there, Dr. W. T. Standiford, about eight times a month during the first half of 1884. Thus, despite his seemingly robust health, he was actually a very sick man.

The end came abruptly. In late December 1884, after hosting a rousing Christmas party, the Earl took to his bed. He had a different doctor now, J. C. Utter, who went more often to the sick man's residence, which for convenience was in the Cosmopolitan Hotel. Utter made two visits daily, charging two dollars each. On January 5 and 6, 1885, he remained from early in the day until late at night, charging ten dollars a visit.

Treatment continued, with Utter then in full attendance. On the thirteenth, the Earl got up and fully dressed, then ate the largest meal he had eaten in several weeks and, conversing freely with friends as they played euchre, appeared well on his way to recovery. At 9:30 that evening, however, he unexpectedly died.

But for a man whose stay at Big Spring was clouded in mystery and who had so charmed townsfolk and cattle people, such an ordinary death invited embellishment. Thus another account claimed that upon learning his time had come, he went to bed, covered himself with a blanket, told his friends "So long, boys," and passed on. Another version was told by a fellow Englishman and early West Texas cattleman, Frank Collinson. Collinson, a "windy" who could stretch the truth with the best, related a more colorful account of the Earl's final moments. Realizing that death was imminent, he told his butler to bring him "another good long drink of good American whiskey, and we'll let her rip." The Earl "took the drink and fell back dead."[31] A remarkable man in life, he had to die in a more dramatic manner than was the reality. Such accounts are typical of the rich body of lore that has survived the man's stay at Big Spring.

Dr. Utter, who brought in the prominent undertaker from Dallas, Ed C. Smith, charged twenty-five dollars to have the Earl's remains embalmed for the return trip to New York City and subsequently to his ancestral home in England. The body left New York on an English war vessel, there at the personal request of the Earl's friend of bygone days, the Prince of Wales, who would become England's next monarch—King Edward VII.[32] The Earl was buried in the family vault at the hamlet of Aylesford, County Kent, just out-side of London. Contrary to the legend at Big Spring, the Earl's body was not returned home in a barrel of brandy, though that might have been a fitting end to his inebriate life.[33]

Newspaper accounts listed the cause of his death as "dropsy and hardening of the liver" and as "inflammation of the bowels."[34] And the *Fort Worth Gazette* castigated the Earl for the "foolish excesses" that shortened his life.[35] He died at just thirty-six, though he looked much older.

At the pioneer West Texas cattle town of Big Spring, England's Seventh Earl of Aylesford spent almost two years in the 1880s. Although he was a failure at the cattle trade, he won wide acceptance among the townsfolk and cattlemen, especially the cowboys. The people of Big Spring now honor the memory of their mysterious, eccentric neighbor of yesterday, naming a street for him. But they misspelled his name "Aylford" on the street sign. That was, perhaps, a fitting way to keep alive the memory of this failed yet successful aristocrat, the legendary figure who became a friend of the American cowboys.

Notes

1. *Chicago Inter-Ocean*, ca. 1884, cited in John R. Hutto, *Howard County in the Making* (n.p., 1938), unnumbered page. For a brief sketch of Gould's life, see Allen Johnson and Dumas Malone, eds., vol. VII, *Dictionary of American Biography* (New York: Charles Scribner's Sons, 1931), 18.

2. Hedley Donovan, ed., "The Cowboys," *Time-Life Books*, 25 volumes (New York: Time-Life, 1993), 18.

3. *El Paso Times*, September 11, 1912, 4.

4. Frank X. Tolbert, *Dallas Morning News*, November 1974.

5. Ed Fisher, "Earl of Aylesford," *Permian Historical Annual* V (December 1965), 41-42.

6. Joe Pickle, *Gettin' Started: Howard County's First 25 Years* (Big Spring, Texas: Heritage Museum, 1980), 84, 282.

7. Mary Price Mosley, *"Little Texas" Beginnings in Southeastern New Mexico* (n.p., 1973), 9.

8. *Times* (London), January 15, 1885, 6d.

9. Mosley, *"Little Texas" Beginnings*, 9.

10. Shine Philips, *Big Spring: The Casual Biography of a Prairie Town* (New York: Prentice-Hall, Inc., 1946), 170.

11. Mercedes Braun, *The Complete Basset Hound* (New York: Howell Book House, 1967), 1, 22-23.

12. John Hendrix, *If I Can Do It On Horseback: A Cow-Country Sketchbook* (Austin: University of Texas Press, 1964), 45-46. For an example of how another Englishman lost big in the cattle trade, see Ellis Douthit, "Some Experiences of a West Texas Lawyer," *West Texas Historical Association Year Book* VIII (October 1942), 41.

13. Mac B. McKinnon, *Lore and Legend: A Compilation of Documents Depicting the History of Colorado City and Mitchell County* (n.p., n.d.), 46; Fisher, "Earl of Aylesford," 40-41.

14. *Fort Griffin Echo*, February 19, 1881, quoted in J. Evetts Haley, *Jeff Milton: A Good Man with a Gun* (Norman: University of Oklahoma Press, 1949), 55.

15. Haley, *Jeff Milton*, 46.

16. Pickle, *Gettin' Started*, 76.

17. Ellis Douthit, "Some Experiences of a West Texas Lawyer," 40.

18. Pickle, *Gettin' Started*, 78.

19. Mrs. T. Lee Jones and Rupert N. Richardson, "Colorado City, the Cattleman's Capital," *West Texas Historical Association Year Book* XIX (October 1943), 46.

20. McKinnon, *Lore and Legend*, 52.

21. *El Paso Times*, September 11, 1912, 4.

22. *Chicago Inter-Ocean*, c. 1884, cited in Fisher, "Earl of Aylesford," 42.

23. Haley, *Jeff Milton*, 46. Milton's life is sketched in Dan L. Thrapp, ed., *Encyclopedia of Frontier Biography*, vol. 2 (Lincoln: University of Nebraska Press, 1988), 992. For his full biography, see Haley, *Jeff Milton*.

24. McKinnon, *Lore and Legend*, 40.

25. Ibid., 40-41.

26. Jones, "Colorado City," 44. Columbus "Lump" Mooney, later, in the fall of 1884, started his own spread, the Cross Tie Ranch, about eight miles southeast of Midland near the old Salt Lake. The first foreman was Dan Watson. See Betty Wingate Orbeck, "Walter Cochran's Memoirs of Early Day Cattlemen," *Permian Historical Annual* I (August 1961), 38. For the biography of Slaughter, see David J. Murrah, *C. C. Slaughter: Rancher, Banker, Baptist* (Austin: University of Texas Press, 1981).

27. Philips, *Big Spring*, 171.

28. *Times (London)*, January 15, 1885, 6d.

29. *Chicago Inter-Ocean*, cited in Fisher, "Earl of Aylesford," 42.

30. *Times (London)*, January 15, 1885, 6d.

31. Tolbert, *Dallas Morning News,* 1974; Frank Collinson, *Life in the Saddle,* edited and arranged by Mary Whatly Clark (Norman: University of Oklahoma Press, original paperback, 1963), 81.

32. For the life of the Prince of Wales, see Giles St. Aubyn, *Edward VII: Prince and King* (New York: Atheneum, 1979).

33. Fisher, "Earl of Aylesford," 42; Probate file 11, 1885, Howard County, Texas.

34. *New York Times,* January 5, 1885; *Texas Livestock Journal,* January 17, 1885.

35. *Fort Worth Gazette,* January 29, 1885.

7 The Cowboy Strike of 1883

ROBERT E. ZEIGLER

In the two decades that followed the Civil War, the open-range cattle industry dominated the Great Plains, and then died and was replaced by enclosed-range ranching and stock farming. In Texas, the movement to enclose the range began in earnest in the early 1880s and was completed by 1890. During this transitional period, there was also a great upsurge in European and Eastern investment in cattle, bringing owners who viewed ranching primarily as a profit-making enterprise rather than as a way of life. Cattle raising, like other businesses in the Gilded Age, was becoming a corporate affair.[1] Growing corporate activity resulted in problems for the cowboy similar to those faced by workers in other industries during the same period.[2]

In addition to having problems in common with other workers, the cowboy, in some rare instances, reacted by forming cooperatives to protect himself from the whims of employers, and on occasion he went out on strike.[3] One such incident was an 1883 cowboy strike in the Texas Panhandle. A group of dissatisfied hands demanded wage increases and launched a protest that lasted two and a half months before ending in failure.

The strike has been treated by other students, who mention that ownership changes that were occurring in the Panhandle region helped to bring on the conflict. These students do not, however, consider the causes of the strike or of its failure in detail. There is a need to examine such factors in the hope that such an examination will help to put the cowhand in better perspective as a working man.[4] Such a study substantiates Kenneth Porter's assertion that the infrequency of cowboy strikes cannot be solely attributed to the

Previously published in *West Texas Historical Association Year Book,* 47 (1971): pp. 32-46.

independent, carefree, and free-spirited nature of the cowhand.[5] Indeed a realistic examination of the various determinants leading to the outbreak of the Panhandle walkout creates serious doubts as to the validity of any romanticized picture of the cowboy.[6]

There were five ranches involved in the 1883 strike—the LIT, T-Anchor, LE, LS, and the LX.[7] All were controlled by large corporations or by individuals whose actions indicate an interest in ranching largely as a speculative venture for quick profit. The LIT was owned by a Scottish syndicate, the Prairie Cattle Company.[8] The T-Anchor was controlled by the Gunter-Munson Company, which was involved in land speculation as well as ranching.[9] The LE was owned by the American-based Reynolds Land and Cattle Company. This firm had some financial backing from John M. Bond of the Alliance Trust, a Scottish Company.[10] The Lee Scott Company owned the LS.[11] The LX brand was brought to the Panhandle in 1877 by two Bostonians, W. H. Bates and David T. Beals, who sold out in 1884 to the American Pastoral Company. All these ranches grew rapidly. Typical of the growth was the LX. After an initial concentration on cattle buying, LX representatives began in 1882 to accumulate land and by 1885 had purchased 123,680 acres.[12] The large ranches had, by 1883, also established the practice of fencing as a means of safeguarding their lands and protecting their cattle.

This trend in the Panhandle ranch industry in the 1880s toward corporate activity and large land holdings had a tremendous effect on the status of the cowboy. The man engaged in working cattle had traditionally viewed a brand mark as the demonstration that property concentrated itself in herds. The cowboy had been devoted to his job of protecting the cattle and there had developed a high sense of group solidarity among the hands. Also, the owners had usually been in constant touch with their employees, thereby establishing the feeling of a common interest between workers and boss. As holdings became larger it proved much more difficult for a cowboy to feel personal fealty to the new symbol of property concentration, the fenced range and the corporation.[13]

As Tascosa Sheriff Jim East explained in 1884, loyalty was breaking down because the cow business is not what it used to be. You take such men as John Chisum or Charley Goodnight. They were real people. . . . Their cowboys would have died in the saddle rather than have complained. See what we have now; a bunch of organized companies. Some of them are foreign and have costly managers and bookkeepers who live on and drink the best stuff money can buy and call their help cow servants.[14]

Not only was there a breakdown of loyalty, there was a definite feeling of resentment toward those outsiders who, out of ignorance or arrogance, failed to recognize that in ranch country titles and inherited wealth meant little, and status had to be achieved.[15] Thus John McNab, a director of the Spur Ranch who insisted that he be driven around his Texas holdings in a buggy with an umbrella to shield him from the sun, very nearly had his hat and buggy shot full of holes by disdainful ranch hands. Likewise Mrs. John Adair, the wife of the eastern partner of the Panhandle rancher Charles Goodnight, had to be protected from a kidnapping planned by sensitive westerners.[16]

This feeling of alienation on the part of the cowboy was, according to Lewis Atherton, intensified by the establishment of stockmen's organizations such as the Panhandle Stock Association, which was formed in 1881.[17] These groups consisted of owners and managers, not of hands. While the meetings of stockmen's associations did not deal with such matters as wages or working conditions, the fact that cowboys were not represented helped to break down the common bond between employer and employee.[18]

The large companies made other changes that adversely affected the position of the small owner as well as the common hand. In the 1870s the ambitious Texas cowboy could become a small owner by taking part of his pay in calves and by branding mavericks (unbranded calves).[19] He was usually allowed to pasture his cattle along with his employer's. As alien and unsympathetic ranchers moved into the Panhandle region, these practices were eliminated. The cattle baron was interested in expanding his own holdings and refused to help in elevating his employees to the status of owner.[20]

Additionally, rising land prices increased the difficulty faced by the hand who desired to become a landowner.[21]

Moreover, working conditions were not improving, even though it was evident the cattle industry was booming.[22] The Panhandle cowhand's work consisted primarily of riding fence line, branding cows, and doctoring sick animals. He usually slept in a dugout or tent, and lacking dishes, often ate out of a common pot. His life was lonely and hard, wholly lacking in glamour and romance.[23] The best-known aspect of ranch work, the roundup, was held twice a year, in September and April, and required a relatively large workforce. Once the roundup was completed, half to three-fourths of the men were discharged, each employer keeping only a small group of regular workers to man line camps and to perform assorted menial tasks around the ranch.[24] The hand's pay varied according to his dependability and skill. An ordinary ranch worker usually made from thirty to forty dollars a month, a top hand received forty to forty-five, and a wagon boss might draw as much as $125 monthly.[25] The pay was definitely earned, for during the busy roundup periods the cowboy worked a grueling one-hundred-five-hour week.[26]

This wage scale compares favorably with that of other workers. For a sixty- to seventy-two-hour week the average industrial employee in Texas earned only twenty-three dollars monthly. In counties that were actively engaged in manufacturing the pay was thirty-three dollars a month. Thus a cowboy who was a permanent and not just a roundup hand earned a relatively good wage.[27]

The lack of year-round demand for employees meant, however, that cowboys were essentially a seasonal workforce with little job security. Adding to the insecurity was the existence of a ranch labor surplus in Texas during the 1880s.[28] Also working against security of employment was the very nature of ranch work itself. While requiring a knowledge of stock and a measure of horsemanship, the job was not so skilled as to be exclusive. This is evidenced by the lack of any effective division of labor and by the fact that employers placed a higher premium on loyalty and reliability than on any particular skill.[29] The tradition of loyalty, combined with the oversupply of labor, made protest on the part of the hand who was fortunate enough to have a steady job both difficult and hazardous. In spite of the difficulty and hazard involved, some Panhandle cowboys did decide to use the only real weapon the working man possessed—the strike.

In late February or early March of 1883,[30] the LIT, the LS, and the LX had "floating outfits," which followed drift cattle. The wagon bosses were Waddy Peacock for the LIT, Tom Harris for the LS, and Roy Griffin for the LX. The three crews were camped at the LS supply depot, near the mouth of the Frio Creek and east of the present location of Hereford, Texas.[31] These men, distressed by the changing conditions in the ranching industry, drew up an ultimatum to be submitted to ranch owners. They demanded a raise in pay to fifty dollars a month for regular hands and cooks, and seventy-five for a range boss. The men set March 31 as their strike date and ended their declaration with the vague warning that "anyone violating the above obligations shall suffer the consequences."[32] The ultimatum was signed by twenty-four discontented hands.[33]

The real leader of the strike was Tom Harris. In addition to being a wagon boss he was the owner of a small herd of cattle, a man of ability who was generally respected.[34] Indeed, most of the original strikers were a stable group of small owners or permanent hands. These men, working in a relatively small area the year round, would obviously be most likely to organize. Also, they were the ones most affected by the increasing difficulty of economic advancement. They hoped to improve their worsening position through higher wages. It is doubtful that the hands accepted their wage earner status as permanent. Money represented a means of becoming an independent entrepreneur, a dream that the cowhands still clung to and that urban workers were only beginning to abandon.[35] Also, higher pay would appeal to all hands and was the concession that the owners were most likely to grant. Therefore, while the cowboys were being threatened by the new business developments, they did not try to halt progress; they simply followed the example of other American workers and asked to share its fruits.

This hard core group of organizers hoped, of course, to unite a sufficient number of hands to force the owners into submission, and various efforts were made in this direction. Harris and his twenty-three companions established a small fund to provide financial aid

A. M. Britton (center, with white hat), and H. H. Campbell (seated next to him on rolled-up tent) of the Matador Ranch with some of their men at a Matador cattle camp, ca. 1883. Britton rode around the ranch in a buggy. Courtesy of Southwest Collection, Texas Tech University. SWCPC File no. 64(e), Envelope 1.

for needy strikers and attempted to convince all cowboys in the area of the five ranches to refuse to work for less than the hoped-for scale.[36] The success of these organizational efforts was limited. Reports on the number involved in the strike vary from 325 down to a handful of twenty-five or thirty.[37] Available evidence indicates that the number involved was never stable; rather it changed as hands joined and deserted the organizers. It is possible that at one time three hundred twenty-five men were involved.[38]

Regardless of the size of the strike, it did pose enough of a threat to cause both ranchers and non-striking hands to fear violence. The strikers had, of course, threatened in their ultimatum that "anyone violating the above obligations [wage demands] shall suffer the consequences."[39] This rather vague statement could well have meant that coercive action would be taken against strike breakers or ranchers—or both. Actually, there were only isolated attempts to intimidate non-complying hands. Individuals argued over the merits of the strike, and on one occasion strike leader Harris warned Kid Dobbs, who had taken a job on the LS, "If you want to keep a whole hide, and know what's good for you, quit the LS at once."[40] Dobbs refused and Harris, due to judiciousness or timidity, did not follow through on his threat. Indeed, there are no reports of injury, and it appears that organized sanctions of hands who violated strike rules were not seriously considered.[41]

Ranchers possibly had more to fear than did non-striking workers. Newspapers reported that the strikers were planning fence burnings, attacks on ranchers, and indiscriminate killing of cattle.[42] The same reports warned, however, that a company of the Texas Rangers under Lieutenant John Hoffer was camped and ready for action in nearby Mobeetie.[43] Since no attacks were made on cattlemen, extreme action was unnecessary. There probably were some threats made by overzealous hands, but Harris in an April 25 letter to the *Texas Live Stock Journal*, denied any plan of organized violence. Harris wrote:

> I will say that it is not the intention of the cowboys to resort to any violence or unlawful acts to get adequate compensation for their services, but to do so by all fair and legal means in their power.[44]

In spite of the absence of violence and the presence of Texas Rangers, Jules Gunter, a T-Anchor owner, prepared for trouble. He had the promise of additional men from Charles Goodnight should it be necessary. Gunter, however, seemed quite able and willing to punish unruly hands. He and some of his non-striking men filled a nail keg with horseshoes and dynamite, connected a fuse, and placed this homemade land mine near a storehouse where it was thought striking cowboys might take refuge to fire upon the house. A strike delegation did come to the T-Anchor to discuss demands but nothing more serious than vehement insults and loud profanity occurred.[45]

This sort of preparation was the exception. The ranchers found means of dealing with the strikers that were less eruptive and more effective than dynamite. The T-Anchor, after its initial preparation, merely replaced striking employees.[46] On the LS, L. E. McAllister, the foreman, offered the hands forty dollars a month. It was refused. W. M. D. Lee, the owner, came in from Leavenworth, Kansas, and reprimanded McAllister for not meeting the strikers' demands. Lee's concern, however, was not with the welfare of the men; instead, he reportedly felt a wage increase would keep the cowboys at work until he could secure new ones at the old prices.[47] Lee did hold a conference with Harris, offered him an opportunity to keep his hundred-a-month job, and promised to pay fifty dollars to all top hands recommended by Harris if they would remain on the job. The organizer refused and was promptly fired, along with every striker on the LS payroll.[48] In the case of the LIT, thirty-five dollars a month for regular hands and a minimum of sixty-five for wagon bosses was the first and only offer made.[49] The LE manager refused the strikers' demands and discharged his hands.

The ranchers' general attitude toward the strike was one of disbelief, a feeling that their men would not refuse to work. If they did, they should be fired. This attitude did not, of course, prevent at least two of the owners, regardless of their motives, from offering some concessions, which were quickly refused by the strikers. Harris and his followers seemed determined to bring all the ranches involved to terms; they were not willing to accept a compromise from only two. This decision was obviously an error; the cowboys were in no position to force further concessions. Moreover, a partial victory might have made organization more attractive to other hands; defeat merely discredited the entire effort.

Press opinion tended to favor the owners.[50] The attitude was that the hands had a right to ask for the wages they wanted and even had a right to resort to group action. There was general agreement, however, to the qualification expressed by the *Texas Live Stock Journal*. The *Journal* argued that some cowboys were worth "almost any money as *faithful servants*"[51] (italics added) and that these cowboys were entitled to all ranchmen could afford to pay.[52] Presumably the owners were thought to be the best and fairest judges of what they could afford to pay.

The news reports spoke of violence being threatened by the hands, referred to Harris as "bold and bad," and viewed the availability of state forces with obvious relief. Yet there was no criticism of the owners and no hint that the hands had just complaints. The consensus was that the employee should seek the goodwill of his employer and that any improvement in pay or conditions should be left to the discretion of the owner.[53]

A rare bit of objectivity was expressed by the *Trinidad Weekly Advertiser*. After reporting that the hands were proposing "to burn the ranches, confiscate the cattle, and kill the owners"[54] the *Advertiser* concluded, "an ordinary cowboy is as explosive as a nitroglycerin bomb, and a good deal more dangerous. We shall watch the war with interest, not caring much which side whips or gets whipped."[55]

The ranchers, even while negotiating, expressed confidence in their ability to secure enough new workers to break the strike and, acting on this confidence, made no change in roundup plans.[56] However, at least one ranch, the T-Anchor, found it necessary to pay new hands fifty dollars a month on the understanding that a twenty-dollar wage cut was a distinct possibility.[57] This qualified submission bears out newspaper reports, which stressed the seriousness of the strike. It also explains a report by the United States Commissioner of Labor that categorized the walkout as a success.[58] Any success was somewhat short-lived—the higher wage lasted only a month.[59]

The owners certainly had the advantage. Only a few hands initiated the strike, and with a transient labor force any type of sustained effort was quite difficult. Moreover, even local cowboys were not unanimous in their views on the revolt, some remaining loyal to the owners rather than joining the walkout. Also, a serious lack of discipline existed among the boycotters. Contemporary accounts indicate that many strikers went to Tascosa, drank up their funds, and then decided to go back to work. Indeed, the pleasures of a town that boasted hospitable saloons and the entertainments of such ladies as "Rocking Chair Emma" were enticing enough to break any strike.[60] The owners' practice of firing recalcitrant cowboys and of hiring those who were sufficiently dutiful certainly helped to break the spirit of the strikers.[61] In the absence of strong class consciousness, effective organization, and sufficient funds, the temptation to seek the owners' favor was understandably strong.

Thus the actual strike failed after approximately two and a half months.[62] This, however, was not the end of the episode. Following the strike, the Panhandle region was plagued with an outbreak of rustling that most contemporary accounts link with the initial revolt. While it is not true that all strikers became thieves, there is evidence that at least some of the discontented hands, after failing in legitimate labor action, gave vent to their frustration in an illegal manner.[63]

The Panhandle Stock Association, in an effort to stop rustling, in July 1883 officially adopted the practice of blackballing any man fired because of "complicity in any illegal branding."[64] Attempts to bring rustlers to trial failed because, according to the Oldham County Grand Jury of 1884, even though crime was prevalent, investigation revealed:

> in some instances that witnesses either from fear or moral perpetude [sic] are exceedingly loath to give any information and from this cause our investigations have not been so satisfactory to ourselves as could be desired.[65]

The Stock Association also hired Pat Garrett to organize a force to stop the thieves, and began to assign inspectors the job of watching markets and trails for stolen cattle.[66] These actions, combined with the election of responsible public officials, made rustling hazardous.[67] Thus, if strikers were involved in cattle stealing, they soon drifted to other ranges or to other occupations.

Even though they failed to attain their objective, the striking hands had responded to the changes brought by incorporation in a manner similar to workers in other areas and in other industries. Cowboys were, to be sure, individualistic and undisciplined—traits that certainly contributed to the walkout's failure. However, the individualism of American workers was not restricted to cowhands; workers in other sections and in other industries also experienced organizational difficulties. Nor does the strike prove that cowboys were more naturally anti-corporation than were other groups in American society.[68] Instead, the hands, like many other American workers, opposed some aspects of corporate activity, but were nonetheless interested in sharing in the economic benefits of a growing industrial society. Higher wages could buy the status and provide the opportunity to advance that the large owners were destroying. However, it must be remembered that the organizers, as permanent employees, were the elite of the workers and as such are not representative of all hands. That all cowboys did not share the strikers' aspirations is partially proven by the failure of the strike.

It is also true that all cowboys were not agreed on the effectiveness of strikes as a method of obtaining benefits. A hard-working hand, by remaining loyal to the boss, could hold his job. The power of the owners, and the labor surplus, made participation in a strike a shortcut to unemployment. These conditions, and not merely the unique character of the cowboy, served to split the ranks of the hands, and were instrumental in making organization difficult and failure imminent. The ranch worker was truly a victim of the progress that was changing the complexion of the cattle business. He was losing his traditional position, yet he was almost powerless to improve his situation by labor action. In this one instance, and probably others, the hand did not effectively organize not simply because he was more independent than other workers, but rather because he was unable to overcome the obstacles thrown up by the conditions that prevailed in the cattle industry.

Notes

. 1. Ray Allen Billington, *Westward Expansion: A History of the American Frontier* (New York, 1967), 684-85; Rupert N. Richardson, Ernest Wallace, and Adrian N. Anderson, *Texas: The Lone Star State* (Englewood Cliffs, New Jersey 1970), 263-66.

2. See Ruth Allen, *Chapters in the History of Organized Labor in Texas* (Austin, 1941), 33-42.

3. See John Clay, *My Life on the Range* (Norman, 1962), 123, 135; Dulcie Sullivan, *The L S Brand: The Story of a Panhandle Ranch* (Austin, 1968), 69; Kenneth W. Porter, "Negro Labor in the Western Cattle Industry," *Labor History* X (Summer, 1969): 364-65; and Clifford P. Westermeier, compiler and editor, *Trailing the Cowboy: His Life and Lore as Told by Frontier Journalists* (Caldwell, Idaho, 1955), 131, 135, for references to other strikes and to the cooperative efforts. The following newspaper citations were taken from the Westermeier collection—*Texas Live Stock Journal, Caldwell Commercial, Trinidad Daily Advertiser, Fort Collins Courier,*

Trinidad Weekly Advertiser. The inaccessibility of the originals made this use necessary. Where it was possible, Mr. Westermeier's reproductions were checked, and they always proved to be correct.

4. For other, earlier accounts of the strike, see Ruth Allen, *Chapters in the History of Organized Labor in Texas;* John L. McCarty, *Maverick Town: The Story of Old Tascosa* (Norman, 1946). For more recent accounts, see Donald F. Schofield, *Indians, Cattle, Ships, and Oil: The Story of M. D. Lee* (Austin: University of Texas Press, 1985), passim; and James R. Gober and B. Byron Price, eds., *Cowboy Justice* (Lubbock: Texas Tech University Press, 1997), especially pp. 48-51.

5. Porter, "Negro Labor in the Western Cattle Industry," *Labor History,* 364.

6. It is necessary to keep in mind that this strike involved ranch hands, not drovers. Drovers might work on ranches between drives as roundup hands, but there were actually three classes of cowboys: regular hands, roundup hands, and drovers.

7. This information is based partially on Ruth Allen, *Chapters in the History of Organized Labor in Texas.* She mentions a list of ranches but gives no citation. Also, John L. McCarty, "The History of Tascosa, Texas" (M.A. thesis, West Texas State Teachers' College, 1945) attributes the strike to the above mentioned ranches. His information was obtained from individuals who were witnesses to the strike. However, an official government report says seven ranches were involved, but does not name them. Where possible, the list has been verified by checking names with lists of ranch workers.

8. Walter P. Webb and H. Bailey Carroll, eds., *Handbook of Texas,* 2 vols. (Austin, 1952), II, 1.

9. Ibid., 250, 686, 748-49; Laura V. Hamner, *Short Grass and Longhorns* (Norman, 1943), 174-75.

10. Harley True Burton, *A History of the J. A. Ranch* (New York, 1966), 28; Hamner, *Short Grass and Longhorns,* 168.

11. Sullivan, *The L S Brand: The Story of a Panhandle Ranch,* 27-33.

12. Webb and Carroll, *Handbook of Texas,* II, 1.

13. Lewis Atherton, *The Cattle Kings* (Bloomington, Texas: 1961), 181.

14. Colonel Jack Potter, *The Lead Steer* (Clayton, New Mexico, 1939), 32.

15. Atherton, *The Cattle Kings,* 120; Frederick Bechdolt, *Tales of the Old Timers* (New York, 1924), 113-14; Interview granted by R. E. Baird to a western history class at West Texas State Teachers' College, August 23, 1933. Class members' notes are in the Panhandle Plains Historical Museum, Canyon, Texas. Hereafter cited as Baird Interview. See also Robert V. Clements, "British Controlled Enterprise in the West between 1870 and

1900 and Some Agrarian Reactions," *Agricultural History* XXVII (1953): 132-41; Charles Goodnight to J. Evetts Haley, interview, July 24, 1925 (Panhandle Plains Museum, Canyon, Texas); Philip Ashton Rollins, *The Cowboy* (New York, 1922), 88.

16. Atherton, *The Cattle Kings,* 120.

17. J. Evetts Haley, *Charles Goodnight: Cowman and Plainsman* (Norman, 1949), 379; Atherton, *The Cattle Kings,* 182.

18. Atherton, *The Cattle Kings,* 182. See Haley, *Charles Goodnight: Cowman and Plainsman,* 364, for an example of large company efforts to take over the Panhandle Stock Association.

19. Fred A. Shannon, "The Farmer's Last Frontier, 1860-1897," Vol. V in the *Economic History of the United States* (New York, 1966), 222; McCarty, *Maverick Town: Story of Old Tascosa,* 108; LaWanda Cox, "The American Agricultural Wage Earner, 1865-1900: The Emergence of a Modern Labor Problem," *Agricultural History* XXII (1948): 104.

20. Baird Interview; *Amarillo Globe News,* 1938 Centennial Edition, Section D, 25; C. May Cohea, "The Cowboy Strike," WPA Project (Panhandle Plains Historical Museum, Canyon, Texas), 1; McCarty, *Maverick Town: The Story of Old Tascosa,* 108; Allen, *Chapters in the History of Organized Labor in Texas,* 36; Sullivan, *The L S Brand: The Story of a Panhandle Ranch,* 64-65.

21. Haley, *Charles Goodnight: Cowman and Plainsman,* 303-4; Lester F. Sheffy, *The Francklyn Land and Cattle Company* (Austin, 1963); Estelle D. Tinkler, "Nobility's Ranche: A History of the Rocking Chair Ranche," *Panhandle Plains Historical Review* XV (1942), 16.

22. For evidence of the striking cowboy's knowledge of this change in relative position see Potter, *The Lead Steer,* 32; Allen, *Chapters in the History of Organized Labor in Texas,* 40; *Caldwell Commercial,* March 29, 1883, *Texas Live Stock Journal,* April 23, 1883, in Westermeier, *Trailing the Cowboy,* 124, 128-29.

23. Billington, *Western Expansion,* 684; Shannon, *The Farmer's Last Frontier,* 207-8; Rollins, *The Cowboy,* Chapter X; Clifford P. Westermeier, "The Cowboy in His Home State," *Southwestern Historical Quarterly* LVIII (October, 1954): 228-29.

24. Harley True Burton, "A History of the J. A. Ranch," *Southwestern Historical Quarterly* XXI (April, 1928): 363; Cox, "The American Agricultural Wage Earner, 1865-1900: The Emergence of a Modern Labor Problem," *Agricultural History,* 103; W. C. Holden, "The Problems of Hands on the Spur Ranch," *Southwestern Historical Quarterly* XXXV (January, 1932): 198.

25. Holden, "The Problems of Hands on the Spur Ranch," *Southwestern Historical Quarterly,* 195; Payroll Ledger, April 1885-November 1893, in

Spur Records (Southwest Collection, Texas Tech University, Lubbock, Texas); Payroll Ledger, 1883-1892, in Alamositas Division, Matador Land and Cattle Company, Ltd., Records (Southwest Collection, Texas Tech University).

26. United States Commissioner of Labor, *Strikes and Lockouts 1887*, Serial No. 2546, Doc. No. 1, 580-83.

27. United States Census, 1880, Vol. 2, *Report on the Manufacturers of the United States*, xvi, 176-79. The Panhandle Counties reported no manufacturing establishments, thus comparison with industrial workers in the Panhandle area is impossible.

28. Holden, "The Problem of Hands on the Spur Ranch," *Southwestern Historical Quarterly*, 198-99; John S. Spratt, *The Road to Spindletop* (Dallas, 1955), 229.

29. Holden, "The Problem of Hands on the Spur Ranch," *Southwestern Historical Quarterly*, 196. The unskilled nature of ranch work is borne out by Holden's study of Spur labor, and it is likely that the situation was similar on other ranches in the same area in the same period.

30. This date is approximate. The United States Commissioner of Labor places the date of the strike at March 23. The first known newspaper report is on March 12 and indicates the hands had already asked for wage increases. See United States Commissioner of Labor, *Strikes and Lockouts 1887*, 580-83; *Texas Live Stock Journal*, March 12, 1883, in Westermeier, *Trailing the Cowboy*, 125-27.

31. McCarty, *Maverick Town: The Story of Old Tascosa*, 109-10; Allen, *Chapters in the History of Organized Labor in Texas*, 37; John Arnot, "My Recollections of Tascosa Before and After the Coming of the Law," *Panhandle Plains Historical Review* VI (1933): 69.

32. "Cowboy Strike Ultimatum," (Southwest Collection, Texas Tech University). The original is located at the Panhandle Plains Historical Museum, Canyon, Texas. There are reports that better food, larger winter crews, and permission to own and run cattle on the owners' land were also strike demands. However, the ultimatum and newspaper reports mention only wages.

33. The list of names is as follows: Thomas Harris, Roy Griffin, J. W. Peacock, J. L. Howard, W. S. Gaton, J. L. Grissom, S. G. Brown, W. B. Borina, D. W. Peeples, Jas [sic] Jones, C. M. Hullett, V. F. Martin, Harry Ingerton, J. S. Morris, Jim Miller, Henry Stafford, Wm. F. Kerr, Juan A. Gomez, Bull Davis, T. D. Holliday, C. F. Goddard, E. E. Watkins, C. B. Thompson, G. F. Nickell.

34. Jack Potter quotes Sheriff Jim East as giving this description of Harris. Potter, *The Lead Steer*, 32. This view is substantiated by Harris' conduct during the strike.

35. *The Western Range Cattle Industry Study: Outline of Project* (Denver, no date), 9. For an astute survey of the workers' dream of entrepreneurship and its abandonment see Gerald N. Grob, *Workers and Utopia* (Chicago, 1961).

36. *Fort Collins Courier*, April 12, 1883, in Westermeier, *Trailing the Cowboy*, 124-25; *Denver Republican*, March 27, 1883; United States Commissioner of Labor, *Strikes and Lockouts 1887*, 580-83, for seriousness of the strike. "Cowboy Strike Ultimatum"; Bechdolt, *Tales of the Old Timers*, 115; Sullivan, *The L S Brand*, 65; Allen, *Chapters in the History of Organized Labor in Texas*, 36.

37. See United States Commissioner of Labor, *Strikes and Lockouts 1887*, 580-83; *Fort Collins Courier*, April 12, 1883, *Texas Live Stock Journal*, April 28, 1883, in Westermeier, *Trailing the Cowboy*, 125, 128-29; *Fort Worth Daily Gazette*, March 29, 1883, for varying reports on the number of hands actually involved.

38. This figure is based on the United States Commissioner of Labor, *Strikes and Lockouts 1887*. This report was derived from newspaper accounts and interviews by government representatives; therefore the figure could be inflated.

39. "Cowboy Strike Ultimatum"; Baird Interview.

40. Quoted in Sullivan, *The L S Brand*, 68.

41. Nowhere in the news coverage or in memoirs is there the charge that strikers actually injured hands who continued to work. *The Houston Daily Post*, April 24, 1883, reported that the Adjutant General had received no reports of "impending serious trouble." There are statements to the effect that trouble was feared or that one individual threatened another. See *Ford County Globe*, May 1, 1883, for an example of this sort of report.

42. *Texas Live Stock Journal*, March 12, 1883; *Caldwell Commercial*, March 29, 1883; *Fort Collins Courier*, April 12, 1883; in Westermeier, *Trailing the Cowboy*, 124-25; *Denver Republican*, March 27, 1883.

43. *Texas Live Stock Journal*, April 28, 1883, in Westermeier, *Trailing the Cowboy*, 127, 129; *Denver Republican*, April 25, 1883.

44. T. B. Harris, Tascosa, Texas, to the Editor, *Texas Live Stock Journal*, April 25, 1883 in Westermeier, *Trailing the Cowboy*, 125-26.

45. "Judge" L. Gough, "Memoirs," typed manuscript (Panhandle Plains Historical Museum, Canyon, Texas, 1935), 203; "The T-Anchor," *Amarillo Sunday Globe News* (clipping, n.d., Panhandle Plains Historical Museum, Canyon, Texas).

46. Gough, "Memoirs"; "Judge" L. Gough, "Reminiscences" typed manuscript (Panhandle Plains Historical Museum, Canyon, Texas, n.d.); *Graham Leader,* April 21, 1883.

47. Sullivan, *The L S Brand,* 66-67; McCarty, *Maverick Town: A Story of Old Tascosa,* 111.

48. McCarty, *Maverick Town: A Story of Old Tascosa,* 111.

49. Sullivan, *The L S Brand,* 65; McCarty, *Maverick Town: A Story of Old Tascosa,* 111.

50. The *Fort Worth Gazette,* March 22, 1883, erroneously reported that smallpox in Mobeetie prevented press coverage, but some papers, both in and outside Texas, did cover the strike.

51. *Texas Live Stock Journal,* March 12, 1883, in Westermeier, *Trailing the Cowboy,* 124.

52. Ibid.

53. *Dodge City Times,* April 26, 1883, in *Panhandle Plains Historical Review,* XL (1967), collected by Lonnie J. White; *Texas Live Stock Journal,* March 12, 1883, *Texas Live Stock Journal,* April 21, 1883, *Texas Live Stock Journal,* April 28, 1883, in Westermeier, *Trailing the Cowboy,* 124-30; *Denver Republican,* April 25, 1883; *Fort Worth Gazette,* March 25, 1883; *Graham Leader,* April 21, 1883.

54. *Trinidad Weekly Advertiser,* April 25, 1883, in Westermeier, *Trailing the Cowboy,* 127.

55. Ibid.

56. *Fort Worth Gazette,* March 25, 1883; *Ford County Globe,* May 1, 1883.

57. Baird Interview.

58. For newspaper accounts, see *Fort Collins Courier,* April 12, 1883, in Westermeier, *Trailing the Cowboy,* 125; *Denver Republican,* March 27, 1883; *Denver Republican,* April 25, 1883. On the success, see United States Commissioner of Labor, *Strikes and Lockouts 1887.* This report calls the strike a complete success, which is obviously in error. However, it does seem logical that the report refers to the period when higher wages were being paid. Since the report puts the dates of the strike from March 23 to April 4, it is probable that the higher wages were decided upon early in the strike.

59. Baird Interview.

60. For the delights of Tascosa, see Baird Interview. Cohea, "The Cowboy Strike," 2. An examination of the 1880-1890 minutes of the District Court of Oldham County reveals numerous convictions for "Vagrancy" and "Gaming," thereby indicating a thriving business in prostitution and gambling. Ladies charged with vagrancy usually pled guilty, were fined five dollars and court costs, and went back to work. See Oldham County

District Court, *Minutes,* vol. 1. These may be found in the County Courthouse, Vega, Texas.

61. Potter, *The Lead Steer,* 31, 75; McCarty, *Maverick Town: A Story of Old Tascosa,* 117, 123; *Dodge City Times,* May 10, 1883. Waddy Peacock, one of the original strikers, is listed as foreman on a list of employees of the L S in 1898. This list may be found in the Panhandle Plains Historical Museum, Canyon, Texas. The Oldham County District Court *Minutes* show J. L. Grissom as being a juror on several occasions in 1884 and 1885.

62. The government reported the strike over on April 4, but newspapers did not mention it ending until late April. The *Houston Daily Post* reported it over on April 25 but the April 28, 1883 *Texas Live Stock Journal* reported the strike sufficiently strong to endanger the success of the May 10 roundup. The *Graham Leader,* April 28, 1883, reported one hundred cowboys on strike. The last press mention was in the *Dodge City Times,* May 10, 1883. Thus the strike ended some time between April 25 and May 10.

63. Baird Interview; Sullivan, *The L S Brand,* 67. Potter quotes Sheriff Jim East as saying the strike was still going on in September of 1884, which indicates some trouble was still occurring. Potter, *The Lead Steer,* 30; Louis Bousman, "Reminiscences" prepared by the personnel of the WPA (Panhandle Plains Historical Museum, Canyon, Texas, 1934); Fred Post, "He Escaped Boot Hill to Help Capture Billy the Kid," *Amarillo Globe News,* 1938 Centennial Edition, Section E, p. 12; Haley, *Charles Goodnight: Cowman and Plainsman,* 376. The "Hoggie" that Haley holds responsible for much of the rustling was Tom Harris's brother-in-law. See McCarty, *Maverick Town: A Story of Old Tascosa,* 112-13. Although some strikers probably resorted to rustling, it hardly seems accurate to assume rustling was actually a part of the strike. Instead, it is more likely that some or many strikers refused to back down, were fired, and became rustlers.

64. "Panhandle Stock Association Constitution and Minutes, 1883" (Panhandle Plains Historical Museum, Canyon, Texas).

65. Oldham County District Court, *Minutes,* vol. 1, 92.

66. McCarty, *Maverick Town: A Story of Old Tascosa,* 129; "Panhandle Stock Association Constitution and Minutes, 1883."

67. Haley, *Charles Goodnight: Cowman and Plainsman,* 380; *Ford County Globe,* June 19, 1883.

68. See Allen, *Chapters in the History of Organized Labor in Texas,* 33, for the assertion that cowboys and corporations were naturally antagonistic. See the essay by John Tipple, "The Robber Baron in the Gilded Age: Entrepreneur or Iconoclast?" in H. Wayne Morgan, ed., *The Gilded Age* (Syracuse, 1963), 14-37 for a good summary of the attempts on the part of American society to adjust to the corporation.

Cowboys lived much of the time in line camps similar to the half-dugout shown here, ca. 1880s. Courtesy of Southwest Collection, Texas Tech University. Unidentified museum photo, File no. 52.36-1a.

8 Work Clothes of American Cowboys: The Pictorial Record

SUSAN KARINA DICKEY

Every time a cowboy put on his pants, buttoned his shirt, and slipped on his vest he did exactly what millions of other working men did each morning. But when he donned his big hat, boots, and leather chaps, he entered the world of the cattle kingdom. In the late nineteenth and early twentieth centuries men who made their living by the sweat of their brow labored mostly in factories and on small farms. Cowboys, in contrast, worked in the Big Sky Country in Montana, the mountain valleys of Utah, the empty ranges of Wyoming, and the sweeping plains of Texas—areas where cattlemen ruled.

Like farmers, waddies (as cowboys were often called) worked outdoors, but the latter were not tied to a small plot of land. Instead of working 160 acres, waddies, during the course of a year, ranged over thousands of acres. The divergence between cowboy and factory life was even more striking. An employee at Carnegie Steel, for example, lived in an entirely different world of sights, sounds, smells, tastes, and textures. But in regard to basic garments, ranch hands, farmers, and factory workers all had much in common.

Cowboys and other working men of the late nineteenth century generally wore drab baggy pants held up by suspenders, an ill-fitting shirt, and a nondescript vest. Ranch hands of the Old West, granted, wore the classic cowboy hat, boots, and chaps, but otherwise their garments were not particularly special.

An examination of 254 photographs, dating from about 1880 to 1910, demonstrates, however, that the old-time cowpunchers did not dress like their modern counterparts.[1] Nowadays in the West, working cowboys tend to wear snug-fitting blue denim jeans and a tooled leather belt, often with the name across the back and a gigantic ornate buckle. They also sport fancy-yoke shirts, pointy-toed boots—some in exotic leather—and a western hat.

To appreciate fully the work clothing of the late nineteenth-century cowboy, one must study the component elements: pants, chaps, shirts, vests, coats, hats, and boots. Smaller, but equally important items, such as handkerchiefs and leather cuffs, should also be considered. Photographs are an excellent vehicle for studying cowboy apparel. Although some western wear is preserved in museums and private collections, most work clothing simply wore out, and is no longer available.

Pictures provide access to a large sample. Furthermore, most photographs show ranch hands in a work environment. Looking at the garments in the historical setting gives insight regarding use and function. The Sears and Roebuck catalogs are another helpful source. The Chicago mail order giant began aggressively in 1896 to target the western market by offering work clothing as well as tack and accessories. In regard to the attire, catalog illustrations give details on construction, fabrics, and sizing not available through other sources (see fig. 1).

The pants worn by the cowboy in figure 1 were typical in cut and style. Pants, often called overalls in the nineteenth century, were made in durable medium- to heavyweight woolens. Cotton duck and denim were available from the 1880s onward. Extant fabric swatches indicate a preference for black, brown, gray, and tweeds. Checks and sober stripes also appeared. Waddies, like other working men, preferred dark colors, what the people of the nineteenth century called "serviceable," because such shades minimized the appearance of dirt and stains.

In its first general merchandise catalog, Sears and Roebuck included pants geared toward cowboy needs. Model 1648, the Western, boasted "high cut inserted gusset seat, good quality of medium heavy duck, 2 pockets [in front], strengthened crotch, 2 ply band and flap." To any man who spent a lot of time in the saddle the reinforced crotch was, no doubt, a selling point. At thirty-five cents a pair, the Western was among the lowest-priced work pants in the catalog.[2] An "inserted gusset seat," by the way, was almost like the back of today's western-cut blue jeans. For another five cents one

Fig. 1. Unknown cowboy in typical garb. Price, Utah, 1890s. Photographic Archives, Harold B. Lee Library, Brigham Young University, Provo, Utah. P-1 George E. Anderson Collection, #4476. Reprinted with permission.

could order Model 1650, Never Rip overalls, which were similar to the Western, but welted side seams increased durability. The Never Rip came in gray denim.

The waistband of men's work pants had a notch at the center back. Below the notch a small belt, like that on the back of a vest, allowed minor adjustments. Most men needed suspenders to keep the pants up; thus, all pants came with suspender buttons (see fig. 1). Ranch hands were no exception, and the photographic record shows that suspenders were widely used. The men pictured without suspenders usually wore a gunbelt or chaps that helped hold the pants in place. Waist belts were not used; they did not become a common men's accessory until well into the twentieth century. Consequently, buckles were not part of the work attire either. The large ornate buckles of the present originated as rodeo trophies, not as a practical accoutrement.

A small number of waddies dressed in apron or bib overalls. Of the hundreds of cowboys pictured in James Beckstead's *Cowboying: A Tough Job in a Hard Land* (1991), however, only four men chose this garment. Near Bluff, Utah, a ranch hand teamed his bib overalls with a dark shirt and hat (see fig. 2). In another photograph, a cowpuncher wore a vest with the bib overalls. A man at the Rees Ranch near Randolph, Utah, donned overalls with natural color woolly chaps. All four pictures date from about 1900 or later, suggesting that overalls had gained acceptance by the turn of the century.[3] Maybe some ranch hands liked overalls for practical reasons. A long-waisted man might select overalls to keep his shirttails tucked in. Or perhaps at higher elevations, overalls made of cotton duck or denim helped conserve body heat.

Chaps, from the Spanish *chaparejos,* were a kind of leather armor. Straight or batwing style, chaps protected the rider's legs against brush, animals, fences, or, in short, anything abrasive. In figure 3 the cowboy on the left wore chaps of smooth leather. Notice his sleeve garter and the turned-up pantleg on the other rider. Smooth chaps were the most common and, in the photographs studied, outnumbered woollies about five to one. Most of the woollies pictured were dark in color.

Fig. 2. Cowboys at Bluff, Utah, ca. 1900. Courtesy of Manuscripts Division, J. Willard Marriott Library, University of Utah.

As for cowboy shirts, figure 1 shows the most common style: a pullover shirt with a turn-down collar, three-button placket, set-in sleeves, and a back shoulder yoke. In the nineteenth century such tops were called "overshirts." Although one style dominated, waddies sported a variety of stripes. Solids were more typical, and even though the photographs are black and white, a discerning eye can infer a wide range of color. Light colors, striped or solid, outnumbered dark about two to one. Extant shirts and catalog descriptions indicate that shades of blue and brown were popular after the mid-1890s.

No photograph shows the fancy yoke we associate with western wear today. And despite their reputation as traditional cowboy garb, not many collarless shirts appear in the photographic record.[4] The most elaborate shirt pictured in *Cowboying,* worn by Utah rancher John Whiting in the 1890s, had a bib of horizontal pleats. Otherwise, in construction it was identical to the other overshirts.[5]

Fig. 3. Taylor Button (left, in chaps) and Edward Lamb, Cane Beds, Arizona, ca. 1892. Courtesy of the LDS Church Archives.

Many ranch hands wore leather cuffs, which kept the shirt-sleeves in place. About five inches long, the cuffs gave some support to the wrists for roping and other strenuous activities. The photographs show tooled leather cuffs embellished with stars, diamonds, and other motifs. Less frequently, gloves, usually of the gauntlet type, show up in the photographic record. Surely, though, most cowboys wore gloves to protect the hands from rope burns and from the cold.

Almost every waddy wore a bandana handkerchief around his neck. He folded the square piece of cloth into a triangle and tied the narrow ends with a square knot. The handkerchief served multiple functions. Placing the knot in the back and pulling the cloth up over the nose made a dust mask. This arrangement also protected the cowboy from sleet or driving wind. Relative to shirt color, about as many light as dark handkerchiefs show up in the photographs. It is impossible to gauge any individual man's intention, but some seem to have deliberately chosen a handkerchief to contrast with the shirt and vest.

Men of all classes in the nineteenth century usually wore a vest, and ranch hands were seldom pictured without the practical garment. Vest pockets typically held matches, cigarette papers, and a tobacco pouch. Often a watch chain or fob is visible in the photographs. On horseback, pant pockets were not accessible, so the vest pockets carried close at hand anything the rider wanted. A few vests had shawl or notched collars, but most had a plain V neck and five buttons. Dark serviceable colors were the norm. Most of the vests appear to be woolen, but Jim Robinson, a Utah cowhand of the 1880s, chose leather. His vest hung open to reveal suspenders and baggy pants.[6]

Sometimes weather conditions necessitated wearing an overcoat or a slicker, but most ranch hands avoided, if possible, the extra layer.[7] Set-in sleeves restricted motion, and the added bulk trapped body heat. A cowboy engaged in even moderately hard work could stay warm enough without a jacket or coat. Every man needed a slicker for stormy weather. The slicker was an oilskin coat designed especially for the equestrian. Also called saddle coats, they came in yellow or black, and in the late 1890s sold for about $2.90.[8]

A picture taken on a crisp day in the 1890s in Salina, Utah, shows several men and boys in front of a saddlemaker's shop.[9] One gentleman wore a full-length coat. A boy on horseback huddled in a long coat as well. A deep center vent allowed the young man to tuck the coattails under his seat, which would have prevented the wind from blasting up his back. Most of the men donned the sack coat.

The sack coat, popularized by England's Prince Albert, was the forerunner of today's suit coat. The three-button sack coat fell to the hip and had a notched collar. In the nineteenth century, men usually fastened only the top button. Wool was the most common fabric. Black and white photographs do not, of course, indicate exact color, but extant suiting swatches are mostly dark solids and tweeds. Cotton duck and denim were available by the turn of the century.

Robert Allen, who worked cattle near Moab, Utah, made a striking appearance in an 1880s photograph.[10] His sack coat was a dark medium-sized check. Striking a manly pose, left hand on hip, he pulled back the coat to display a six-shooter and suspenders. The shirt was also a medium-sized check, a dark grid on a light background. Mixing patterns was not uncommon in the nineteenth century. Even the pants were a small-scale check in a dark color. Like most of the men photographed, Allen turned up the bottom of the pant legs. A broad-brimmed hat and cowboy boots completed the outfit.

Headwear and footwear, more than any other part of the outfit, evolved in response to the needs of working cowboys. In time, others adopted both items, but in the nineteenth century, as today, functional considerations determined both form and appearance.

The wide brim of the cowboy hat shaded the eyes, thus improving clarity of vision. In the rain, the brim protected the face and the back of the neck. Cowhands also used their hats to fan a new campfire and carried water in the crown to douse the embers later. Some men rolled the soft felt hats into a sort of pillow, and when waved in the air the hat was a signaling device.[11] In sum, the hat was the most versatile article of clothing in the cowboy wardrobe.

Rarely was a waddy seen without a hat. Jim Robinson, the man with the leather vest, wore a black hat with an unusually high rounded crown for an 1880s studio picture.[12] A few years later Sears and Roebuck sold an almost identical hat. "The Texas Chief . . . [a] high crown Mexican style sombrero hat."[13] It had a five-inch brim and a six-and-one-half-inch crown. In buckskin tan, the Texas Chief sold for $3.75 and included a fancy braided silk band. The "never flop brim" turned up slightly.

The authentic Mexican sombrero had a high conical or cylindrical crown with a saucer-shaped brim. Many Southwestern cowhands wore these embroidered hats, which were made of plush felt. On the Northern Plains, men like Robinson preferred an anglicized version.[14] The Mexican sombrero inspired John B. Stetson, an eastern hatmaker, to design a western hat, and within a few years he dominated the cowboy market.

Stetson, born in 1830 in New Jersey, learned the hat business from his father. The young Stetson had planned to continue in the family business, but had a falling out with his brothers. By this time Stetson had developed tuberculosis and, like so many "lungers," he decided to go west. When the Civil War began, Stetson was in Missouri. He tried to enlist for the Union, but the medical officer turned him down. Stetson then joined a small group of men heading for Colorado to prospect for gold. The route took the men across the Great Plains, and the dry fresh air restored Stetson's health.[15]

One winter day while encamped in the Rockies, Stetson showed the others how to make felt from rabbit fur. He matted the fibers by sprinkling them with water. He shrank the small mat by dipping it into boiling water. By repeating the process, Stetson worked the felt into a hat. A short time later, a Mexican horseman rode into the camp. Spying the hat, he offered Stetson five dollars. The New Jersey hatmaker had sold his first cowboy hat.[16]

Stetson returned east. With about a hundred dollars start-up money he opened a hat shop in Philadelphia. Initially, he made hats for the local wholesale trade, but remembering his time in the West, the entrepreneurial Stetson decided to try selling cowboy hats. He shipped samples of the Boss of the Plains with its four-inch crown and four-inch brim to several western dealers. Orders poured in for the Boss, which retailed at five dollars. Soon, Stetson designed other models and made them in different grades.[17]

The best quality nutria (beaver) hats sold for thirty dollars, about a month's wages for a waddy. To break in a stiff new hat, a cowboy might douse it in a water trough or a nearby stream. He could then place the dripping wet hat on his head and shape it to his fancy. This process also resulted in a custom fit.[18]

Owning a Stetson carried a certain cachet; other makers sold less expensive models. Sears and Roebuck sold a variety of cowboy hats. Colors included tan, brown, jet black, and nutria (a very dark brown). The company displayed, in the Fall 1897 edition, two pages of cowboy hats ranging in price from $2.25 to $5.25. Sears also advertised made-to-order hats, which required an additional three weeks to manufacture and ship. Model names reflected the western emphasis: the Reservation Sombrero, the Pine Ridge Sombrero, the Mustang, the Vaquero, the Cow Puncher's Sombrero, the Pride of the Platte, and the Cattle King.[19] By now (1897) the term *sombrero* referred to any cowboy hat; none of the hats pictured were true Mexican sombreros.

A photograph dated about 1910 of three cowboys at the McIntyre Ranch, Leamington, Utah, shows some of the variations in hat styles.[20] Burt Johnson turned up the brim a bit and poked dents into the crown. Del Bradfield wore a lighter colored hat, perhaps gray or tan, with the peak shaped like a mountain summit. The brim was flat. The third hand, Cal Brimley, turned up the left side Australian style.

Boots were just as important as hats, but for different reasons. Rising to just below the knee, boots protected the legs and feet of the mounted herder. Bootmakers avoided lacings, hobnails, or other protrusions that might catch on vegetation. Tapered high heels prevented the foot from slipping through the ox-bow stirrup. Better to be thrown from a horse than to be dragged. The heels also helped the cowboy to "dig in" when throwing a lariat from the ground or holding the lead of a skittish horse. Made of soft leather, the typical boot had a thin sole so the rider could "get the feel" of the stirrup.

In the photographs studied, the high heel is visible, but several other features are not. For the most part, nineteenth-century cowboy boots were rather plain. Decorative stitching enhanced some, but the colorful inlays, complicated embroidery, and extremely pointed toes we associate with cowboy boots date not from the nineteenth century, but from the silent movie era. Film directors wanted something flashier than the workaday cowboy boot.[21]

Legends to the contrary notwithstanding, no single maker invented the cowboy boot. Some writers credit Herman Joseph Justin or Charles Henry Hyer as the earliest bootmakers of the Old West. These two were among the more successful, to be sure, but the historical record proves they were not the first to set up shop. Investigation also reveals that the cowboy boot probably evolved from northern European prototypes. Boots of the late nineteenth century resembled both English Wellington and Hessian military boots. Census data also supports the likelihood of European models. Numerous shoe and bootmakers in the West, as throughout the United States, were themselves immigrants from England or Germany or were the sons of such immigrants. Interestingly, both Justin and Hyer were first generation German Americans.[22]

Wellington boots, named for the Duke of Wellington, first appeared during the Napoleonic wars. These below-the-knee boots with piped side seams, one-inch stacked heel, wide toe box, and pull-on straps remain popular with equestrians. The Hessian boot is today less well known, but was in the nineteenth century very popular in the United States. A notch under the knee, adorned with a small tassel, distinguished the Hessian boot.[23]

The 1880 census recorded 535 shoe- and bootmakers in California. East of the Rockies, Kansas, home of 212 makers, dominated the trade. Texas ran a distant second with 163. The practice of wearing the pantlegs stuffed into the boot top may have originated in Kansas. At the completion of the trail drives, cowboys could have purchased fresh boots and tucked in the legs so as to show off the new acquisition.[24]

By 1900, Justin and Hyer had emerged as the largest manufacturers of factory-made cowboy boots, but the best boots were custom-made. These two companies, as well as other bootmakers, ran a mail order division for the custom work. Upon request, the manufacturer sent back a kit so the cowboy could obtain the necessary information. He outlined each foot and measured the feet and legs in six places. The manufacturer then carved a pair of lasts from which to make the boots.[25]

Headwear and footwear were among the cowboy's most important and most cherished possessions. Not only did hat and boots facilitate the ranch hand's work, but they also shaped his identity. More importantly, the special apparel molded the popular image of the cowboy. "Clothes make the man" is a cliché, but one with a grain of truth. In recent times we tend to dismiss the symbolic functions of apparel. We think clothing makes a personal statement, and indeed, it does. But in the heyday of the cowboy, clothing was more an indicator of social status. The pants, shirt, and vest identified the ranch hand as a manual laborer, but his hat and boots telegraphed to others that he was a cowboy.

The later widespread adoption of western headwear and footwear is unusual in the history of work clothing. Painter's overalls, for example, occasionally appear as casual wear in women's fashion magazines, but only as a passing fad. Cowboy clothing is another story. Even in Chicago's Loop business district one spots businessmen in Stetsons and Tony Lamas.

Why did cowboy clothing spread to the mainstream and endure for over a hundred years? The broad-brim cowboy hat might appeal to anyone who spends time outdoors, and well-fit boots are comfortable, but there is more to the unshakable popularity of the clothing. The cowboy was unique among American laborers. Ranch work, relative to most industrial occupations, was self-directed. Cowhands punched no time clock and often worked in settings of spectacular beauty. Writers and early filmmakers recognized that these unusual attributes would appeal to the growing urban population.

No other occupation has been so dramatized and romanticized as that of the western ranch hand. Dime novels and silent movies capitalized on the self-reliance and courage of the waddy and made him into a cowboy of mythic proportions. After World War II, Americans born in the 1920s and 1930s, watching movies featuring John Wayne, Gary Cooper, and other stars, fell under the spell of the cowboy mystique. Baby Boomers, too, through television Westerns—*Wagon Train, Maverick,* and *The Wild Wild West,* to name just a few —responded to the allure. Ask children of the 1950s or early 1960s what games they most enjoyed, and they are likely to include "cowboys and Indians." To wear a cowboy costume and use a toy six-shooter added to the excitement.

Today we recognize the stereotypes of Indians, Mexicans, women, and cowboys presented in the popular media. Still, the mystique persists. As long as it does, cowboy clothing will continue to be worn by people thousands of miles from the ranches of the West, the vast majority of whom have never roped a cow from horseback, branded a calf, or lived by choice on bacon and beans through all seasons of the year.

Notes

1. The photographs studied appear in *Cowboying: A Tough Job in a Hard Land* by James W. Beckstead (University of Utah Press, 1991). Most of the pictures are from Utah, but comparison to images from other states showed only minor differences. Catalog illustrations studied are from the Sears and Roebuck Catalog, 1896-1910 (microfilm editions).

2. Sears and Roebuck Catalog, 1896, 234.

3. Beckstead, *Cowboying,* 55, 76, 221, 223.

4. Philip Ashton Rollins, *The Cowboy: An Unconventional History of Civilization on the Old-Time Cattle Range*, rev. ed. (New York: Charles Scribner's Sons, 1936), reprint with foreword by Richard W. Slatta (Norman: University of Oklahoma Press, 1997), 108.

5. Beckstead, *Cowboying*, 210.

6. Ibid., 94.

7. Rollins, *The Cowboy*, 109.

8. Sears and Roebuck Catalog, Spring 1897, 188.

9. Beckstead, *Cowboying*, 128.

10. Ibid., p. 82.

11. Rollins, *The Cowboy*, 105.

12. Beckstead, *Cowboying*, 94.

13. Sears and Roebuck Catalog, 1887, 210.

14. Rollins, *The Cowboy*, 106.

15. Jack W. Pilley, "El Sombrero," *Noticias* 9 (1963), 24. Pilley's account is based on the recollections of Stetson sales representative Harry Walker Jr.

16. Ibid., 25-26.

17. Ibid., 26.

18. Ibid., 27.

19. Sears and Roebuck Catalog, Fall 1897, 264-65.

20. Beckstead, *Cowboying*, 54.

21. Barbara Brackman, "Legacy Posing as History: Hyer, Justin, and the Origin of the Cowboy Boot," *Kansas History* 18 (1995), 40.

22. Ibid., 41.

23. Ibid., 40. Brackman's article contains a photograph of Billy the Kid showing his Hessian boots.

24. Ibid., 40.

25. Ibid., 46.

9 Cowboys and Sheepherders
PAUL H. CARLSON

After the Civil War, cowboys and sheepherders both worked through the large, empty ranges of the American West. Indeed, herders and Spanish-trained vaqueros had worked the southwestern ranges for more than a century before that time. But after the Civil War, during the great age of the American cowboy, when cattle moved out of Texas and scattered over the larger West, cowboys and sheepherders presented marked contrasts in their dress, habits, work, and general lifeways. Their differing styles, attitudes, and approaches to life help to define the cowboy of the late nineteenth century and reveal something about cowboy culture of the period.

The images that cowboys and sheepherders have left us are different, too, although, of course, both are stereotypical, mythic, and dominated by folklore. Charles Russell, the famed cowboy artist, for example, fashioned a painting of some cowhands who had lassoed an angry bear, an event in Montana that actually occurred. Bursting with energy, the painting shows cowboys, on excited horses, engaged in a dangerous, lively activity. The young men, with guns on hips, are neatly dressed in Levi's, vests, and broad-brimmed hats.[1]

N. C. Wyeth, who also painted western scenes, produced one of a western sheepherder. In it, as described by Winifred Kupper, is an old man, rugged, tattered, and alone under the stars with his sleeping flock in the distance, a watchful dog nearby, and a dying fire in front of him. In one weathered hand is a well-worn pipe, in the other is an open book. But the man is staring into the fire, not into the book. Here is quiet and solitude and none of the energy of the youthful cowboys.[2]

A different version was published in *West Texas Historical Association Year Book* 58 (1982): 19-28.

Consider the two paintings. They suggest that the cowboy was a man of action; the sheepherder a philosopher. Philosophers in America seldom attract attention. Few, if any, movies have been made about them. Americans admire people of action over people of intellect, and if modern-day advertising and fashion modeling mean anything, Americans prefer young people, too.

Cowboys were young. Their average age, according to William H. Forbis, was twenty-four, and they remained cowboys an average of only seven years. In the West between 1860 and 1890 there were about forty thousand of them. The majority were Anglo Americans, but a good number were Hispanic, African American, or Native American. Many cowboys in the Southwest were southern Civil War veterans looking for new stakes or perhaps some rugged action that would help them work out frustrations of a lost cause. A number of them were ex-Union soldiers who were unwilling to return to quiet farm life in the Midwest or dairying in New England. A rare few were former sailors, and a few more were ne'er-do-well Englishmen. Some were educated; some were outlaws on the run.[3]

Sheepherders were older and on the average remained in the occupation longer. There were about twelve thousand of them in the West between 1870 and 1900. A minority of them were Anglo Americans. Most sheepherders were Mexicans and Mexican Americans, or such newly arrived immigrants as Germans, Basques, and Englishmen who were hoping to carve out a home in the West. The few American settlers among them grazed sheep because of the low overhead and quick profits. If he were an Englishman or Basque, the sheepherder was a potential millionaire. Many of the Mexican and Mexican American sheepherders were *pastores,* proud and idealistic but often in perpetual debt to the sheep's owner.[4]

There were, then, ethnic, age, and other factors that distinguished cowboys from sheepherders. Cowboys displayed such qualities as restlessness, impatience, impulsiveness, and self-confidence. Compared to sheepherders, they were more outgoing, open, and extroverted. Sheepherders displayed such qualities as stability, innovation, open-mindedness, and a tolerance of uncertain situations. In a mythic sense, both manifest such characteristics as honor and hard work and a rural, agrarian, open-range way of life. Cowboys, in addition, symbolize youth, freedom, strength, and independence; sheepherders symbolize experience, duty, durability, and fealty.

In the American West in the late nineteenth century, sheepherders, when compared to cowboys, have been called better educated, steadier, quieter, and more careful. Folklore in the Southwest had it that "you could class a sheepman as a man who had money, and a cattleman as one who hadn't."[5] Many more sheepherders than cowboys, for instance, became local pillars of society or citizens of substance and wealth. In Texas, more sheepmen than cattlemen became governors of the state.[6]

Of the two, the cowboy was younger, often African American or Hispanic, occasionally Native American or Chinese, and the popular view held that he dreamed of wealth, craved action, and cared little for anything else. Edward Everett Dale, once a ranch hand himself, wrote that many a "cowboy was a wild, reckless type who rode hard, swore hard, and feared neither God nor man." He "was lavish with his money," and contemporaries regarded him "as a swaggering swashbuckler, who carried a gun, had little regard for horse flesh, and who seemed at all times to be jealous of honor, sudden and quick in quarrel."[7] Although that was unfair, the cowboy did have an exalted, heroic image of himself as a hard-riding, fast-shooting hombre who broke wild horses, shot it out with Indians, and lassoed bears for fun. Often he tried to live up to that idealized version of himself. He lived, according to William Forbis, "by a code compounded of hardfisted frontier desperation and Victorian-era social values."[8]

Sheepherders were different. A sheepherder went on foot with a dog or two, lived in a tent or wagon, seldom carried a weapon on his hip, and dressed miserably. Rarely was he African American or Chinese. Few were outlaws on the run.[9] Sheepherders have been pictured as peaceful, docile, and law abiding, but more often a sheepherder's life was virile, adventurous, and full of obstacles.[10] Like cowboys, many of them held dreams of owning their own spread, living in a big house, and employing large numbers of others.

The cowboy's work involved action, danger, and spectacular skill on horseback. The cowboy sat high on his horse with the rein tight, and gauged another's worth by the way he could ride. He looked down (figuratively and literally) on anybody who (like the sheepherder) walked the range instead of riding it, or who, if he did ride, rode with a slack rein that called for nothing spectacular or dangerous. He believed that "a man on foot was no man at all."[11] His recreations were boisterous and carefree. His songs, writes Winifred Kupper, contained "the innocuous words and the simple rhymes of a Mother Goose jingle; his stories were tall tales recalling the lore of dragons and giants." He was, according to Kupper, "the perennial adolescent of the West, the Peter Pan of the range."[12]

The sheepherder was more mature. He was careful, conscientious, hardworking. He was alone with his sheep for weeks at a time. His solitary life called for intelligence and inner resources. If he fought his aloneness, he was likely to go insane—so popular opinion held. But Robert Maudslay, a Texas Hill Country sheepherder, laughed at such opinions. "There was too much to do," he writes, "too much to look at, too much to read, too much to think about, for loneliness." Nonetheless, the western phrase "crazy as a sheepherder" embodies the idea that loneliness drove herders insane.[13]

Insane or not, the reading, thinking, philosophizing sheepherder was often utterly unintelligible to the cowboy. Cowboys, for example, thought Martin H. Kilgore a bit strange. In 1887 in the Texas Big Bend Country, a delegation of three cowpunchers rode up to Kilgore, a recently arrived sheepherder. They wanted to know what he was doing there in cattle country with his sheep. Outnumbered and concerned about rough treatment he might receive from the cowboys, Kilgore resorted to guile. He had been reading *From the Earth to the Moon*, Jules Verne's 1865 science fiction novel about a space flight, and he began paraphrasing the story to the cowboys. He and his partner, Kilgore told the cowboys, had built a machine for a trip to the moon and got it loaded with ballast, supplies, and their old sheep dog. They had risen to over five thousand feet by midnight, he said, when they discovered that they were losing altitude. They began throwing out ballast. They threw out supplies. They finally threw out the old dog. After that, they traveled without difficulty. In the morning, upon looking out the window, said Kilgore, they saw their old dog flying along beside them in space. At this point in the yarn, one of the cowboys said, "Let's go. This damned fool is crazy!"[14]

Despite Kilgore's experience, there is something elusive, paradoxical, and contradictory about the matter of personality and character differences between cowboys and sheepherders, for cowmen often ran sheep themselves. Richard King, for example, in the 1860s ran some thirty thousand sheep on his giant cattle ranch in South Texas. Dominicker Hart, one of America's largest wool growers, in the 1890s grazed on his upper Big Bend ranges more than six thousand cattle with his enormous flocks of sheep. In the Texas Hill Country, the earliest German settlers profited from raising both cattle and sheep (sometimes in the same pasture), and Charles Schreiner of Kerrville insisted that if they wanted to borrow money from his bank, cattlemen in the 1880s must use part of the loan to buy sheep.[15]

Still, between the early 1870s and late 1880s, cowboys and sheepherders argued over grazing rights, water supplies, fences, and other matters. Contrasts in lifestyles, background, and equipment helped to produce the hostility. Fundamental differences between sheep and cattle meant that the animals required different amounts of water, different types of food, and different manners of herding. Sheep were thoroughly domesticated, subsisting on short grass and weeds. Except in hot weather or when being trailed, they needed little water, but got by on the morning dew. They needed a constant attendant, however, preferably one who went on foot. The longhorn of the post–Civil War era, on the other hand, was a wild creature, requiring long grass and plenty of water. His very nature required men on horses who could cope with his feral constitution.

Cowboys met the requirement. A good cowboy, writes William Forbis, was "a dirty, overworked laborer who fried his brains under a prairie sun or rode endless miles in rain and wind to mend fences or look for lost calves."[16] His work was difficult. It consumed his waking hours through three seasons. In winter, about half the cowboys, unemployed on the ranch and with no long drives to join, stayed in town and took odd jobs, like painting houses. (Some people complained that they painted the town red.)[17]

Cowboys and visitors spread canvas on the ground for this 1918 dance at a Spur Ranch camp. While most of the men are wearing big hats and tall boots, the fiddle and guitar players are in "button" shoes, and one of the dancers is wearing bib overalls. A Frank Reeves photo. Courtesy of the Southwest Collection, Texas Tech University. SWCPC File no. FR 265.

Sheepherders did not look like the cowboys, but they worked as hard, and all year long. A sheepherder cared for two thousand highly dependent sheep, and of necessity he stayed with them at all times. Sheepherders could swear as hard as cowboys, play poker and other games of chance as well, and during the few times when they were in town (or so folklore maintains) drink as much. But they seldom made as much noise. (Philosophers usually don't paint a town red.)[18]

But the two were hardly similar. Cowboys and sheepherders differed in their dress and appearance. Although there are few photographic records of him ever looking much like the characters from old western movies or even recent television productions, a cowboy might get himself dressed in expensive clothes when he could afford it. A cowboy might also spend as much as a month's wages on a pair of custom-made boots, and, writes Kupper, "he could [often] present a pleasing picture of a utilitarian sort."[19]

A sheepherder never looked pleasing in any sort of way. Of one, John Clay writes, "I met old English Joe, an old, weather-beaten man. There he stood on a knoll, unshaven, a greasy hat on his head, his clothes worn and ragged, watching a flock of sheep as they kept slowly trailing a bench of rich grass."[20] The sheepherder, according to western folklore, wore a hat without a brim, mesquite thorns for buttons, shoes that were shapeless masses of leather, and pants that were the color of the ground around him. For the sheepherder there was rarely anything to dress up to. As Kupper writes, "the hours spent with greasy sheep, walking in dust and mud, and living out-of-doors" were reflected in his appearance. Rarely was there a wife to help keep him neat; indeed, "he often spent months without seeing a woman," and in general he "had less incentive" than a cowboy "to make himself attractive."[21]

The basic ingredient of the sheepherder's function, and his identity, was the wagon. Sometimes covered by canvas, sometimes an all-wood structure, the wagon carried all the man's worldly possessions and the tools of his trade. It was built for efficiency of storage, and it was a marvel of compactness. In harsh weather, a kerosene lamp and a stove kept the wagon warm. The end gate made into a table, the tongue in front, to which horses or mules were hitched while traveling, made a seat or bench upon which to sit before the fire. The wagon was seldom swept out, never cleaned. Over much of the West it was both characteristic of and indispensable to the sheepherder.[22]

The basic ingredient of the cowboy's function, and his identity, was the horse. Although the relationship between rider and horse was a practical arrangement, rather than a love affair, the horse nonetheless became the major element in the self-image of the cowboy. The horse provided transportation and mobility, power and strength. Veteran cowmen, proud of the feeling of height and power that came from being mounted, displayed (suggests western folk tradition) an aversion to walking any distance greater than, say, from the bunkhouse to the barn.[23]

The cowboy has become part of America's western tradition. Far more has been written about him than about sheepherders, and he has appeared in more movies too. His chroniclers and poets have perpetuated the fiction. With longhorns and mustangs, writes Joe B. Frantz, the cowboy has become part of the "Holy Trinity of Texas," and the legend is enjoyed from one end of the modern globe to the other.[24]

No such romance is associated with the sheep industry. Essentially, the sheepherder has been forgotten; he has few chroniclers and fewer poets. He is the central part of few, if any, movies. "The sheepherder," writes Arthur Gilfillan, "ain't got no friends." Even such leading environmentalists as John Muir took the sheepherder and his flock to task: "hoofed locusts" Muir called sheep. A lonely and, if the poets of the West are correct, lowly character, the sheepherder makes up no part of the western myth. (This book, for instance, is not about herders; it is about cowboys.) For sheepherders there was no trinity of any kind.[25]

Although in fact they had much in common, clearly cowboys and sheepherders of the nineteenth century American West often presented sharp contrasts in dress and appearance, in character and personality, in work and leisure. They rarely got along, purchased supplies when possible at separate mercantiles, ate and drank at different establishments, and in general avoided one another.

But after the close of the open range in the 1890s, changes came. Many cattlemen, especially in the Southwest, added sheep to their operations, and sheepmen often acquired cattle and added Angora goats to their pastures. Cowboy-sheepherder antagonism declined.

In the twentieth century, the sheepherder survived, but he retreated in the face of enclosed pastures, net-wire fences, and new ranching techniques. Sometimes, such as the immigrant herder John Moore Shannon, who became a prominent banker, he prospered as a "sheepman," a rancher unafraid of raising exasperating, timid little animals.

The cowboy changed, too, but he did not disappear. Rodeos, pulp fiction, movies, radio shows, and modern cowboy culture gatherings kept his nineteenth-century image alive and well. Indeed, unlike the sheepherder, the cowboy and the cowboy way are celebrated the world over.

Notes

1. The painting hangs in the Amon Carter Museum in Fort Worth. A reproduction of it is in William H. Forbis, *The Cowboys* (New York:

Time-Life Books, 1973), 7–8. For a brief discussion of the evolution of cowboy garb as presented by painters and photographers, see Richard W. Slatta, *Social History in the Saddle: Trailing the History of the Cowboys of the Americas* (Lubbock: The International Center for Arid and Semiarid Land Studies, Texas Tech University, Publication Number 98-2, 1998), 5–8.

2. Winifred Kupper, *The Golden Hoof: The Story of the Sheep of the Southwest* (New York: Alfred A. Knopf, 1945), 78; a copy of the painting faces the title page.

3. Forbis, *The Cowboys,* 17.

4. Ogden Tanner, *The Ranchers* (Alexandria, Virginia: Time-Life Books, 1977), 103; Kupper, *The Golden Hoof,* 69-70; Paul H. Carlson, "Bankers and Sheepherders in West Texas," *West Texas Historical Association Year Book* 61 (1985): 5-14.

5. Kupper, *The Golden Hoof,* 70.

6. See Carlson, "Bankers and Sheepherders in West Texas," 5-14.

7. Edward Everett Dale, *Cow Country* (Norman: University of Oklahoma Press, 1942; 1965), 223.

8. Forbis, *The Cowboys,* 7, 17. See also John Clay, *My Life on the Range* (Chicago: privately printed, 1924), 56; Dale, *Cow Country,* 218-23.

9. Tanner, *The Ranchers,* 85, 87, 97, 102-5.

10. Kupper, *The Golden Hoof,* 45-47; *Corpus Christi Caller,* July 19, 1925.

11. J. Frank Dobie, *On the Open Range* (Dallas: The Southwest Press, 1931), 102. See also Dale, *Cow Country,* 218-23.

12. Kupper, *The Golden Hoof,* 72.

13. Cited in Ibid., 73.

14. Florence Fenley, *Old Timers, Their Own Stories* (Uvalde, Texas: Hornby Press, 1939), 19-20.

15. See Paul H. Carlson, *Texas Woollybacks: The Range Sheep and Goat Industry* (College Station: Texas A & M University Press, 1982), 178-80.

16. Forbis, *The Cowboys,* 7.

17. Kupper, *The Golden Hoof,* 78; Dale, *Cow Country,* 122.

18. Kupper, *The Golden Hoof,* 78, 101.

19. Ibid., 74. See also Slatta, *Social History in the Saddle,* 5–8.

20. Clay, *My Life on the Range,* 56.

21. Kupper, *The Golden Hoof,* 77.

22. Tanner, *The Ranchers,* 96.

23. Richard W. Slatta, *Comparing Cowboys & Frontiers* (Norman: University of Oklahoma Press, 1997), 91-92.

24. Joe B. Frantz, *Texas: A History* (New York: W. W. Norton & Company, Inc. 1976), 130, 135. See also William W. Savage Jr., ed., *Cowboy Life:*

Reconstructing an American Myth (Norman: University of Oklahoma Press, 1979), 3-14; and several articles in Buck Rainey, guest editor, *Red River Valley Historical Review* 2 (Spring, 1975): 9-65.

25. Arthur Gilfillan, *Sheep* (Boston: Little, Brown & Company, 1929), 85.

10 Stockyards Cowboys
J'NELL L. PATE

Some of our earliest images of cowboys include herding cattle to the "cowpens" in the Carolinas in colonial days, transporting a few head eastward for the Confederate army in the Civil War years, and especially in the 1870s, trail-driving thousands of Texas longhorns northward to railheads in Kansas. The working men in these instances were not really called cowboys; they were drovers, herdsmen, or waddies. Another category of livestock workers with various names and different job descriptions who were not really called cowboys but who fit the broader definition of the term were stockyards cowboys, employees of the urban livestock markets.

From approximately 1860 to 1960, a unique system for marketing livestock existed, and it used workers we can call "stockyards cowboys." The system developed almost completely as a consequence of the expansion westward of railroad lines. Where railroads connected in large cities, terminal market centers developed as several smaller stockyards united. Eventually, a few large meatpacking and slaughtering facilities accepted nearly all the livestock brought to the market.[1] In fact, the large stockyards companies, railroads, and meatpackers worked together so well that during the last two decades of the nineteenth century they made meat slaughtering the largest industry in the United States.[2]

The general public did not think of workers or employees in such terminal markets as "cowboys," in part because a dozen or so specific job descriptions applied to the various tasks the men performed. When men work with livestock, however, some incidents are universal. If an angry bull began chasing a handler, the man generally did the quickest thing possible to get out of the way. A cowboy on the trail, for example, got on his horse fast; a ranch cowboy climbed a fence, and so did a stockyards worker or a rodeo cowboy.

Work in the outdoors in the sweltering heat of summer, as another example, or the numbing cold of blue-northers in winter, was for stockyards workers not unlike the work of trail-driving cowboys or cowboys on ranches. "Horsing around" was also typical of cowboys anywhere. A prankster might unsaddle a friend's horse that had been tied to a fence, put a mule in the same spot with the cowboy's saddle on it, and hide the horse elsewhere. Such an incident happened at a stockyards; it could just as easily have happened on a ranch.

Job titles for workers at stockyards make up a long list, as cows are not the only animals involved and many people at a stockyards are in business for themselves. Animals generally bought and sold at a stockyards included cattle, calves, hogs, sheep, goats, horses, and mules. In stockyards of Ohio, Illinois, Missouri, and Kentucky, hog receipts predominated, but cattle dominated in western centers like Kansas City, Denver, Fort Worth, and Omaha.

A terminal market or stockyards center existed to facilitate the buying and selling of livestock. Commission companies, large and small, sold livestock for producers, those farmers and ranchers who sent them to the market. Order buyers bought lots of animals for someone else. Meat packers hired experts who bought only cows, calves, hogs, or sheep. Other order buyers might have requests to buy calves to stock the northern ranges, while traders and dealers generally bought and sold for themselves to make a profit. The stockyards company did not buy and sell but only provided the place for others to do so.

Nevertheless, the company employed sometimes hundreds of workers to unload animals from railroad cars, drive them to pens, feed and water them, and clean up the pens. Folks working for any of such businesses were simply called "yard hands." Other livestock workers included muleskinners or wranglers who worked with the horses and mules,[3] cattle raisers' brand inspectors, men dipping animals for ticks (after the turn of the century), veterinarians (or folks who knew animal care without the formal schooling involved), weighmasters, telegraph operators, and reporters for livestock market newspapers. During the dry season, somebody served as fire marshal and patrolled the stockyards with a crew with buckets. In hot weather with plenty of hay around, and when manure got dry as powder, for example, someone might accidentally flick the ashes off a cigar in the wrong place. The ashes might smolder for hours and then flare up and soon be out of control.[4]

Big stockyards at the rail terminals began to emerge rapidly in the 1870s and 1880s as investors multiplied and the livestock industry expanded. After the 1890s and by the turn of the century, a larger workforce became available as immigrants of the "new" immigration from Central and Eastern Europe arrived in the United States seeking perhaps political, economic, or religious freedom. The immigrants, arriving by the millions, often traveled directly to stockyards cities for available jobs. Most worked in meatpacking plants in conditions perhaps not quite as bad as Upton Sinclair described in *The Jungle*, but almost. Only a few managed to acquire jobs in the stockyards.[5]

All of the job titles and the various activities of workers—those who handled the cattle, anyway—could in a broad sense be called "stockyards cowboys." Their duties depended on which business within the yards employed them.

A president of the Wichita Union Stock Yards once used an analogy long accepted in the livestock industry. He described a stockyards as a "hotel for livestock." "For instance," he said, "we rent 'em a room—a pen; if they want to eat, we feed them. We furnish branding service as a hotel furnishes a manicure; we have a veterinarian on duty offering a complete service, just like a hotel employs a house physician."[6] The length of time the animals generally stayed in a "livestock hotel" before being sold was twenty-four hours.[7]

A typical sixteen- to eighteen-hour workday at the yards began when a shrill whistle blew at one of the packing plants as a wake-up call for the broader stockyards community—usually at 5:30 A.M.[8] In the days before U.S. Department of Agriculture and Packers and Stockyards Administration rules, buyers were able to buy livestock any time there was enough daylight to see the cattle. "You started when you could [see to bid] and quit when you couldn't."[9]

A condensed version of a routine day at a stockyards, a day from the time livestock arrived until they left, involved multiple stockyards "cowboys." Workers unloaded cattle from railroad cars at the

Swenson Land and Cattle Company stockyards workers on horseback, early 1930s. Courtesy of Southwest Collection, Texas Tech University. SWCPC File no. 123(b), Envelope 39.

unloading dock, and a stockyards employee called a key man received them and locked them in catch pens. A gate man from a commission company to which the livestock had been assigned filled out a weigh bill and moved the cattle to the pens of the commission company. Cattle salesmen from the commission company brought buyers around to see the animals, listening to bids and accepting the highest. Then the salesman told the yard man of his commission company to take the animals just sold to the weigh scale. The weighmaster wrote the weight, owner, purchaser, and price on a scale ticket. Once the animals stepped off the scales, they belonged to the new owner. A stockyards company employee drove the animals to a pen for holding until picked up by an agent for the owner, which might be a packing company, an order buyer, a speculator, or someone else. The commission company finished the paperwork, deducted its fee, and sent a check to the owner before business closed that day.[10]

The men worked hard all day, especially if the railroad cars kept coming in steadily. The boss of a commission company or order buying outfit worked just as hard and got just as dirty as his lowest paid worker. The poorest paid workers were the stockyards employees who scooped up the manure from the pens, hosed the animals down, put out fresh hay and water, and drove the animals to and from the weigh scales. They all worked long hours. Usually the men brought a lunch pail and ate on the run. Sometimes an enterprising person made the rounds with a food wagon and sold individual items reasonably. On the Fort Worth market someone sold tamales.[11] If the runs of livestock kept coming for days, often the workers did not stop unloading until 1:00 or 2:00 A.M. Some kept a bedroll on the dock and slept between carloads.[12]

In the days when most animals arrived by rail, steers sometimes had horns so wide they could not walk straight out of a railroad car opening—although they tried. Stockyards workers had to turn a steer's head so that one horn came out at a time. In this type of ticklish situation, clothes got snagged or men got jabbed and hurt by the sharp horns. Cattle often were wild, having grazed on quiet ranges for a few years before being loaded in the railroad car where the animals remained penned for hours.[13]

One of the major chores in handling cattle at the terminal market was sorting the animals by size, shape, color, weight, sex, age, or any other characteristic the buyers and sellers deemed important. Stockyards cowboys sorted the animals into various pens, often riding on horseback as they did so. (Stock handlers in the nineteenth century would never have imagined that driving cattle down the alleys at a stockyards could be accomplished a century later by long-haired kids on motor scooters, but it was.. Cowboys have indeed changed!)[14] Some workers drove and sorted on foot, however, for nearly everyone carried a heavy walking cane, sorting pole, or buggy whip to use when driving cattle.[15]

Fortunately for the workers, there were lulls and slack periods. On cold days, workers gathered around the potbellied iron stoves in the scale houses to warm, catch up on the news, and spit tobacco in the spittoons. On most other days out in the open on the yards, they whittled. Wooden pens, in a day when all men carried pocket knives, proved an irresistible temptation to men trying to kill a little time. If he had made an offer to a calf salesman of a large commission company, an order buyer might whittle a while in the silence while the other man thought over the offer. Actually, whittling proved to be a good technique in making a deal.[16] Some stockyards cowboys printed up flyers and addressed and mailed several hundred, or perhaps a thousand, notices to customers of the commission company to quote prevailing prices. In a day before female office help, either the owner of the company or his staff handled such duties.[17]

Stockyards personnel—cowboys—used terms unique to the stockyards or to livestock marketing experience. They would not be the only persons ever to use the terms because folks who raised livestock on the ranch or farm used them too, but the terms originated at the market. For businessmen earning a living from the cattle, or other livestock trade, the terms "speculator," "order buyer," "trader," "dealer," or "cattle broker" were used rather broadly. The meatpacking stockyards lingo that developed generally referred to the animals. The following terms offer an example:

> *catch pens*—any small pen or a group where cattle were held after being weighed
> *clearinghouse*—an organization that furnished money, office space, and clerical help to traders, order buyers, or commission men on a percentage basis
> *draft*—any group of cattle weighed at one time
> *fill*—the weight that cattle gained from feed and water while in the yards
> *grassy*—cattle grazed on range or pasture alone
> *hat racks*—Nellies, old thin cows
> *killing cattle*—cattle in the condition to be slaughtered
> *runs*—going to market, especially in rather large quantities
> *scalawags*—stock that was thin and emaciated
> *she-stock*—cows and heifers
> *shipper*—the owner of cattle or livestock shipped to the yards

shrink—the amount of weight that cattle would lose in shipping or standing around not eating

springer—a cow due to calve

stale—an animal kept on the yards for too long

walkway—wooden walkways that were built on top of fences
to allow the men to walk above the pens

yardage—the fee paid for use of stockyards' facilities and
food

Men who worked around cattle or livestock seemed to love their
work as a way of life; they had a code of honor uniquely theirs and
were often tough men who used tough language—except around
ladies. If promised so much for a carload or a trainload of cattle, a
buyer was honor bound to pay that price "regardless of the consequences." If he backed out of a deal, a person might as well leave the
country because he wouldn't be trusted henceforth.[18] Surprisingly,
the men were polite to each other. "If you walked up on someone
making a trade, you politely walked off and left them alone. And if
you heard what they were talking about, you left the deal alone."[19] In
addition, the men had incredible memories. The dealers and commission agents could recall the weight, cost, quality, and condition
of cattle they were selling and had sold in times past. For the most
part, the yards were peopled by those who appreciated opportunity,
hard work, and integrity. Others sometimes dealt on the yards, but
they did not last long.[20]

Men became so accustomed to the bidding process, which was
the real feature of private treaty sales on the yards long before auctions, that when they shopped in stores adjacent to the yards to buy
hats, shoes, or other clothing, they generally bid a little lower than
the marked price. Merchants did not especially like the practice.[21]

Reasons for being a stockyards cowboy varied, but in general
one made more money as a commission man or order buyer than
cowboying on a ranch or running a few head oneself. Even when
prices fluctuated, thus hurting the producer, the dealer on the yards
got his cut, and his job stayed reasonably steady. Often one could be
his own boss on the yards as a speculator, agent, buyer, or other
dealer. One got to deal with cattle, after all; one was in the business

of livestock. Only the hands who worked for either the stockyards company or other agents on the yards earned lower wages and always performed the dirtiest chores. Theirs was a job; in fact, they likely did not think of themselves as cowboys or cowmen, even. Several years ago when this writer interviewed more than a hundred people from the Fort Worth Stockyards in their homes, she noticed that retired commission company owners, packer buyers, and self-employed dealers lived in fine brick houses, but the retired stockyards company laborers lived in small frame houses in poorer sections of town.

Quite often being a "stockyards cowboy" resulted after opportunities for driving herds up the trail in the 1870s or 1880s had ceased. One was a cowboy at heart and wanted to stay in the business. George W. Saunders was an example. Saunders had hired on in 1871 at age seventeen for his first trail drive. After a dozen years of going up the trail, he bought a livestock commission business in San Antonio, but sold it and drove horses up the trail. Finally, in the spring of 1886, Saunders returned to San Antonio and went into the commission business at the stockyards for good, even incorporating and selling shares of stock in a business that grossed between five and six million dollars annually. He also managed some ranches on the side.[22]

Another former cowboy who completed his working years on a livestock market was Charles C. French. As a young man in 1878 he hired on with a small trail drive north from Austin, Texas. His older brother Horace was the trail boss, and they took the cattle north to Dodge City, Kansas. Then in 1879 they drove a herd for a different cattleman all the way to Ogallala, Nebraska. Later they took four thousand steers from that same herd on northward to the Cheyenne agency, crossed the Missouri River when it was frozen twenty-eight inches thick and the temperature, according to French, was 72 degrees below zero.[23] By 1888, French was traveling in Texas buying cattle for such commission companies as McIlhany, James H. Campbell, and Evans-Snider-Buel. In 1896, he held a job as promotional agent for the Fort Worth Stockyards. Over the next forty years, with only a couple of jobs elsewhere, French promoted the Fort Worth market. He helped create a fat stock show that survives over a century later as the Southwestern Exposition and Livestock Show, Fort Worth's largest single annual event. French also helped to create pig, beef, and corn clubs for youngsters throughout Texas, Oklahoma, and Louisiana—clubs that predated, and later contributed to, the expansion of the 4-H movement.[24]

People may have an image of a cowboy as a lone, silent, and bashful man, but get the species in groups, at a stockyards, and someone begins to tell good stories. To be a successful salesman one had to be an "inveterate talker," someone like Charles C. French. He was said to know all the cattlemen in Texas.[25]

Typical of the ranch manager–cattleman whose entire working life revolved around cattle but who ended up at a stockyards was Scotsman John Clay. Born in the Border Country of Scotland in 1852, Clay came to the United States as a representative of fellow Scots who were investing in America's booming livestock industry. He managed the Swan Land and Cattle Company for eight years, served his turn as president of the Wyoming Stock Growers' Association, and eventually turned to the commission business. His John Clay Commission Company was still going strong on several stockyards half a century after his own death in 1934.[26]

While a few of the big stockyards' terminal markets still operate, most of them have ceased their trading. Activity moved to rural and small-town auctions, or packers purchased directly from feedlots in the grain country. Many sales occur these days after buyers, at home or in their offices, have seen the herd on a videotape that shows the cattle on a ranch. The big stockyards era, which lasted over a century, effectively came to an end for many stockyards people in the 1970s and 1980s. Some young men who began by shoveling manure in the pens worked their way through various jobs to the top, owning the commission company or order-buying company for which they had worked; or they started their own firms. Others moved around a lot, working for most of the companies on the yards sooner or later.[27]

Some retirees from the old days doubt if the stockyards they knew could ever operate under today's regulations of OSHA, constitutional rights, overtime restrictions, civil rights, animal rights, affirmative action, or minimum wages. For them, a forty-hour, five-day work week was unknown. "Cattle had to be tended whether it was Christmas, Thanksgiving, or Sunday."[28] Cowboys on the range knew that situation too. It is no wonder, then, that the old time stockyards cowboy is pretty much a cowboy of the past.

Notes

1. J'Nell L. Pate, "Livestock Hotels: America's Historic Stockyards," unpublished manuscript, 4.

2. Harmon Mothershead, "The Stockyards, A Hotel for Stock or a Holding Company," *Nebraska History* 64 (Winter 1983): 519.

3. "Horse and Mule Notes," *The Cattleman* 1 (June 1915): 37.

4. Bill Sloan, *In and By: Items Sorted Out of My Experiences on the Fort Worth Stockyards 1945-1955* (Austin: Nortex Press, 1987), 85.

5. "Committee Tells Packers Bohunks Are Not Wanted," *The Texas Stockman-Journal,* December 25, 1907, 6.

6. Ralph Hinman Jr., "Stockyards Functions as Hotel: Room, Meals, Even Manicure," Clipping in Wichita Meat Industry and Trade Stockyards folder, n.d., Wichita Public Library, Wichita, Kansas.

7. Pate, "Livestock Hotels: America's Historic Stockyards," 98.

8. J'Nell L. Pate, "Livestock Legacy: A History of the Fort Worth Stockyards Company 1893-1982," Ph.D. dissertation, Denton, Texas: University of North Texas, 1982, 172.

9. Interview with Johnnie Stubbs, Fort Worth, Texas, September 15, 1983.

10. Interview with Gary Allen, Fort Worth, Texas, February 22, 1982.

11. Sloan, *In and By,* 17.

12. Interview with Johnny Adams, Fort Worth, Texas, June 7, 1983.

13. Interview with Claude Marrett, Fort Worth, Texas, September 12, 1983.

14. Paul W. Horn, "Ponies with Wheels," *The Cattleman* 63 (June 1976): 66-68.

15. Sloan, *In and By,* 43.

16. "Whittlers Are Rough on Posts and Benches at Stockyards," *Fort Worth* 22 (January 1948): 14.

17. Interview with William E. Jary Jr., Fort Worth, Texas, July 12, 1983.

18. Sloan, *In and By,* 2, 10.

19. Wade Choate, *Swappin' Cattle* (San Angelo: Newsfoto Publishing Co., 1990), 118.

20. Ted Gouldy, editor and publisher of *The Weekly Livestock Reporter,* in foreword to Sloan, *In and By,* v.

21. Transcript of oral interview by W. H. Barse Jr., Fort Worth, Texas, 1977, in Junior League Oral History Collection, Oral Histories of Fort Worth, Inc., 7-8.

22. George W. Saunders, "Reflections of the Trail," in *The Trail Drivers of Texas,* compiled and edited by J. Marvin Hunter, 1924; reprint Austin: University of Texas Press, 1985, 430, 441, 449-50.

23. C. C. French, "When the Temperature was 72 Degrees Below Zero," in *The Trail Drivers of Texas,* 742-43.

24. Rossie Beth Bennet, "History of the Cattle Trade in Fort Worth, Texas" (Master's Thesis, George Peabody College for Teachers, 1931), 4; "Arousing Interest in Hogs," *The Fort Worth Daily Live Stock Reporter,* April 11, 1911, 1; "The Cattlemen's Convention," *Fort Worth Gazette,* March 10, 1896, 4; and C. C. French, "The History of a Great Fort Worth Institution: The Southwestern Exhibition and Fat Stock Show," C. C. French Manuscripts, as cited in *Research Data,* Vol. 62, *Federal Writers Project,* Fort Worth Public Library, Fort Worth, Texas, 24582.

25. "Personal Mention," *Texas Live Stock Journal,* September 5, 1891, 10; and Pate, *Livestock Legacy,* 77.

26. John Clay, *My Life on the Range,* with an introduction by Donald R. Ornduff (1923; reprint; Norman: University of Oklahoma Press, 1962), ix-xv.

27. Sloan, *In and By,* 45.

28. Ibid., 53.

11 Cowboy Humor
KENNETH W. DAVIS

The humor of the American cowboy is distinctive, yet traditional and universal. Like most folk occupational groups, American cowboys find humor in their work. This humor is situational, earthy, spare, and often abrasive. It reflects many aspects of the really strenuous work that is cowboying. Despite the romanticized view of the cowboy's life and work as being glamorous and fraught with high adventure—a view fostered in dime novels and in cinema—in reality the work of a cowboy is today as it was in the nineteenth century: difficult, challenging, and demanding. To succeed as a cowboy, a man has to be strong physically and mentally. As humorist Curt Brummett has observed, cowboys also must have a sense of humor to survive their difficult and often dangerous occupation.[1]

Cowboys use humor to relieve the sometimes monotonous routines of long days spent in the saddle herding cattle or equally long days branding and castrating calves, tending fences, painting barns, or doctoring sick animals. In such work situations, humor can take the edge from boredom or enable cowboys to cope with serious physical challenges. If, for instance, a cowboy has a "wreck" (a fall from a horse, or a bad encounter with a reluctant cow that causes potentially grave physical injury), his co-workers will more often than not laugh at his misfortunes rather than be sympathetic. When he is jeered for not being a better horseman, a cowboy learns to be more careful, more attentive. The seemingly harsh, even callous humor at the expense of a cowboy who has had a bad fall or a run-in with an angry cow can also serve to relieve tension or defuse sincerely felt but seldom expressed anxiety about the welfare of a companion.

In their times away from their demanding labors, cowboys enjoy socializing just as do other folk groups. They like good parties, dances, barbecues, storytelling, and of course, weekend rodeos in which they compete for prizes for their occupational skills, such as milking wild cows, roping, and bulldogging. Whatever their recreations may be, cowboys enjoy jokes at each others' expense. Traditional elements are common in much of the humor that occurs in recreational situations. In folk humor, a motif as old as ancient Roman comedy is that of the proud or "uppity" individual being brought down or taught a lesson. An example from the short grass country of north central Texas illustrates this principle. On a ranch near Throckmorton, Texas, a rather boastful, vain cowboy worked diligently—sometimes at two jobs—to get the money to buy a brand new pickup. Cowboys take pride in their horses and in their pickups. They are prone to brag about the superiority of whatever horse they ride or whatever brand of pickup they happen to own. The Throckmorton cowboy could hardly wait to take his new pickup to a dance in nearby Albany. When he arrived, he began showering attention on an attractive woman, who had recently broken up with another cowboy who happened to drive a battered old pickup whose original paint job was barely discernible. The proud owner of the new pickup finally persuaded the woman to go with him for a ride out in the country. When they left the dance hall and got in the sparkling new vehicle, it would not start. The soon angry owner tried every thing he knew to do in his efforts to get the motor to start, but he had no success. In a while, the woman's ex-boyfriend came out and with exaggerated concern for the new pickup's owner, volunteered to tow it to a service station that had a mechanic on duty. In just a few minutes, a tow chain was attached to the new pickup and the journey to the service station began. But the driver of the battered old pickup did not go the direct route to that station; instead, he drove around the courthouse honking his horn so that every one would notice that he was towing a new vehicle. The owner of that new vehicle was powerless, of course, to do anything but endure the laughter of the cowboys and townspeople who watched the small but comic parade. The following Sunday morning at the Dairy Queen, the coffee drinkers chuckled about the spectacle and commented on its significance: "Well, that just serves him right for being so uppity and all about having a new twenty-thousand-dollar truck. For all that money he spent, he had to be towed by Tommy's old beat-up Ford. I don't care none if he does have his nose out of joint." For weeks after the towing incident, the once vain owner of the expensive new pickup was teased mercilessly.

An interesting corollary to cowboy humor is also in the grand tradition of folk humor. The interaction of the oral narrator with his audiences can be an instance of humor that augments the story's comic quality. In the story about the defective new pickup that had to be towed, variants developed as it was repeated by different narrators. Interruptions such as "But that's not the way I heard it," or "You forget to mention the look on that woman's face when her old boyfriend came up and offered to drive the two of them to the mechanic's," or "You reckon that old boyfriend had messed with the wires in that brand new truck?" Each of these comments or queries by members of audiences hearing the account would be incorporated into the narrative. Cowboy tales thus continue to grow by accretion as folk tales have done for centuries. When a tale such as the one about the pickup is being told, the interaction between the teller and the one who interrupts him can be a source of wit, also. An interrupted narrator will sometimes say, "Well, who is telling this story, me or you?" The interrupter may retort, "Well, you are tryin', but you shore need all the help you can get."

In their times away from their work, cowboys keep alive another ancient tradition in folk humor: the telling of tall tales, or as the cowboys call them, "windies." Sitting around the modern equivalent of an eighteenth-century British coffee house—a Dairy Queen in a small town—cowboys sometimes delight in trying to outdo each other in telling of phenomenally intelligent animals or prodigious feats of physical strength. Cattle dogs are found on many modern ranches. These highly trained dogs can in reality do amazing things in working cattle. Quite naturally, in the tradition of folk humor their skills become greatly magnified in stories told about them. And over the coffee cups a tale of one such talented cattle dog, an Australian blue heeler, described this amazing creature's tremendous skills. One cowboy who had worked with the dog said it could make a herd of frightened and angry steers calm down and then separate them into even numbered groups, and could make them file into the

133

branding pens one at a time. As the tales of this dog's competence spread, its talents grew immensely. Soon in some stories, this prodigy of a dog could sort the steers from the heifers. Then, it was claimed that it could be told to get all the two-year-old cows from a herd and bring them in. After it had done that chore, it was told to bring in the three-year-olds, and so on. In some versions of the brilliant dog stories, it could be told to separate the brown cows from the red ones, the spotted ones from those with solid colors, and those with horns from those who had been dehorned. One creative yarn-spinner, after hearing various accounts of how intelligent this Australian dog was, came up with a boast to top all the previous tales. "My old cow dog is so smart he can cull out the just-bred heifers from the old cows and it can put the older one in bunches according to how many calves they have had. It will have a bunch of mama cows over here who have had three calves, and on the other side of the pen, it will have another bunch who have had only two. And . . ." But at this point, the cowboys at the table with him shook their heads, laughed, and left to go back to their labors. As they left, one cowboy who had told a milder yarn about a cow dog lamented between laughs, "Hell, the first liar in this outfit don't have the chance of a snowball in July."

Cowboy humor has a strong element of machismo in it. In the nicknames they give each other there is a stout undercurrent of admiration as well as of depreciation of physical attributes. An overweight cowboy may be nicknamed "Tubby" or even "Lard" (shortened version of "lard-assed"), but with wry irony, the obese cowboy may be called "Slats" or "Tiny" or "Skinny." Far up in the Texas Panhandle, a cowboy who had a tremendous amount of body hair was called "Ape" by some and "Hairless" by others. Sexual humor is common enough also in the cowboy world. On a large ranch south of Wichita Falls, Texas, a cowboy and his wife had eight children in the first nine years of their marriage. This man's fellow cowboys expressed their admiration for his fecundity by calling him "Stud." On another ranch was a young Texas Tech University College of Agriculture graduate whose successes with good looking women were a cause of major envy in his fellow cowboys. This amorous cowboy was nicknamed "Lucky" by his friends who wished secretly or openly that they, too, could be as successful with women. Other nicknames reflect various physical attributes as well. On the Pitchfork Ranch between Dickens and Guthrie, Texas, a six-foot eight-inch cowboy was called "Big Dog." On a neighboring ranch one of the cowboys who was as strong as any two men but who had a peculiar voice was called "Squeaky Steel" or sometimes merely "Squeak." A tall, spindly cowboy who was described as being so thin he could not cast a shadow was called "Yard," for as one of his companions said of him, "He's as thin as a yardstick."

A type of humor common in the world of the cowboy is the practical joke. Usually practical jokes are done for vengeance or "getting back" at an individual who has played a joke earlier. Playing practical jokes often becomes highly competitive. Part of the humor in this activity comes from seeing who can outdo whom in inventiveness. One hapless cowboy was victimized regularly by his buddies in the ranch bunkhouse He was quite gullible and totally trusting, so he was fair game for his companions, who thought themselves ever so much more intelligent than their seemingly "innocent" companion. These other cowboys delighted in short-sheeting their friend's bed, or in filling his pillow case with rocks or corn cobs, or in filling his boots with wet socks, or with telling him that the manager or foreman wanted him to be at the corral at 4 A.M. saddled and ready to ride. Of course, when the cowboy got to the corral and saddled his horse, he had to wait an hour or so until the long day actually began. This much-teased cowboy accepted the taunts and tricks of the other men all through a long winter. When the warm days of spring came, he began to leave the chuck house soon after finishing the evening meal. He would sometimes not return until after 10 or 11 P.M. When asked what he did when he was out so late, he would say either "nothing" or he would say "I've been looking for a pet." But he would say no more. The cowboys persisted in their playing of jokes on the man, and he continued to do nothing to retaliate. Finally, in the first week of June, when high temperatures were common, he came in one night from his mysterious after dinner jaunts with a canvas sack hung over his shoulder. The other cowboys asked him what was in the sack. "Oh," he replied, "Ain't nothing much, just a little pet, a sort of guard, I guess." He then opened the sack and poured out a half-grown rattlesnake. "You reckon this here snake can guard my bed and my boots?" Cowboys of all ages, despite whatever infirmities they had, jumped out of windows, ran for the doors, and screamed or cursed in their complete terror. "God-amighty, you sumbitch," one frightened cowboy said to the man who had endured countless practical jokes, "Have you gone out of your damn mind? You got to be crazier than an outhouse rat to bring a rattler in here. GIT RID OF THAT—" But at this point, the snake coiled as if to strike. When it opened its mouth, the frightened man saw that the snake had no fangs. The butt of jokes had removed them. "Well, I'll be double damned if you ain't pulled a good one on us." After that, there were no more practical jokes played on the one-time victim.

As old as the comedies of Roman playwrights Plautus and Terence are situations in cowboy humor in which individuals of a lower socioeconomic status manage through native wit to outdo or "best" their wealthy, better educated employers or any group that holds itself above supposed lower classes. In what is probably B. M. Bowers's best known novel, *Chip of the Flying U* (1905), there is an account of a practical joke played by a cowboy on a group of archeologists from an eastern university who came to Montana in search of the bones of ancient hominids. These scholars were not having any success in their search, so one of the Flying U cowboys decided to help them out. He donned a partially cured cowhide with most of the hair still in place and ran through the group's field camp. He screamed wildly, then ran out of sight. The searchers for evidence of hominid bones were traumatized fully. Although the cowboy who pulled this trick was severely reprimanded by the ranch's owner, the other cowboys felt that the proud easterners got what they deserved for behaving as if they were superior to the lowly cowboys on the ranch.

In contemporary cowboy humor, the cartoons of the late Ace Reid often depicted the mighty brought down by the humble. In Reid's cartoons, there is a Hogarthian realism in the depictions of scrawny, wiry cowboys and their haggard wives struggling to survive in a harsh, unforgiving environment in which human beings vie with mean cattle and spooky horses, and struggle with drought,

First old cowboy reunion, 1897, held at Plainview. Courtesy of Southwest Collection, Texas Tech University. SWCPC File no. 64(b), Envelope 1.

floods, snakes, and dangerous varmints. Reid sometimes allowed his cowboys to win out verbally with a stereotypical flint-hearted banker or lawyer or government official.

In the real world of the modern day cowboy, life often imitates art. Cowboys and ranchers alike must cope with difficulties that seem insurmountable. Here, too, humor becomes a device to make coping with problems endurable. Celebrated novelist Elmer Kelton wisely inserted a lengthy comic interlude in his classic novel *The Time It Never Rained* (1973). This section complements the novel's grand theme of endurance and survival. The interlude recounts a coyote chase in which one skinny, half-starved coyote outwits men on horseback, in pickups, and even a man in an airplane. The coyote's survival skills are counterparts to those of Charlie Flagg, the novel's protagonist, who has to battle drought, famine, government officials, and even his erstwhile friends and fellow ranchers. The humor in this part of *The Time It Never Rained* is not only a complement to the novel's themes; it is also grandly comic in its sheer physical action.

The speech of the cowboy reflects in its imagery the world of nature in which cowboys live and work. Particularly noteworthy are metaphors and similes. From the lips of Festus Hagan, Marshall Dillon's friend in the popular TV series, *Gunsmoke,* came this one: "He was covered all over with ugly like an ape." In the real world, cowboys have a similar natural poetry in their verbal images. Some of their language is bawdy, even seemingly crude, but it has about it a genuineness that makes it noteworthy. Of a job or chore easily done, a cowboy may say "A short horse is quickly curried." Of something that is well-constructed, he may say that it's "built like a brick outhouse," or of something shoddily or awkwardly built, that it "looks like a widder woman's henhouse." Or of bone-killing fatigue after a long and difficult day's work, a cowboy can say "I feel like I've been rode hard and hung up wet." Admiration for feminine beauty evokes the poetic as well: "She's one slick heifer," or "What a mane of hair," or "She's got the eyes of a doe," or a suggestive "She'd do in my stall anytime." Such examples may be more accurately labeled examples of wit or sapience rather than instances of humor, but the distinctions between wit and humor are fine ones at times.

Cowboy speech is often subtle or wry. Of a long-winded preacher's sermon one elderly cowboy in the Big Bend of Texas area said "I never heard a man ignore so many good places to stop his braying. That feller in the pulpit just kept on like a lonesome jackass in a whirlwind." And of a ranch owner well-known for his fixed opinions on every topic, one of his cowboys said "He's just got to be the head stud on everything." A cowboy who lacks character or who vacillates in his opinions or conduct can be described as "anybody's dog who will hunt with him."

The humor of the contemporary cowboy includes not only traditional elements from the Anglo-American heritage; it also reflects the post–World War II explosion of high technology. Cowboys sitting around a table in a restaurant having coffee or having beers on a hot day in a bar may tell the latest jokes from television comedians along with older forms of humorous tales. High-speed transmission of jokes via satellite to television and radio and even e-mail—which also allows rapid dissemination of jokes throughout the world—equips cowboys with some of the same examples of humor found in other parts of the world in vastly differing occupational groups. Today's cowboys sometimes have battery-powered small radios in their shirt pockets or clipped to their belts and listen to news and music through earphones. They may have weather radios so that they can have the latest forecasts to guide them in the making of plans for the day's work. A contemporary cowboy no longer has to rely on sometimes witty sayings about weather, such as "You know it's going to rain when it's raining all the way around and straight down in the middle."

Despite the ready availability of rapid communication devices, cowboy humor tends to remain grounded in its reliance on the immediate world of the modern West. In its most honest forms, cowboy humor reflects genuinely the character of the men and women who live on ranches where they must be in daily contact with all that nature has to offer, good or bad. To survive in so challenging a world, the cowboy or "Westerner" really needs a sense of humor. In times of flood and drought, for example, the cowboy must somehow cope, lest the extremes of weather drive him quite mad. One cowboy saying that was common in Benjamin, Texas, in the first decade of the twentieth century illustrates well the quiet humor of the cowboy: "In Texas, when it rains, it never stops, and when it stops, it never starts." Or consider this statement of what faced a bunch of cowboys who had more work to do in one day than they had time or manpower to accomplish: "Well, boys," the foreman said, "it's gonna be a tough fight today with a short stick."

Without humor, the life of any individual can be difficult if not impossible. Members of the one authentic American folk occupational group, cowboys and ranch owners, enjoy a richly diverse body of humor that illuminates their world and enriches the world of other Americans. Indeed, it enjoys international recognition and admiration. Whether that humor is of the taciturn sort of the one-liner or of the more loquacious, such as a tall tale or "windie," it reminds us of both our frailties and our strengths as human beings who share with the cowboys the need to survive with the aid of laughter. The humor of today's cowboys appeals to us, for in its varied forms it has something to offer to even the most jaded sophisticate as well as to individuals whose agrarian western heritage is still fresh in their psyches.

Note

1. Curt Brummett, Interview with Lawrence Clayton, Abilene, Texas, 1986.

12 Cowboy Songs and Nature in the Late Nineteenth Century

ROBERT G. WEINER

The cowboy songs of the mid to late nineteenth century play a vital role in recording the history of western ranch hands. Such songs contribute greatly to understanding the life of cowhands and the day-to-day activities they experienced, including the hardships and the joys of a life herding cattle. The songs also contribute to the popular image of the cowboy in the twentieth century, as some of these songs have become part of the western music canon ("Home on the Range" and "Git Along Little Dogies") and remain well-known today.

It is important, however, to point out that the waddies (cowboys) of the 1800s, who sang many of the songs now published in songbooks, bear little resemblance to the "singing cowboy" of the movie or television screen, or to the modern country and western singer. Although singers like Garth Brooks or the late Gene Autry might perform traditional cowboy songs, their dress, image, and musical instrumentation have little to do with the original cowhands who sang those songs. Much of the romantic image of the cowboy was fostered through literature, such as Owen Wister's 1902 novel *The Virginian* and the genre of popular twentieth-century western novels.

Cowboy songs are not the same as modern country music.[1] Many of the early cowboy songs took melodies from popular tunes of the day and adapted them with new lyrics. "The Cowboy's Dream," for example, is actually set to the melody of "My Bonnie Lies over the Ocean," and the well-known "Dying Cowboy" was adapted from an earlier British song, "The Ocean Burial."

There are many studies about the history and folklore of cowboy songs.[2] Most early cowhand tunes were not written down, but were passed on through oral tradition. By the late 1800s, however, cowboy songs were occasionally published in newspapers and magazines, and some were discussed in cowboys' memoirs.[3] Ultimately, it was the publication of N. Howard Thorp's *Songs of the Cowboys* (1908) and John Lomax's *Cowboy Songs and Other Frontier Ballads* (1910) that truly brought cowboy songs to the eye of the American public.

While there have been many other collections of cowboy songs, these two volumes (of which several editions have since been printed with various titles) remain the most widely studied and influential. Again, although the "singing cowboys" of the screen, radio, and stage bear little resemblance to the cowboys who first composed and sang the songs in Thorp's and Lomax's volumes, the two books indirectly influenced them. Works like Lomax's book helped to "define . . . the romantic nature of the singing cowboy figure that was to follow in mass media."[4] A major difference, of course, is that the cowboys who originally sang the songs found in Thorp's and Lomax's books were not professional entertainers, as were the "singing cowboys."

While many tunes discuss real figures and events (e.g., Jesse James, the Chisholm Trail, or Billy the Kid), like most other folksongs and tales, the cowboy songs cannot always be relied on to tell historical truth about those people and events. They can, however, be relied on to tell about the real experience of the day-to-day living of the waddy who sang them.[5] As one writer noted, "There is a conscious realness, a vivid touch of life in the cowboy ballads that make them more than mere bits of 'rag muffin' verse. They are an expression of the . . . lives of those who composed them. . . ."[6] The world described in cowboy songs "is more actual than imaginary."[7]

Cowboy song lyrics, perhaps, suggest how cowboys viewed nature and the environment. Perhaps, too, cowboy songs shaped their views of the natural world. Cowhands composed and sang songs for many reasons. Because life on the cattle trails and ranches was dangerous, isolated, and lonely, songs provided cowboys some comfort, security, and entertainment. Sometimes, after a long cattle trail or a hard day's work, herders would get together and put on sing-out dances.[8] They learned their songs while on the job or during chow time at the "mess wagon,"[9] and very often at night they sang to the cattle they tended, mainly to soothe them, as any unexpected noise could start a stampede.[10] Cowboy "Teddy Blue" Abbott states in his memoirs that some cowboys would sing to their cattle and that songs were sometimes shared. ". . . [I]f it was a clear night and the cattle [were] bedded down and quiet," he said, "and one man would sing a verse of a song, and his partner on the other side of the herd would sing another; and you'd go through a whole song that way. . . ."[11]

The lyrics to many of the songs reflect the cowboy's tough life and lack of stability, and they often painted pictures of the environment around him. Some songs, however, delighted in the free and unencumbered life away from society that a herder's existence provided.

When one examines the lyrics to many of the cowboy songs in relation to the natural environment, two distinct ideas and one overall theme emerge. There are songs that describe nature as a dreadful place, and there are songs that describe it as a wonderful place. The one theme that appears throughout the songs is that nature must be conquered, either by the waddy himself or by the slowly encroaching civilization and economic development.

The real life of the cowboy was a far cry from the romanticized image portrayed in many western novels and films. It was a lonely, isolated life with few comforts. Cowhands had to live in, and make the best of, what many of them viewed as a hostile natural environment, and some of them did not relish their hard life, long work hours, and little pay. (Indeed, some went on strike to improve their pay.) Folklorist Ina Sires points out that "the grayness of the prairie" is a common theme throughout the range of cowboy songs.[12] The prairie is often seen as a barren wasteland with no real redeeming characteristics.

Perhaps the most famous song that exemplifies the idea is "The Dying Cowboy" (also known as "The Lone Prairie"). The lyric is supposedly sung by a dying cowboy who asks that his body not be buried on the ". . . lone prairie, where the wild coyotes will howl o'er me . . . where the buzzard waits . . . [and] the rattlesnakes hiss and the crow flies free. . . ."[13] This song goes on to describe how the cowboy wishes to be buried in the same churchyard as his father. The freedom that nature offers is no comfort to this dying cowhand. Another version of the song shows the herder reminiscing about his "home, and cottage in the bower" and pleading with his comrades who will soon be burying him, "For the sake of the loved ones who will weep over me/Oh bury me not on the lone prairie. . . ."[14] The waddy takes no consolation in the idea that his grave would be alone in the wilderness. Bernard Buie writes that the dying cowboy in this song "tells his story straight and his words are picturesque and realistic with the details of the range over which he had made his last ride."[15] Despite the dying man's pleas, the other cowboys buried him on the prairie:

> But we took no heed of his dying prayer
> In a narrow grave just six by three
> We buried him there on the lone prairie . . .
> Where the wild coyote and winds sport free
> On a wet saddle blanket lay a cowboy-eee[16]

Not all versions of the song have this ending, but the ending illustrates that even the dying cowboy's friends—now his undertakers and sole mourners—saw nature in a dismal light.

Songs such as "The Dreary, Dreary Life" (sometimes called the "Cowboy's Dreary Life"), state that all the waddy does is round up cattle for hours on end "on the bald prairie so bare." This song describes the hands' harsh working conditions and how the "weather being fierce and cold" almost freezes the men to death. In particular, the Pecos River is a horrible place in which to camp because "the wolves and owls (have) . . . terrible howls." The narrator dispels the idea that the herder's life is romantic and free from care; in fact, he advises cowboys to "Sell your bridle and your saddle, quit your roaming and your travels" and settle down and get married.[17] Irwin Silber points out in *Songs of the Great American West* that "there's more truth in this one song" than in many other cowboy ballads. The "working conditions were frequently abominable," and nature sometimes threw an "onslaught of bad weather" into the ranch hand's lap.[18] In "The Dreary, Dreary Life," the cowboy narrator

Cattle of the Matador Ranch grazing on open range still free of mesquite and juniper in the early 1880s. Courtesy of Southwest Collection, Texas Tech University. SWCPC File no. 64.

dreams of going back to civilization, finding a wife and house, and leaving the prairie, as the prairie is far too dangerous to deal with.

The song "Poor Lonesome Cowboy" describes the waddy's life as isolated, with no family around to allay his loneliness. The narrator longs for the comforts of town life, pointing out that he is "a long ways from home."[19] In "The Cowman's Prayer" the singer asks God to bless the cattle and prays for rain to stop the "prairie fires."[20] Again, the image of the prairie here is a negative one. The lyrics of "The California Trail" describe the prairie as "bleak," a place where there is no wood for fires, and where the cowboy has to stand up to eat.[21]

"Doney Gal" describes a situation where the cowboy narrator has only his horse, Doney Gal, to help him do his work through rain, sleet, and snow. In this song, the life of the herder is a "weary thing," working day and night through "rain or hail." By the end of the trail, the narrator is able to laugh at the "storms, the sleet, and snow/ When we reach the little town of San Antonio."[22] When the cowboy comes back to civilization and no longer has to battle with the environment, he is happy. He no longer has to deal with the hardships of his work and fighting the elements; his weary life is transformed, even if only for a short while, when he enters the mainstream of city life.

One of the more common negative themes expressed in cowboy songs is the comparison of the herder's environment to hell, or the dwelling place of Satan, forsaken by God. "Hell in Texas" describes "a supposed deal between the Lord and the Devil by which the latter acquired some land that the Lord couldn't use. . . ." The Devil put the "thorns on the cactus and horns on the toads . . . [and] poisoned the feet of the centipede" and "called [the land] Texas." A similar song, "Eastern Shores of the Rio Grande," describes how the Devil sought out a piece of land to call his own. He chose the "eastern shores of the Rio Grande." In order to keep his evil reputation, he created a "vegetation of thorns . . . with Spanish daggers stiff, sharp pointed and tall." The indigenous American Indians were "direct from hell"; the rivers were dried up and the temperature fixed at a scorching 107 degrees. According to the song, Texas was a dreadful place until the white man came along and fixed it up; the Devil may have gloated that before then "on earth there is no hellisher land . . . [but] should the devil visit his little hell in West Texas now, he wouldn't know it." Man has remade a once-hellish nature into a "little paradise";[23] he was able to tame it and make it suit his needs. The song goes on to describe how irrigation techniques, the sinking of wells, and the ability to turn cactus into cattle feed put Satan in his place. By subjugating the environment, the cowboys and farmers have prevailed not only in the physical, but also in the spiritual realm.

In another song, "A Cowboy Alone with His Conscience," a cowhand on the prairie describes the emptiness he feels about being alone with "nobody near him 'ceptin' God." Even the most sinful person, he claims, would be no match for nature:

> Take the very meanest sinner that the nation ever saw
> One that don't respect religion more'n he respects the law,
> One that never does an action that's commendable or good
> An' immerse him fur a season out in Nature's solitude,
> An' the cog-wheels o' his conscience'll be rattled out o' gear.[24]

Being alone in the remoteness of the wilderness is too much for anyone to bear; the silence becomes maddening and the surroundings are grim. Again, the theme of the prairie being a place not for human habitation comes into play. The song gives the impression that the narrator would rather be anywhere else than forced into this isolation.

There is a similar example in the song "Buffalo Skinners." The narrator describes crossing the Pease River, where there is no "worse hell on earth than the range of the buffalo." After the cowboys finish their work and start on their homeward journey, they declare that they are never going back to that "hell-fired country . . . (because) . . . God's forsaken the buffalo range and the damned old buffalo."[25] Throughout "Buffalo Skinners," the narrator faces hardship after hardship and ultimately decides he doesn't really see the value of chasing after and skinning the buffalo and encourages his listeners to tell others not to go there, but rather to stay with their wives and sweethearts. Here again, nature, at least in the Southwest, is described as being evil and abandoned by God. Little of the modern-day romance of the Old West is found in such contemporary cowboy songs.

Nonetheless, not all cowboy songs espouse a negative view of nature. Some songs laud the freedom that the cowboy's life provided and express a fondness for the prairie. One of the most famous cowboy songs, "Home on the Range," for example, describes a view quite opposite from that of the negative "Buffalo Skinners." The first line, ". . . give me a home where the buffalo roam," presents a positive view of the life of herders. The song also describes the fresh air, "wild flowers," and "glittering skies." The narrator enjoys his life away from the city where "the deer and the antelope play."[26] For some cowhands, there was joy to be found on the open range and prairie; the wonderful experiences and views that the environment offered made up for the loneliness of the herder's life.[27]

Cowboy songs sometimes express a very romantic view of the natural environment. Just as some of the songs describe the environment as being godforsaken, there are those, such as "The Cowboy's Prayer," that view nature as a place where God lives. The song thanks God for the freedom away from civilization and "where the churches grow":

> *I like creation better as it stood*
> *That day You finished it so long ago . . .*
> *I thank you Lord that I am placed so well,*
> *That you have made my freedom so complete;*
> *That I am no slave of whistle, clock or bell. . . .*[28]

The narrator expresses a love for living life on his own terms away from the monotony of living by schedules. Here, nature is analogous to freedom and finding God. Other songs, like "Lasca" and "Old Time Cowboy," both describe the positive aspects of freedom in the cowhand's life.

The wide and open landscape was seen as a representation of a life exempt from constraints of civilization.[29] "Where the Grizzly Dwells" describes an idyllic West where "nature's handiwork lies in virginal beauty."[30] The narrator describes losing himself within the macrocosm of the wilderness and his reflective experiences while herding cattle. Nature is pure precisely because it has not been touched by society.

Working with the cattle is also sometimes described as a positive experience. The song "Cowboy's Life" describes how the "bawl of a steer, to a cowboy's ear/Is music of the sweetest strain."[31] Other animals are also seen in a friendly light: the "yelping of the coyotes" is a "glad refrain." The song goes on to state that the life of the cowhand is equal to that of royalty. The boundless plains and free life of the cowboy close to nature taught him directness and simplicity.[32]

Other cowboy songs portray the harshness of nature as something that one should not just get away from (as the negative songs suggest), but rather as something to learn from and adjust to, something useful and good. "Ridin' up the Rocky Trail from Town" celebrates herders as "children of the open" who despise the "haunts of men," and refers to city life as a "constant round of strife." The ranch hands could learn how to be rugged and stern from the most fearful animals; the song describes the centipede, rattlesnake, bobcat, and bronco steed as friends who helped the cowboy become tough and fearless. One example is the line, "From the rattlesnake we learned to guard our rights."[33] In this song, nature represents freedom, and to a certain degree it can be internalized. Those fierce aspects of nature that townsfolk feared are adopted by the hands to make them fierce as well.

In "I'm Bound to Follow the Longhorn Cow," the narrator says that his destiny is to herd cattle through the "desert heat . . . rain and sleet . . . and winter's icy snows."[34] Despite having a loving family and girlfriend, the herder must roam; he wants to be with cattle even more than in the loving arms of a girl. He is able to overcome nature's discomforts within himself.

Perhaps no other cowboy song expresses both inner and outward conquering of nature better than "Westward Ho." The song discounts living in Colorado, Montana, Kansas, Arizona, Nevada, and New Mexico. Instead, the narrator wants to live "where the grizzlies wander." In this song, the herder welcomes what some would consider the dangerous aspects of the environment. He knows he can conquer anything that comes his way:

I'll chase the wild tarantula
And the fierce coyote I'll dare
And the locust grim, I'll battle him
In his native wildwood lair.[35]

While they do not really present nature in a romantic way, the songs do indeed exemplify a lack of fear. They suggest that cowhands were not afraid to go into, and make a home in, the wilderness. Thus, many songs express aspects of nature that some feel are dangerous, but which can be overcome and even used to benefit the herder's life.

Other songs describe the prairie as a positive place. The song "Jolly Cowboy" calls ranch life on the prairie "always fun" for the herder and his horse companion. "I love the rolling prairies, they're free from care and strife."[36] Even when nature throws a rainstorm his way, the cowhand is content to be with his horse. Eventually, the cowboy gives up roaming the range to be with his wife, but he has no regrets and remembers fondly the life he had as a waddy. Life on the prairie was equated with the independence and spirit behind it. "The Call of the Plains" is perhaps the best example of this idea:

Ho' wind of the prairies
Free as the waves of the sea . . .
I dream of the wide, wide prairies . . .
The call of the Spirit of Freedom
To the spirit of freedom in me
My heart leaps high with a jubilant cry
And I answer in ecstasy.[37]

The song is filled with romantic notions of the environment and the life within it. The narrator wants to be a part of nature's free life; the waddy can't escape it.[38]

Texas also is portrayed in a positive light. Unlike "Hell in Texas" and other songs noted above, some cowboy songs describe Texas as a wondrous place that no other state can match. "The Texas Cowboy," for example, describes working in other states, such as Montana and Nebraska, and finding the environment less than enjoyable. The narrator longs to come back to Texas, where there is always work and happiness to be found.[39] While it does not portray the natural environment kindly, the song expresses the romantic love for the life the cowboy left in Texas. Another song, also entitled "The Texas Cowboy," describes a different scenario, where the cowhand leaves his homestead and lives on the rolling prairies to forever be a "roving cowboy."[40] "The Cowboy's Christmas Ball" describes Texas as a place where the "Clear Fork's waters flow. . . monstrous stars are winkin' o'er a wilderness profound."[41] In all of these songs, Texas is a place that is loved. Other songs that express a similar sentiment include "I'd Like to Be in Texas" and "We Love the Name of Texas."

In the late nineteenth century, the open range was disappearing. The long cattle drives to Kansas, Montana, and elsewhere were becoming a thing of the past; the expansion of civilization, railroads, and the farming-ranching industry with the use of barbed-wire fences confined much of the open range. The open, free, and often dangerous life of the cowhand was dying. Cowboy songs also speak to this death and the "progress ideal" behind it. Clark Stanley, writing in 1897, pointed out how the cowboy's way of life was coming to an end: "As the Indian and buffalo are fast disappearing before advancement of civilization," he wrote, "so ere long the cowboy will be a thing of the past, but his part in the development of the country, . . . his songs and legends deserve a place in its history."[42]

It is precisely this advancement that certain songs lament. "I'm Going to Leave Old Texas Now" describes how "they plowed and fenced my cattle range." The narrator refuses to give up his free life on the wide range, deciding to go someplace where he will not be confined (in this case, Mexico).[43] In "To Hear Him Tell It," a "relic of [a] Texas Cowman" describes his life as a cowboy trailing the herds north. The world now, he says, is very different, and those days are gone forever.[44] "The Old Cowman" expresses a similar sentiment from the perspective of a cowhand who went across a range on which he used to work, finding it "spoilt and strange . . . [with the] stretchin' of the wire." The old waddy complains about the "progress" that has divided the wide range where once there was "no fence or fuss . . . [where nature] was lovely and free."[45] The narrator is grateful to have been born before the encroachment of civilization destroyed his way of life, saying, "I hate to see the wire fence a-closin' up the range."[46]

"The Tough Longhorn" tells a story from the point of view of a dying cow. All the grass and water are now gone and the "nester came with his . . . barbed wire fence." The cowboys and longhorns, who were once partners, are no more, phased out respectively by land development and other breeds of cattle. "The time has come when longhorns/And their cowboys are no use."[47]

Other songs describe the railroads covering the land and shutting out the waddy.[48]

Clearly, the cowboy songs selected here express some of the different ways that herders felt about their lives and the natural environments in which they lived and worked. While none of them can be taken as depictions of specific historical events, the songs depict the daily lives of cowhands, their trials, tribulations, joys, and triumphs. They illustrate the cowboy's love for his life in nature without the man-made constraints on his coming and going. Finally, they show how the waddy felt about the "conflict between the wilderness ideal and the cult of progress."[49] In other words, they refer to the historical changes that caused the late nineteenth-century real-life cowhand and his way of life to be sacrificed on the altar of progress. Cowboy songs of the late nineteenth-century reflect an oral tradition among western ranch hands that both glorified and condemned the natural environment in which they lived.

Notes

1. Anne Dingus, "Cowboy Songs," *Texas Monthly* 17 (November 1989): 120-25.

2. Julia M. Hirsch, *Cowboy Songs: A Select Bibliography of Books: With Library of Congress Call Numbers* (Washington, D.C.: Archive), LC 1.12/2:C 83/2; Guy Logsdon, *The Whorehouse Bells Were Ringing* (Chicago: University of Illinois Press, 1987); Guy Logsdon, "The Cowboy's Bawdy Music," in *The Cowboy: Six Shooters, Songs, and Sex,* edited by Charles W. Harris and Buck Rainey (Norman: University of Oklahoma Press, 1976), 127-38; Austin E. Fife and Alta S. Fife, eds., *Cowboy and Western Songs: A Comprehensive Anthology* (New York: Clarkson N. Potter, 1969); N. Howard Thorp, *Songs of the Cowboys,* Austin E. Fife and Alta S. Fife, eds. (New York: Clarkson N. Potter, 1966). This collection also reprints Thorp's original 1908 version of *Songs of the Cowboys.*

3. One example includes Clark Stanley's "Cowboy Songs and Dances," *The Life and Adventures of the American Cowboy: Life in the Far West* (privately published in 1897).

4. Mark Fenster, "Preparing the Audience, Informing the Performers: John A. Lomax and Cowboy Songs and Other Frontier Ballads," *American Music* 7 (Fall 1989): 261.

5. Lawrence Clayton, "Factuality Versus Artistic License in Western Folk Songs," *West Texas Historical Association Year Book* 58 (1982): 176-80.

6. Bernard Buie, "Cowboy Ballads of West Texas," (M.A. thesis, Simmons University, Abilene, 1929), 25.

7. Martin A. Cannon, "Cowboy Song Adaptations" (M.A. thesis, Texas Technological College, Lubbock, 1966), 76.

8. Ibid., 6; Buie, "Cowboy Ballads of West Texas," 42.

9. Clark Stanley, "Cowboy Songs and Dances," *Cowboy Reader,* Lon Tinkle and Allen Maxwell, eds. (New York: Longmans Green and Co., 1959), 213.

10. Walter Prescott Webb, *The Great Plains* (New York: Ginn and Company, 1931), 459.

11. E. C. "Teddy Blue" Abbott and Helena Huntington Smith, *We Pointed Them North: Recollections of a Cowpuncher* (Norman: University of Oklahoma Press, 1978), 223. Originally published 1939.

12. Ina Sires, "Songs of the Open Range," in J. Frank Dobie, ed., *Texas and Southwestern Lore* (Dallas: Southern Methodist University Press, 1967), 192. Originally published 1927.

13. "The Dying Cowboy," in Charles A. Siringo, ed., *A Lone Star Cowboy: Old Favorite Cow-Camp Songs* (n.p.: Norwood Editions, 1975, originally published 1919), 5-8.

14. "Lone Prairie," in Margaret Larkin, ed., *Singing Cowboy: A Book of Western Songs* (New York: Alfred A. Knopf, 1931), 22-23.

15. Buie, "Cowboy Ballads of West Texas," 22.

16. "The Dying Cowboy," in Siringo, ed., *A Lone Star Cowboy*, 7.

17. "The Dreary, Dreary Life," in N. Howard Thorp, *Songs of the Cowboys* (Lincoln: University of Nebraska Press, 1984), 61-62. This version was originally published in 1921.

18. Irwin Silber, ed., *Songs of the Great American West* (New York: Macmillan, 1967), 192.

19. "Poor Lonesome Cowboy," in Larkin, ed., *Singing Cowboy*, 108-10.

20. "The Cowman's Prayer," in Jules Verne Allen, *Cowboy Lore* (San Antonio: Naylor Company, 1971), 138-39. Originally published 1933.

21. "California Trail," in Thorp, *Songs of the Cowboys*, 18-20.

22. "Doney Gal," in John Lomax, *Our Singing Country* (New York: Macmillan Company 1949), 250-51.

23. "The Eastern Shores of the Rio Grande," in Siringo, ed., *A Lone Star Cowboy*, 40-42.

24. "A Cowboy Alone with His Conscience," in John Lomax, *Songs of the Cattle Trail and Cow Camp* (New York: Duell, Sloan, and Pearce, 1950), 184-85. Originally published in 1919.

25. "Buffalo Skinners; Range of the Buffalo," in Allen, *Cowboy Lore*, 115-17. There are many variations of this song. See Thorp and Fife, *Songs of the Cowboys*, 195-218.

26. "Home on the Range," in Siringo, ed., *A Lone Star Cowboy*, 32-33.

27. Katie Lee, "Songs the Cowboys Taught Me," *Arizona Highways* 36 (February, 1960): 34-39.

28. "A Cowboy's Prayer," in Thorp, *Songs of the Cowboys*, 47-48.

29. "Lasca," in Lomax, *Songs of the Cattle Trail*, 23-26; "Old Time Cowboy," in Thorp, *Songs of the Cowboy*, 121-22.

30. "Where the Grizzly Dwells," in Lomax, *Songs of the Cattle Trail,* 174-75.

31. "The Cowboy's Life," in Thorp, *Songs of the Cowboy,* 44-45.

32. Cannon, "Cowboy Song Adaptations," 71-72.

33. "Ridin' up the Rocky Trail From Town," in Lomax, *Songs of the Cattle Trail,* 179-81.

34. "I'm Bound to Follow the Longhorn Cow," in Harold W. Felton, *Cowboy Jamboree: West Songs & Lore* (Eau Claire, Wisconsin: E. M. Hale, 1951), 102-3.

35. "Westward Ho," in Thorp, *Songs of the Cowboys,* 161-62.

36. "The Jolly Cowboy," in ibid., 86-87.

37. "The Call of the Plains," in Lomax, *Songs of the Cattle Trail,* 172-73.

38. Other positive prairie examples include "The Cowboy," which describes cowboys as "sons of the prairie," in Clifford P. Westermeier, ed., *Trailing the Cowboy* (Caldwell, Idaho, Caxton Printers, 1955), 260-61; "A Prairie Song" also expresses the freedom of life on the plains in Thorp, *Songs of the Cowboys,* 129-30.

39. "The Texas Cowboy," in Larkin, ed., *Singing Cowboy,* 54-55.

40. "The Texas Cowboy," in Arbie Moore, ed., "The Texas Cowboy," in Dobie, *Texas and Southwestern Lore,* 196-97.

41. "The Cowboy's Christmas Ball," in Siringo, ed., *A Lone Star Cowboy,* 27-31.

42. Stanley, in *Cowboy Reader,* 212.

43. "I'm Going to Leave Old Texas Now," in William A. Owens, *Tell Me A Story, Sing Me a Song: A Texas Chronicle* (Austin: University of Texas Press, 1983), 107-8.

44. "To Hear Him Tell It," in Lomax, *Songs of the Cattle Trail,* 36-39.

45. "The Old Cowman," in ibid., 165-67.

46. Lee, "Songs the Cowboys Taught Me," 39; "Bronc Peeler's Song," in Thorp, *Songs of the Cowboys,* 11-12.

47. "The Tough Longhorn," in Siringo, ed., *A Lone Star Cowboy,* 21-23.

48. One example is the "The Camp Fire Has Gone Out," in Thorp, *Songs of the Cowboys,* 20-21. It describes how the railroad companies covered the West and there was nothing left for the cowboy.

49. E. Martin Pedersen, "The Dreary Life of the Cowboy: Memoir and Myth in Cowboy Ballads," *Social Education,* 61 (March 1997): 134.

13 Rodeo Cowboy: "Booger Red" Privett and the Origins of Rodeo

J. BOYD TROLINGER

The popularity of rodeo in the late twentieth century represents a continuing fascination with the American cowboy. Informed largely by popular tradition, many people hold fast to a view of the rodeo and its cowboy heroes as symbols of the bravery, skill, and rugged individualism long associated with America's westward expansion. Despite the infusion of legend and folklore into the popular image of the cowboy, a closer look into the origins of rodeo reveals a genuine connection with the working cowboy of the past.

The modern sport of rodeo evolved out of cowboy contests held by rival range cattle operations in the nineteenth century. By the 1870s, many towns in the cattle country of the West, such as Deer Trail, Colorado, and Pecos, Texas, included cowboy exhibitions in 4th of July and other celebrations.[1] But the emergence of rodeo as a popular spectacle, and the rodeo cowboy as a symbol of the American West, began in the early 1880s with Buffalo Bill and his Wild West Exhibition.[2] Buffalo Bill, by showcasing the talents of cowboys, helped transform the popular image of the cowboy from that of a common laborer who lived a life of monotony and endless drudgery to that of a colorful frontier hero. The success of Buffalo Bill's exhibition led to the creation of a number of other western-themed shows. By 1885, there were more than fifty Wild West shows touring the United States.[3]

Buffalo Bill and other operators of western shows paved the way for the emergence of rodeo as a popular entertainment and a respectable sport. Rodeo provided a stage for competitors to display the qualities and special skills of cowboys. And though modern-day rodeo competitions include a number of events that showcase "cowboy skills," few events with clear ties to the past remain.

Saddle bronc riding, however, is one of the modern day rodeo events with a direct connection to the open range cattle industry of the nineteenth century. Indeed, many people regard saddle bronc riding as the "cornerstone of rodeo" and the "'classic event' of the [rodeo] program."[4] Unlike bareback bronc riding or bull riding, events that have no antecedents in real ranch work, the ability to break an unruly horse to the saddle was and is a valuable skill for the working cowboy.

One of the brightest stars of the early years of rodeo was a short, slight Texas cowboy who earned recognition as perhaps the greatest saddle bronc rider of his time. His name was Samuel Thomas Privett, but in his day most people knew him as "Booger Red."

Booger Red Privett combined renowned cowboy skills with a natural sense of showmanship to become one of the most famous saddle bronc riders of the late nineteenth and early twentieth centuries. His skills as a bronc buster brought him work with some of the largest ranches in Texas and led ultimately to a career as a performer in Wild West shows and rodeos. Before the days of an organized rodeo circuit and a sophisticated ratings system for competitors, Booger Red established himself as one of the first champion rodeo cowboys. "For more than a quarter of a century," wrote renowned rodeo announcer Foghorn Clancy, "Booger Red was regarded as the greatest bronc rider in the world."[5]

Booger Red is all but forgotten today, but he was a significant force behind the establishment of rodeo as a popular entertainment and the creation of the popular image of the cowboy and the American West. He served as a direct link between authentic cowboy work and rodeo performances. Indeed, his career spanned a period in history that witnessed the glory days of the range cattle industry and the dawn of the modern rodeo. He earned his reputation as a skillful horseman working for some of the largest ranches of West Texas in the late nineteenth century. He established a Wild West exhibition as a showcase for his talents, and in the process enhanced the growing popularity of the cowboy and rodeo. He became a popular rodeo champion who, even as an old man, could attract a crowd of "madly cheering thousands" to see him ride.[6] Booger Red—cowboy, show-man, and rodeo champion—helped to create, and indeed became, the cowboy hero of the American West.

Samuel Thomas Privett was born on April 29, 1862, on the TP Ranch in Williamson County, Texas.[7] Privett grew up on his father's ranch, the SP, in Erath County, where at an early age he discovered a love for the ranch life and the work of a cowboy. He acquired a repu-tation locally for his talent for breaking horses. So great was his abil-ity that by the age of twelve he was known as the "Redheaded Kid Bronc Rider."[8]

The name "Booger Red" came to Privett after a childhood acci-dent left him badly injured. On Christmas Day, 1877, Privett and a friend filled a hollow tree stump with gunpowder to create a large explosion. The gunpowder ignited prematurely, and Privett's friend was killed. Privett sustained serious burns on his face, and doctors repeatedly had to cut open his eyelids and nostrils as the tissue healed. A child who saw Privett's injuries remarked "Gee, but Red sure is a booger now, ain't he?"[9] Privett's siblings began to call their brother "Booger Red." The injured boy liked the nickname, and went by the name Booger Red for the rest of his life.

After the death of his parents, Booger Red purchased property near Sabinal, Texas. Following a brief but unsuccessful stint run-ning his own ranch, he found work on the Billy Guest ranch near Sonora.[10] As a cowboy on the Guest ranch, Privett became well-known in West Texas for his skills as a bronc buster. Ranchers in the area brought their "bad ones," or unbreakable horses, for him to tame. He contracted his services out to some of the biggest ranchers in West Texas, including the Shannons, Hendersons, and Harrises.[11] Not staying long on the Guest ranch, Privett worked for a number of West Texas ranches through the 1880s, including the Sugg, Door Key, Broome and Farr, and Fayette Tankersley ranches.[12]

Booger Red ultimately moved to San Angelo, Texas, where he bought a wagon yard and broke horses for ranchers in the area. He also contracted his services out to the Texas Rangers, who made a gift to him of a chrome-plated .45 revolver. For a time, Privett trained horses for the U.S. Cavalry at Fort Concho. Offered a choice between a monthly salary and a by-the-head fee by the U.S. govern-

A cowboy rides a bucking horse in a rodeo, ca. 1920. Photo by Frank Reeves, courtesy of Southwest Collection, Texas Tech University. SWCPC File no. FR 265.

ment, Privett opted for the latter. But when it learned that he could break sixty to eighty horses a day, the government placed Privett on a monthly salary.[13]

Privett's skills on horseback became an attraction for the people of San Angelo. Booger Red discovered that he could charge admission to the people who wanted to see him ride. Working in an empty farmers' merchandising lot in downtown San Angelo, he began his career as a showman by passing around his hat to gather a small fee from those who wanted to see him work or, as was often the case, wanted to bet that Booger Red could not ride a particular outlaw horse. Booger Red's exhibitions drew increasingly larger audiences, and he began to hang ducking on rope fences around his arena to keep out non-paying onlookers.[14]

Though he increasingly capitalized on his local fame and bronc riding abilities, Booger Red remained a working cowboy. He normally took the horses he rode for the crowds in San Angelo to his ranch to complete their training. Though the crowds wanted to see a spectacular ride, Privett's real work was breaking horses for the saddle and the plow. The unruly horses he contracted to break demanded not only skill but also perseverance. Booger Red's determination to train horses properly was such that he often spent his days chasing horses on the plow across his fields until the horses exhausted themselves.[15] He made a successful living for many years in San Angelo breaking horses for West Texas ranchers. But the call of the show life eventually took its hold on him.

In 1901, Booger Red started his own Wild West show. At first he performed in San Angelo and the towns of the surrounding area. In its early years, Booger Red's Wild West Wagon Show consisted of little more than Privett, a wagon, a few horses, and a helper or two. The show season lasted from April through October, and Booger Red began and ended each year's tour in San Angelo.[16] But as his fame spread throughout West Texas, the show quickly grew beyond the bounds of Privett's home town.

Booger Red took his Wild West exhibition all over Texas and the Southwest. At its peak, the traveling show had thirty-two wagons, twenty-two horses, twelve steers, twelve saddle horses, and fifteen cowboys and cowgirls.[17] Despite the size of Booger Red's show, it always remained a family affair. The Privett children worked for their father, and Privett's wife Mollie acted as the cook, seamstress, and stand-in mother for Booger Red's employees. She also sold tickets before each show, and it was her responsibility during the show season to exercise the horses that Privett and his cowboys rode.[18]

Booger Red was the principal attraction, but his children were also skilled riders, ropers, and performers. All six Privett children worked for or performed in the show. After Booger Red's retirement, a few of his children went on to achieve fame in their own right. His son Bill was a championship rider in the 1920s and 1930s, and worked as a trainer and jockey for Bing Crosby at Crosby's stables in California. Ella Privett, skilled and courageous on horseback, worked as a trick rider with numerous circuses, and for a time she rode in the Tom Mix Circus. Privett's younger son Tommy was a noted trick roper and spent some time on the Ringling-Barnum circuit. Alta, the youngest of the Privett children, performed as a rider for the Hagenbeck-Wallace Circus until she retired in 1930.[19]

Booger Red's Wild West Wagon Show was also the starting place for some of the most famous cowboy performers of the early twentieth century. Bill Pickett, a renowned show cowboy who won fame for his unique style of bulldogging steers, got his start with Booger Red's show. Privett discovered Bill Pickett as he stood by watching a group of boys trying to ride a burro. Pickett, the lone African American in the group, was the only one of the boys to ride the animal successfully. Booger Red recognized the young man's talent and asked Pickett's parents for permission to take Bill on as a performer in the Wild West show. Pickett stayed with Booger Red's show until 1905 when he joined the Miller Brothers 101 Ranch Real Wild West show.[20]

"Texas Jack" Lewis and "Oklahoma Curley" Roberts, both of whom became world champion riders, also gained their first experience with Booger Red's show, as did Hugh "Huckleberry Slim" Johnson, a bronc rider with a wooden leg.[21] Foghorn Clancy, a longtime rodeo announcer of some fame in the Southwest, worked for Booger Red in 1903 and 1904.[22] Many of the cowboys in Privett's employ achieved fame, but in his day Booger Red was always the star of the show.

Booger Red managed his traveling show much as a trail boss on a cattle drive during the golden years of the range cattle industry. Life on the road for the people who worked for him was much as it must have been for cowboys on the trail during the heyday of the cattle trailing industry. Booger Red's troupe traveled by wagon and camped at nearby streams where the livestock could be watered and fed. Privett held a "kangaroo court" in the cowboy fashion every Saturday night, during which those guilty of petty offenses were tried, and when convicted given "lashes" with a pair of chaps.[23] Though he had become a full-time showman, he maintained the lifeways of a working cowboy.

Booger Red achieved fame not only for his Wild West exhibitions but also for his championship performances at some of the first rodeo contests in Texas. He took first prize at the first bronc riding contests in both San Angelo and Fort Worth.[24] Privett established a national reputation at the 1904 World's Fair in Saint Louis, Missouri, where he was named the best all-around cowboy after winning twenty-three awards in a six-week contest. In 1915, at the age of fifty-three, Booger Red won honors as the world's champion bronc rider at the World's Fair in San Francisco.[25]

Booger Red continued to perform until late in his life, but his level of participation in the show life lessened as he grew older. Tired of the ceaseless travel that the Wild West show demanded, and burdened with financial difficulties, Booger Red sold out to the Miller Brothers 101 Ranch show in 1913.[26] He did not end his involvement with the Wild West shows entirely, however, and continued as a performer with the Al G. Barnes troupe, Buffalo Bill's Wild West Exhibition, the Hagenbeck-Wallace Circus, and others into the early 1920s.[27]

After he retired as a performer, Booger Red maintained an important connection with the Wild West shows. On his ranch near Miami, Oklahoma, he raised livestock for some of the biggest shows of the time, including the Miller Brothers 101 Ranch Real Wild West Show.[28]

Booger Red's celebrity faded after he retired from Wild West exhibitions and rodeo. Further, the emergence of the western movie and its cowboy heroes overshadowed his fame. Booger

Red's reluctance to be filmed or photographed limited his exposure to the relatively small number of people who made up the audience at his Wild West shows and at the rodeos in which he performed. Booger Red, because of his appearance and his short stature (five feet four inches) does not seem to fit the mold of the movie cowboy, but in fact Hollywood moviemakers took an interest in making a star of the Texas bronc rider. "Many were the times," remembered his wife Mollie, "he would start into the arena and see a machine [movie camera] set up in some obscure place, but they never tricked him." Booger Red might have achieved lasting fame, as many of his contemporaries did, had he entertained the cameramen who tried to capture a picture of him on a wild horse.[29]

Booger Red's exploits became the source of numerous stories that transformed him into a legend. Embellished stories of Booger Red's life blur the distinction between fact and fiction. Yet the embellishments attest to the stature of a man who, even before his death, was larger than life. One such story describes Booger Red's last performance. As the story goes, Booger Red traveled to Fort Worth for the Fat Stock Show in 1924, intending to take in the events as an anonymous spectator in the crowd. Despite his effort to remain inconspicuous, some of the audience members recognized the famous bronc rider. When a particularly ferocious bronco threw and seriously injured a rider, a call went up in the crowd to "Give us Booger Red!" Over sixty years old and suffering from Bright's disease, Booger Red rose from his seat and descended the grandstand to the arena floor. He rode the bronco until the animal was exhausted, and as the audience thronged to get a closer look at him, Booger Red slipped out of the arena and returned to his home in Oklahoma. Just two weeks later, Booger Red died.[30]

Booger Red did ride a bronco in Forth Worth in early 1924 to the thrill of a large crowd; and he died two weeks later. But beyond these facts the story becomes more the stuff of legend than of history. The *Fort Worth Record* announced Booger Red's ride a day in advance, and Privett rode in his exhibition a "flea-bitten" bronco that had been the star of his Wild West show.[31] Even without the embellishments, the story reflects the popularity of Booger Red with the fans and his importance to the sport of rodeo.

The *Fort Worth Record,* in announcing the exhibition, described Booger Red as "one of the greatest cowboys in the history of rodeo sports," and his performance at the 1924 Fort Worth Fat Stock Show became front page headline news.[32] The story, "Veteran Rider and Broncho [sic] United in Arena," began as follows:

> Veteran cowboy and veteran broncho [sic], both great ones in their time—and both retaining more than a vestige of that greatness—they plunged from the chute, the band struck up "Dixie," and the big crowd literally went wild.[33]

Even though Booger Red did not suddenly emerge from the crowd to tame a wild horse, it is clear that the fans appreciated the significance of the event they witnessed—the last ride of one of the greatest rodeo champions of the time.

He did not achieve lasting national fame, but Booger Red played an important role in the creation and growth of the myths of the cowboy and the American West as well as the establishment of rodeo as a popular entertainment. His Wild West show, though small in comparison to exhibitions such as Buffalo Bill's, was an important starting place for many circus and rodeo cowboys. As trick riders and ropers, Booger Red's children extended his influence on the popularization of the cowboy and rodeo beyond the bounds of West Texas. Some of the cowboys who worked for Privett achieved the enduring national celebrity that he never realized.

Individually, Booger Red had a significant impact on the history of West Texas and the growth of the sport of rodeo. While in many ways typical of the cowboy/showman of his era, the breadth of his career, the singularity of his skill, and the colorfulness of his character set Booger Red apart from his contemporaries. As a working cowboy, he broke many horses that helped to herd the cattle and participated in the pursuit of American Indians in West Texas. As a showman, he used his skills to entertain, and the Wild West shows he helped to popularize were important in creating the myth of the Old West. As a champion bronc buster, he participated in and brought recognition to the emergent rodeo industry.

Samuel T. "Booger Red" Privett died in Miami, Oklahoma, in 1925. For his contributions to early rodeo, he was inducted into the National Cowboy Hall of Fame in November, 1975. Few people who knew Booger Red or saw him ride are alive today, and the paucity of records of his life, both written and photographic, have left him almost entirely forgotten. But in his day he was a hero and a champion, and he gave his audiences a chance to see a real cowboy in action. His life holds interest today because of the role he played both as a model for and a creator of the image of the cowboy in the popular imagination. The men and women of the modern rodeo owe a debt to the people who in the late nineteenth and early twentieth centuries popularized the sport, including the man who called himself "Booger Red, Booger Red, the ugliest man alive or dead."[34]

Notes

1. Kristine Fredriksson, *American Rodeo: From Buffalo Bill to Big Business* (College Station: Texas A & M University Press, 1985), 10; Elizabeth Atwood Lawrence, *Rodeo: An Anthropologist Looks at the Wild and the Tame* (Knoxville: The University of Tennessee Press, 1982), 80.

2. Fredriksson, *American Rodeo,* 10; Lawrence, *Rodeo,* 44-45.

3. Fredriksson, *American Rodeo,* 11.

4. Lawrence, *Rodeo,* 25.

5. Foghorn Clancy, "Memory Trail," *Hoofs and Horns* (October 1937), 8.

6. "Veteran Rider and Broncho Reunited in Arena," *Fort Worth Record,* March 14, 1924, 1A.

7. Booger Red's exact birth year is difficult to determine, and is recorded in various sources as having been some time between 1858 and 1864. See Lloyd Linford, "Early Rodeo's Booger Red," *The Quarter Horse Journal* 27 (February 1975): 358; Tom Mulvany, "Booger Red's Last Ride," *Southwest Review* (Autumn 1944), 31; Charlsie Poe, *Booger Red: World Champion Cowboy* (n.p.: Quality Publications, 1991), 1.

8. Clancy, "Memory Trail," 8; Linford, "Early Rodeo's Booger Red," 358; Mulvany, "Booger Red's Last Ride," 31-32.

9. The origin of the name Booger Red is the best known and most frequently retold story of Privett's life. See Dean Chenoweth, "'Booger Red' was King of Texas Bronc Riders," *San Angelo Standard-Times,* February 19, 1963, 7A; Clancy, "Memory Trail," 8; Linford, "Early Rodeo's Booger Red," 358; Mulvany, "Booger Red's Last Ride," 32; Poe, *Booger Red,* 2.

10. Mulvany, "Booger Red's Last Ride," 32; Poe, *Booger Red,* 3-4.

11. Poe, *Booger Red,* 4.

12. *San Angelo Morning Times,* April 25, 1933.

13. W. C. Beck Collection, Southwest Collection, Texas Tech University, Lubbock, Texas; Poe, *Booger Red,* 34.

14. Interview of Tommy and Mollie Privett, W. C. Beck Collection.

15. Interview of Tommy Privett, W. C. Beck Collection.

16. Elizabeth Doyle, WPA interview with Mrs. (Mollie) Samuel Thomas ("Booger Red") Privett, in Jim Lanning and Judy Lanning, eds., *Texas Cowboys: Memories of the Early Days* (College Station: Texas A & M University Press, 1984), 222; Mulvany, "Booger Red's Last Ride," 33.

17. Doyle, WPA interview with Mollie Privett, *Texas Cowboys,* 225; *San Angelo Morning Times,* April 25, 1933.

18. Doyle, WPA interview with Mollie Privett, *Texas Cowboys,* 221, 224.

19. Grady Hill, "Booger Red, Greatest Bronc Peeler of the West, Hits Saddle Again as his Widow and Four Famed Riding Children Reunited," *San Angelo Evening Standard,* January 5, 1945.

20. Interview of Tommy Privett, W. C. Beck Collection; Poe, *Booger Red,* 47-48.

21. W. C. Beck, Author's notes, W. C. Beck Collection.

22. Ibid; Foghorn Clancy, *My Fifty Years in Rodeo* (New York: The Naylor Company, 1952), 20; Poe, *Booger Red,* 48-49.

23. Interview of Tommy Privett, W. C. Beck Collection; Doyle, WPA interview with Mollie Privett, *Texas Cowboys,* 223.

24. "Booger Red Privett," in Willard H. Porter, ed. *Who's Who in Rodeo* (National Cowboy Hall of Fame, 1982), 99.

25. Interview of Tommy Privett, W. C. Beck Collection; Doyle, WPA interview with Mollie Privett, *Texas Cowboys,* 228; Poe, *Booger Red,* 30, 49.

26. *San Angelo Morning Times,* April 25, 1933.

27. Mulvany, "Booger Red's Last Ride," 34; Doyle, WPA interview with Mollie Privett, *Texas Cowboys,* 225.

28. "Booger Red Privett," *Who's Who in Rodeo,* 99.

29. Interview of Tommy and Mollie Privett, W. C. Beck Collection; Doyle, WPA interview with Mollie Privett, *Texas Cowboys,* 228.

30. Doyle, WPA interview with Mollie Privett, *Texas Cowboys,* 228; Mulvany, "Booger Red's Last Ride," 30-31; Poe, *Booger Red,* 74-77.

31. "Veteran Rider and Broncho Reunited in Arena," *Fort Worth Record,* March 14, 1924, 1A.

32. "Thousands Marvel at Escapades Made by Stars of Rodeo Here," *Fort Worth Record,* March 13, 1924, 4A.

33. "Veteran Rider and Broncho Reunited in Arena," *Fort Worth Record,* March 14, 1924, 1A.

34. "The Singing Prairies," *Avalon Dispatch,* September 1984, manuscript in Tom Green County Library, Biography Files.

14 French Cowboys: The *Gardians* of the Camargue and Buffalo Bill

JUDY GREAVES RAINGER

In 1905, Buffalo Bill's Wild West and Congress of Rough Riders traveled through the cities and towns of France. The Wild West Exhibition, or Wild West as Buffalo Bill himself preferred to call it, began its tour in Paris for three months, then traveled through 117 French provincial towns.

There in Paris to enjoy the entertainment and to marvel at the Wild West was a young aristocrat from Avignon, a city located in southern France. The Marquis Folco de Baroncelli-Javon had an unusual affinity for the oppressed populations of the world, especially Gypsies, Albigensians of the Middle Ages, and local speakers of the Provençal dialect. Now, by lingering behind the tents and befriending the participants of the show, Baroncelli came to empathize greatly with the American Indians of the show and was so enamored of their culture that he expressed the thought that he should have been born an Indian. He struck up a close friendship with Pedro (Joe) Esquival, Jacob White Eyes, and Sam Lone Bear, some of Buffalo Bill's Indian stars. Not only did he attend as many Paris shows as possible, he was also able to spend considerable time with his new friends when the Wild West traveled into southern France. What is especially significant about the new friendships is the fact that Baroncelli's family domain lay just outside of the Camargue in southern France and that he himself was a *manadier*—a "ranch owner" and representative of the French cowboy culture.[1]

It is important to examine the *gardians'* history and culture in the Camargue and the changes that occurred as a result of Buffalo Bill's Wild West Exhibition and its impact on Baroncelli. Superficial

similarities bind the two cowboy cultures together, but more important is the fact that the cultures were revitalized on both sides of the
Atlantic at approximately the same time. Buffalo Bill in the United
States and Baroncelli in France both worked in their own way at
mythologizing and idealizing a nineteenth-century tradition that
survived into the twentieth century. Another Frenchman, Joë
Hamman, knew both Buffalo Bill and Baroncelli, and made his contribution to both cowboy cultures in French cinema.

The Camargue is one of the most unusual areas in France.
Located on the coast of the Mediterranean Sea at the mouth of the
Rhone River Valley delta, it is a vast area that is home to famous
black bulls and white horses. Measuring two hundred thousand
square miles, the Camargue resembles a treeless plain, spotted with
freshwater and salty lagoons. For centuries it was almost uninhabit-
able except for the wild horses, bulls, and the *gardians* that lived in
isolated, primitive circumstances in order to "guard" the wildlife.
The lifestyles of the *gardians* and the existence of the horses and
bulls have been documented since the 1500s, when a confederation,
la Confrèrie, was created in response to threats of war to France and
its southern region of Provence, where the Camargue is located.
Recognizing that their skills with horses made them especially vul-
nerable to conscription, the horsemen of the vast area created an
association that would allow them to maintain contact with each
other and provide financial help to needy *gardians*. When the
confederation was created in 1512, the *Confrèrie* numbered only
twenty-three members; in 1527 it numbered 130. Mainly the
Confrèrie required its members to pay certain sums of money each
year and to observe and celebrate the saint's day of St. George (the
patron saint of the confederation) every year with masses and
parades.[2]

Although there are few historic documents that chronicle the
gardians' activities, it is clear that the *gardians* were enmeshed,
unlike American cowboys, in the religious and political turmoil of
the late eighteenth century. Those were years of divided loyalties
throughout France. People loyal to the throne and the prerevolu-
tionary days of France were known as Loyalists or Royalists; those
more liberal and loyal to the Republic of France were Republicans.

Provence and the *gardians* were not spared these civil conflicts. A document of 1817, addressed to the mayor of Arles, asked for the reorganization of the *Confrèrie* and reaffirmed its humanitarian, religious, and social missions. Considering the events of France in that epoch, particularly the reemergence of the monarchy following the rule of Napoleon, the document seems to be a declaration of loyalty to the king, Louis XVIII.[3]

Such turmoil divided the *gardians* into two camps, one of Catholic against Protestant and another pitting Royalists against Republicans. By the end of the nineteenth century the *gardians* faced other problems. They lived scattered throughout the Camargue, many of them in very isolated circumstances with no chance to socialize. Even more critical was the ominous threat of advancing technology, which would replace the use of the Camargue horse for the thrashing of the harvest and turn the swampy Camargue country into arable land.

It was into this rich provincial tradition that Buffalo Bill and the Wild West entered, and it was from this rich, threatened tradition that Baroncelli brought his interest and influence. Buffalo Bill (William Frederic Cody) had been to France once before, in 1889, but only to the cities of Paris, Lyons, and Marseilles. On that trip the Wild West numbered 140 cowboys, 75 Indians, and 250 animals, including buffalo, long-horned steers, and many horses.[4] The show arrived in Le Havre in northern France in May 1889 by the steamship *Persian Monarch* and then proceeded directly to Paris to participate in the Universal Exposition, which was to celebrate the centennial of the French Revolution. There was no more room under the recently built Eiffel Tower, so the Wild West set up camp on the western edge of Paris, near the Bois de Boulogne in Neuilly.

In the arena some of the participants of the Wild West included Annie Oakley; William "Buck" Taylor as "King of the Cowboys"; Johnny Baker, another dead-eye shot; Joe and Antonio Esquival, famous ropers from Mexico; and Lillian Smith, one of the famous cowgirls of the era. Also present were famous Sioux, including Featherman, Red Shirt, and Rocky Bear.[5] Another important element of the Wild West was the Deadwood Coach. Cody had learned from his previous trip to England in 1888 that the coach served, for the aristocracy, as an important visual reminder of the American West.

There was no denying Buffalo Bill's influence on the French culture. Rosa Bonheur, a distinguished artist of the time, painted Buffalo Bill's portrait and a dozen other paintings of the Indians and the animals. Her fascination with the members of the show was indicative of the *bourgeoisie's* attraction for exotica at the end of the nineteenth century, and her repeated presence at the Wild West was also indicative of the approval her social class bestowed on Cody. Reflecting this approval, social columns in the newspapers wrote of the "*chapeau* (hat) Cody" and the "*chemisier* (shirt) Cody."[6] Long before the far-reaching effects of the global advertising power of the twentieth century, Paris had succumbed to its first wave of "*le look cowboy.*"

Not only was the influence felt in fashion, but in all manifestations of the popular culture of the epoch: children's toys, chocolates, candies, cigar boxes, cabaret songs, operettas, books, comic books, postcards—anything that could be printed or designed—were soon transformed into free publicity for the show. Not all of this was by chance. The publicity team that preceded the arrival of the Wild West had covered the city with enormous posters and had discovered the ingenious gimmick of advertising through cheap postcards.

The Wild West left Paris in November 1889 to spend ten days in Lyons in central France and two weeks in Marseilles in southern France. From there the successful show went to Spain, Italy, Belgium, and England before returning to the United States in 1892.

In 1905, Cody reorganized his show under the name Buffalo Bill's Wild West and Congress of Rough Riders of the World and found a new partner in James Bailey. As a result of Bailey's experience with the Barnum and Bailey Circus, new, more modern, and quicker means of transportation and organization were available to Cody. A new system of coupling for railroad cars permitted an even larger troupe of men, women, and animals to be transported efficiently, thus enabling the Wild West to schedule one-night stands in distant, provincial towns of France.[7] This new Wild West arrived in France on March 10, 1905, debarking in Dunkirk and then moving on to Paris to set up camp under the Eiffel Tower in an area that covered forty thousand square meters. The eight hundred men and women, five hundred animals, and sixty wagons required vast accommodations and enormous quantities of food: per day, eight hundred kilograms of meat and one thousand kilograms of bread were consumed.[8] After two months of twice daily shows the Wild West moved on, using sixty-seven railroad cars and a system of loading and unloading materials that was so precise that the Prussian Guards of the Kaiser came to take notes for future use.[9]

The route took the Wild West through the north to cities such as Rouen, Cherbourg, and Caen; through the south in Montpellier, Toulouse, Nîmes, and Arles; through the west in La Rochelle and Bordeaux; and in the east through Nancy, Sedan, and Verdun. One hundred and seventeen French cities and towns were entertained with parades, twice-daily shows of the Congress of Rough Riders, and sideshows, all carefully crafted to exhibit American showmanship and ingenuity. In the larger towns, the show stayed for two days, sometimes more, but usually the troupe followed a grueling schedule of one-night stops for eight months. In November 1905, the show stopped to spend the winter in Marseilles, then recommenced the tour in March 1906, and finally left the country on March 12, 1906, when it went on to Italy, Croatia, Austria, Hungary, Germany, Luxembourg, and Belgium. It finally sailed for the United States in September 1906.[10]

The Wild West and Congress of Rough Riders presented a thrilling show in all of these countries. A typical show included most of the following activities: music from the Cowboy Band directed by William Sweeney; a Grand Review introducing the Rough Riders of the World and featuring riders from America, England, Germany, Japan, Russia, Arabia, and Mexico; reenactments of the Pony Express and an emigrant train crossing the American plains included an attack by Indians; the entrance of Buffalo Bill and his exhibit of shooting from horseback; a reenactment of the Battle of Summit Springs when Buffalo Bill killed Tall Bull; lassoing by the Esquival Brothers of Mexico; a scene of the Great Train holdup of the Union Pacific; cowboy, Indian, and Cossack games on horseback; and finally, as a climax, a "holiday" at the T-E Ranch in Wyoming interrupted by an Indian attack that was, in turn, repulsed by the

Virgil Leonard at Croton Breaks on Matador Ranch branding a calf, ca. 1906. Courtesy of Southwest Collection, Texas Tech University. SWCPC File no. 64(B), Envelope 1.

cowboys.[11] This program was an interesting mixture of American Manifest Destiny and international camaraderie that brought in the crowds and high gate receipts.

At the Paris show in March 1905, the Marquis de Baroncelli-Javon, while loitering around the Wild West tents, met some of the members of the show. Joë Hamman, one of the interpreters for the Wild West, introduced the Marquis to Jacob White Eyes and Sam Lone Bear at a Paris restaurant near the Champs de Mars.[12] After this initial encounter, Baroncelli engaged a friend to write a letter inviting Buffalo Bill to use the services of the Camargue *gardians*. The friend, Alfred Runel, wrote back: "On the subject of the special horsemen in your locale, Colonel Cody has begged me to say that we will study this thing when we have the occasion to send one of our representatives to the Gard [the political division where Nîmes is located]: . . . I told them to write to you for complete and detailed information. I also told them (to Buffalo Bill) that if he wanted to come himself, you would be happy to receive him."[13] No reply to this letter exists and one is left to speculate on the possible outcome.

From this time, however, there sprung up an unusual friendship and exchange of many gifts. When the Wild West stopped in Toulouse on October 13, 14, and 15, Baroncelli met his Indian friends for a long afternoon visit. On the banks of the Garonne River, Baroncelli related the injustices of French history; Lone Bear and White Eyes translated the history to the other two Indians present, Blue Shield and Iron Tail. "Suddenly, old Iron Tail put his hand on the shoulder of Baroncelli and in Sioux addressed the Marquis while looking at him with a look of inexpressible sadness. White Eyes translated: 'The Chief says, Let my friend forget his Pale-Face name. For all Indians, this man, who after seven hundred years still remembers the great sorrow of his nation and has come to despair on the very soil that drank the blood of warriors, will be forever known as Loyal Bird, *Zihtkala-Waste*.'"[14]

Soon after, the Wild West arrived in Nîmes. There, Baroncelli brought a group of *gardians* with him to see the show. A couple of days later, Baroncelli invited Iron Tail, Lone Bear, and two of the cowboys from the Wild West to his *manade* (ranch) near Caylar to see a roundup of the wild Camargue bulls. A report in the newspaper, written anonymously by Baroncelli, described how the Indians applauded the skillful maneuvering of the Camargue ponies. Then the group proceeded to Gallargues to see the release of the bulls at the local arena. "When they arrived at the village, the two Indians suddenly stood up. Letting their cloaks fall to their feet, they put their hands on their heart and sang enthusiastically the war chant. . . ."[15] The next day the Indians sent the Marquis a pair of moccasins and a beaded purse.

Exchanges continued over several weeks. On October 30, the Wild West set up the tents in Arles for a two-day show. In November, the winter camp was set up in Marseilles, where Baroncelli and the cowboys spent many winter evenings visiting and comparing the American West with the Camargue, according to Jeanne de Flandreysy, a personal friend and biographer of Baroncelli.[16]

There is no definitive evidence that Buffalo Bill met Baroncelli. There is a book, however, at the *Palais* of Roure in Avignon, *The Last of the Great Scouts,* dedicated by Cody to Baroncelli and a book by Baroncelli, *Babali,* dedicated to Cody for "all the kindness he had toward me."[17] Other than these two books, one can only theorize about the actual meetings between the two men. One can read today the ensuing correspondence between the Marquis and the Indians in the *Palais* of Roure in Avignon.[18] Mostly the letters relate proposed meetings that failed to take place and requests from both sides for authentic items. The correspondence with White Eyes; Pedro (Joe) Esquival, the Mexican roper; and Ould Ziza, an Algerian horseman, continued when the Wild West left France for the rest of the European tour in 1906; in September 1906, when the Wild West boarded the steamship in northern Belgium to return to the United States, Baroncelli was there to say good-bye. "He spent with them the last days of their European stay. He stayed with them until the last minutes and from the quay cried to them, *'Chekpa a oue enachevi,'* which means 'Brothers, may we see each other again soon!' At the moment when the boat pulled away from the quay, Jacob White Eyes threw a package which landed at the feet of Baroncelli. It contained Red Chief's beautiful costume, all in leather, decorated with painted feathers and colored beads."[19]

These gifts are guarded today by the descendants of Baroncelli, the family of Henri Aubanel that lives in Saintes Maries-de-la-Mer. The family insists that Baroncelli met Buffalo Bill, and shows visitors a personalized, dedicated portrait of Buffalo Bill. They, and others in the region, also insist that the American cowboy traditions were borrowed by Buffalo Bill from the Camargue *gardians.* It is an interesting theory to consider if one is intent on establishing a transatlantic cultural borrowing, but it seems more probable that the Marquis de Baroncelli and Buffalo Bill were simply equals in popularizing, packaging, and publicizing a fading provincial tradition of men on horseback.

Today the *gardians* of the Camargue speak little of Baroncelli. Although buried in Saintes Maries-de-la-Mer, relatively far from his aristocratic home in Avignon, Baroncelli never has received the local distinction of having a street named after him, and his gravesite is hard to find by outsiders. Despite such neglect, Baroncelli's foresight at the turn of the century has guaranteed a distinctive look and culture for the *gardians* that set them apart from other Frenchmen.

It is interesting to note that Baroncelli's first attempt to give the *gardians* a certain look occurred during the same time period that he had established contact with the Wild West. In the summer of 1905, he created the *Coumitat Vierginen* (the Virgin Committee) to estab-lish the Camargue costume and traditions.[20] That costume for spe-cial festivities, then and today, consists of suede pants with a single dark braid down the outside of the leg, a brightly printed shirt with *Provençal* designs, a velvet vest, and a felt hat, formed into a shape identifying the *manade* the *gardian* represents.

Baroncelli's committee changed its name to *Nacioun Gardiano* (Nation of Gardians) in July 1909. It was described by its first cap-tain, Alphonse Arnaud, as ". . . having the goal of maintaining and glorifying the Arlesian costume, the uses and traditions of the land of the bulls . . ."[21] This group, in contrast to the *Confrèrie*, was only for the amateurs or those not owning land. Following the inspira-tion of Baroncelli, the *Nacioun Gardiano* began the tradition of "games" presented in arenas and featuring feats of horsemanship.

Another change initiated by Baroncelli was the insistence that the bulls used in the arenas should be pure Camargue bulls, not those interbred with Spanish bulls for stronger qualities. At stake, Baroncelli and others believed, was the entire local economy and culture based on the bull. If Spanish bulls were to predominate, so would the Spanish bullfights and the entire panoply of ritual that surrounds the Spanish bullfight.

A French bullfight, more appropriately called a "race" or "run," pits the agility and speed of the bull against a dozen "razeteurs"—young men dressed in white who try to capture a small red ribbon and strings from the bull's horns. Such a competition would not have survived long in the twentieth century if Spanish bulls had been allowed to dominate in the arena. Today, however, this is the culture and entertainment one enjoys in the Camargue because of Baroncelli's work at the turn of the century. Just as he standardized the costume and the equestrian games performed at the yearly festi-vals, he preserved the most critical element of the *gardian* culture and economy, the Camargue bull.

Baroncelli's work in the Camargue, just as Buffalo Bill's work with the American West, codified the look and traditions of the *gardian* perhaps too much. In the latter part of the twentieth century, one sees today too much of the "drugstore" cowboy in the Camargue. The complaint heard now is that the *gardian* is too much a caricature of folklore and that the true value of the *gardian's* role is lost in the tourism of the Camargue's economy. Similar complaints can be heard in the American West.

One Frenchman has worked the similarities of French and American cowboys to his advantage. Joë Hamman, mentioned before as one of the interpreters to the Wild West, spent several years in Montana in the early twentieth century. He learned first-hand the Indian culture, the English language, and bronc riding. When he returned to France, he was instrumental in establishing the strong bonds between Baroncelli and the Sioux of the Wild West. Not long after the Wild West left France and Europe, Hamman turned his attention more fully to the Camargue, painting, and the cinema. His paintings are evocative of the American West, although they almost always portray the Camargue. Purposefully or not, he painted the scenes so that the *gardians* have their faces turned away or their hats worn low, and the impression is that the man on horseback might be anyone, anywhere.

More ironically, Hamman became a director and producer of western movies in France. In 1907, he produced *Cowboy,* one of the very first of the genre in any country, and in 1909 he produced *Les Aventures de Buffalo Bill.* Both of these films are lost today, but one may still see *Arizona Bill, The Railway of Death, Hundred Dollars Dead or Alive,* and others.[22] Hamman always wore cowboy dress, and one could easily mistake him for another nationality. He also occasionally posed in authentic Indian dress, and a friend remarked of him that he actually looked like an Indian in his later years.[23] He died in 1974.

Clearly, it seems more than coincidence that two Frenchmen, the Marquis Folco de Baroncelli-Javon and Joë Hamman, and one American, William Frederic Cody, had a common vision of cowboy cultures as the way to preserve a link to rural traditions and to assure the survival of these cultures into the twentieth century.

Cody, Baroncelli, and Hamman, all three, had similar missions creating images and myths through traveling shows, cinema, and regional rituals. The men, realizing that appearance was everything, standardized the costumes and physical look to which one is accustomed today. Though the American cowboys had other myth-makers and institutions to credit their success, the French *gardians* of the Camargue owe their existence to the unique confluence of time and place that brought together Baroncelli, Hamman, and Cody.

Notes

1. Some words in this article are not French but of Provençal origin. This is the dialect of the Provence region of France, the area that almost entirely includes the Camargue.

2. Remi Venture, *La confrèrie des gardians et sa fête annuelle* (The Confederation of the gardians and its annual festival) (Marguerittes, France: Editions de l'Equinoxe, 1992), 17.

3. Ibid., 22.

4. *New York Sun,* April 29, 1889.

5. *Le Figaro,* August 9, 1889.

6. Musée des Arts et Traditions Populaires, *Petit Echo de la Mode* (Paris) June 25, 1905.

7. *Barnum & Bailey En Route: Leur Système Spécial de Transport* (Barnum and Bailey on the road: Their special system of transportation) Paris, 1902.

8. *Côte Libre* (Paris), March 30, 1905.

9. On Russell, *The Lives and Legends of Buffalo Bill* (Norman: University of Oklahoma Press, 1960), 371.

10. Christina Stopka, "A list of dates for Buffalo Bill's Wild West" (photocopy) Buffalo Bill Historical Center Archives, Cody, Wyoming.

11. Charles E. Griffin, *Four Years in Europe with Buffalo Bill* (Albia, Iowa: Stage Publishing Co., 1908), 91-94.

12. Thierry Lefrançois, ed., *Les Indiens de Buffalo Bill et la Camargue* (Indians of Buffalo Bill and the Camargue) (Paris: Editions de la Martinière, 1994), 119.

13. Ibid., 49.

14. Ibid., 52.

15. Ibid., 36.

16. Jeanne de Flandreysy, "Mistral et les Peaux Rouges" (Mistral and the Red Skins) *Les Tablettes d'Avignon et de Provence* 227 (September 7, 1930).

17. Lefrançois, 59.

18. Jacob White Eyes, different cities in Europe, to Marquis de Folco Baroncelli-Javon, Stes. Maries-de-la-Mer, February 1906 to September 1906, Baroncelli Collection, Palais du Roure, Avignon, France.

19. Lefrançois, 69.

20. Sabine Barnicaud, "Quelques dates importantes de la vie du Marquis de Baroncelli" (Important dates in the life of the Marquis of Baroncelli) (photocopy) Baroncelli Collection, Palais du Roure, Avignon, France.

21. Pierre Dupuy, *La Guide de la Camargue* (Guide to the Camargue) (Besançon: La Manufacture, 1991), 83, 89.

22. Musée Camarguais, *Camargue, terre de cinéma* (The Camargue, land of movies) (Parc Naturel Régionel de Camargue, 1984), 6.

23. Lefrançois, 109.

15 Reel Cowboys: Cowhands and Western Movies

ALBERT B. TUCKER

Typically on any Saturday morning in the summer of 1947, boys and girls across rural America completed their morning chores, changed into go-to-town jeans, and scraped together twenty-five cents in pennies, nickels, and dimes they had squirreled away in an old cigar box. Then they walked or hitched a ride into town, where for nine cents they bought a ticket to the matinee movie at the Majestic Theater. A dime bought a cup of cola and a bag of popcorn.

Then, they settled down into a soft theater seat to watch a Three Stooges short subject. They grinned with embarrassment when a bouncing-ball cartoon character invited them to sing, but no one ever did. Finally, the serial started. This Saturday's serial was episode number 10 of *Billy the Kid*, starring Johnny Mack Brown. The kids in the movie theater knew that Billy was really a U.S. Marshal just pretending to be an outlaw in order to catch some bad guys. Episode 10 ended with Billy the Kid plunging down a cliff, tied up inside a burning stage coach. During the previews the boys and girls returned to the concession stand to spend their last six cents on a piece of bubble gum and a *Holiday* all-day sucker. Some of them strolled up and down the aisle looking for coins dropped by careless youngsters.

Soon, everyone was back in their seats ready to ride the range with their favorite cowboy hero starring in what adults called the B Western. The B stood for budget, although some liked to say it stood for bad. Sometimes the movies starred Hopalong Cassidy, Wild Bill Elliot, or Roy Rogers, but on this Saturday it was Gene Autry in *The Old Corral*. Hardly anyone noticed the guitar player in the background, who looked like a very young Roy Rogers. The kids watched the movie with rapt attention, for the "reel" cowboys were their

Roy Rogers and Trigger topped box office sales from 1940 to 1949. When Trigger died, Roy had him stuffed and now displays him at the Roy Rogers Museum. Courtesy of Albert B. Tucker.

heroes of the silver screen. From a kid's point of view, the "reel" cowboys never let you down. They always did the right things, even to their own discomfort. They always helped the oppressed, fought the bad guys, and never doubted their ability to win. After saving the ranch from the evil rustlers or cleaning up the town from the crooked gamblers, the cowboy hero waved goodbye to the pretty girl and rode off into the sunset. (The boys silently wondered why the cowboy left that pretty girl.)

The Drifter

That cowboy was my hero
'Cause he always stood for right,
And he was lightning with a six-gun
When he was forced to fight.
But when he rode off into the sunset,
I'd have turned that horse around,
'Cause I loved that girl he left there
In that dusty western town.[1]

They rode off into the sunset. That means that after great accomplishments, cowboys did not hang around living off their reputation. They did not bask in the praises of people, but went in search of other wrongs to be made right. They always roped better, rode better, sang better, and shot better than anyone. They were strong, handsome, and intelligent, and they never smoked, cursed, or drank whiskey. They always had coins to pay for their food, and they never had to change clothes, brush their teeth, or go to the restroom. A mother rarely showed up in the cowboy movies unless she was the mother of the pretty girl. Cowboy heroes did not need mothers to look after them.[2] The pretty girl was never helpless, squeamish, or whiny. She was independent unless she chose, for her own reasons, to let the cowboy hero help her.

Cowboy heroes always won fist fights, six-gun duels, and the admiration of the townspeople. They were looked up to, even worshiped, for their integrity, clean living, and impeccable judgment about right and wrong. The bad guys always lost and the cowboy heroes always won.

Sound ridiculous? Maybe. But in the post-World War II era, to the front row kids—boys and girls alike—that such a world might exist gave hope that their own miserable life of doing thankless chores, riding swayback old mares, and wearing hand-me-downs would someday transform into the world of the B Western. Somewhere out there was a place where everyone respected and cheered for cowboys.

The B Western Genre

Jon Tuska wrote in *The Filming of the West* that he was determined to avoid the stereotype of the Hollywood cowboy, claiming that "nostalgia has no place in film history."[3] Yet Tuska could not bring himself to destroy the movie cowboy's image by including photographs of them long after their prime. To picture a B Western hero as aging is a sacrilege. No place for nostalgia? Well, when it comes to cowboy heroes, logic and consistency can take a walk.

The world of the B Western is a vaguely described West that began in the 1830s and lasted for about a hundred years. For the most part, the B Westerns portray the West as a concept that does not change over time. The front row kids saw nothing amiss with having covered wagons, six-shooters, flashlights, wood stoves, refrigerators, radios, telephones, automobiles, and airplanes, all in the same movie. These items were invented somewhere between 1830 and 1930. If the West is perceived as a constant concept, historical sequence is inconsequential. The B Westerns are a composite and predictable West, authentic-looking in overall settings, but with details blended and shuffled for action and drama.

The oral legends, dime novels, and Wild West shows that claimed to reflect the cowboy culture created composite fictional heroes out of Wyatt Earp, Butch Cassidy, Wild Bill Hickok, and Buffalo Bill Cody. Even the bad guys underwent transformation. Johnny Mack Brown, who used one of Billy's guns in *Billy the Kid*, the serial, started the trend to sanctify outlaws.[4] Billy the Kid was later played by such cowboy heroes as Roy Rogers, Bob Steele, and Buster Crabbe. When Billy could no longer be passed off as an undercover good guy, Alfred "Lash" LaRue inherited the role.[5] He had the persona of Humphrey Bogart, with hints of a shady past, now reformed and ready to literally whip the bad guys into line.[6]

Young cowboys went to the movies to learn how good cowboys were supposed to act. Good cowboys shot straight, rode hard, and always found ways to "head 'em off at the pass." The good cowboy was a combined lawman, counselor, preacher, doctor, civil rights worker, and all-around best cowboy who could outshoot, outride, and outfistfight any of the bad guys. The cowboy hero was twenty feet tall, or so it appeared on the silver screen.[7]

The Working Cowboy and the Westerns

All wannabe cowboys have a hero cowboy after whom they model their own lives. Alex Gordon, Gene Autry's press agent and former president of the Gene Autry Fan Club, stated that Autry would appreciate it if something might be written about Buck Jones, because Jones was Gene's cowboy hero.[8] Here was a case where a leading cowboy hero had his own cowboy hero.[9]

In a survey conducted at the 1998 Cowboy Poetry Gathering in Alpine, Texas, working cowboys registered their vote for three of their favorite "reel" cowboys from the 1930-to-1950 era. These same cowboys, for the most part, had participated in a nationwide vote in 1947 in which kids voted between Roy Rogers and Gene Autry for the title, "King of the Cowboys." Roy Rogers won that election. Now, as adult working cowboys, they were once again given a chance to vote, only this time they could choose from a long list of movie cowboys and they could vote for three. From a sampling of eighty-three working cowboys, the top four "reel" cowboys were Gene Autry (thirty-seven votes), John Wayne (thirty-six votes), Roy Rogers (thirty-four votes) and Hopalong Cassidy (twenty-five votes). The vote tally differences among the top four are not statistically significant, but fans may take secret delight in knowing that this time Gene Autry came out on top. The votes ranked the cowboys as follows:

Gene Autry and Champion enjoyed top box office draw from 1930 to 1939, when Gene enlisted in the Army. He returned in 1946, but had to share top cowboy honors with Roy Rogers. When Champion died, Gene had him buried. Courtesy of Albert B. Tucker.

Votes	Movie Cowboy
37	Gene Autry
36	John Wayne
34	Roy Rogers
25	Hopalong Cassidy (William Boyd)
21	Randolph Scott
14	Tom Mix
10	Tex Ritter
9	Rex Allen
8	Wild Bill Elliot and Lash LaRue
7	Johnny Mack Brown and Buck Jones
6	Bob Steel and Clayton Moore (Lone Ranger)
4	Duncan Reynaldo (Cisco Kid)
3	Alan "Rocky" Lane and William S. Hart
2	"Sunset" Carson and Charles Starrett
1	Ken Maynard, Ben Johnson, and Gary Cooper

Interestingly, no one voted for Colonel Tim McCoy, "Crash" Corrigan, or Eddie Dean. McCoy and Corrigan were often seen in a series called "The Three Mesquiteers" or "The Range Busters." Eddie Dean was a good singer and actor, but apparently he has been largely forgotten.

Several cowboys wrote comments on the survey form or spoke into a tape recorder to express their opinions about the movie cowboys of their childhood. Howdy Fowler voted for Alan "Rocky" Lane because he once had his picture taken with Lane at Truth or Consequences, New Mexico. Many of the movie cowboys traveled with their movies across the country, making personal appearances at the theaters. Having seen a cowboy hero in person was cause enough to vote for him. Some cowboys had actually worked cattle and rodeos with the cowboy heroes, and one was proud that he had appeared with Gene Autry on stage in 1943.

Many of the cowboys commented that they preferred the "authentic" movie cowboys over those "fancy dans" who became cowboys only by wearing a cowboy outfit. One such working cowboy was especially vocal about the fake cowboys. He preferred the

Wild Bill Elliot played Red Ryder from 1944 to 1946, filming sixteen features. He later played Wild Bill Hickok, a peaceable man, wearing his guns with the handles facing backwards. Courtesy of Albert B. Tucker.

heroes who could ride, used authentic gear, and could throw a loop. When asked which ones he considered to be authentic cowboys, he pointed to Monte Hale, Lash LaRue, and Roy Rogers.

Joaquin Jackson, a retired Texas Ranger, quickly voted for Wild Bill Elliot, who is the universal role model for all Texas Rangers. Mike Wohleking, a Texas Department of Public Safety Trooper, modeled his talk and mannerisms after Wild Bill Elliot, although he stopped at wearing his guns in reverse.[10]

A group of working cowboys at the Alpine Cowboy Poetry Gathering got into an argument as to who was the best Red Ryder: Wild Bill Elliot or Alan "Rocky" Lane, both of whom co-starred with a young Robert Blake playing the part of Little Beaver. Wild Bill Elliot won the Red Ryder contest when one cowboy noted that Alan "Rocky" Lane ended up his career as the voice of television's *Mr. Ed*.

Henry Ogletree, a former sheriff of Brewster County in the Big Bend of Texas, said, "We need some more of these type heroes for the young people to look up to." If a consensus could be found among the working cowboys, Mr. Ogletree expressed it well. Most of the working cowboys lamented the lack of high level cowboy role models for the youth of today.

John Kuenster is a former line rider and hunting guide for the 06 Ranch. He knew as a child that the B Western was not real, but hoped that it did exist, at least somewhere between imagination and real life. When he returned home from the Saturday matinee, he tried the same things the movie cowboys did, but they never worked. That's when reality set in, although he never gave up trying, hoping that somehow better days awaited the hard-working cowboy.

Billy Hill is a day worker for the area ranches. He was raised on a ranch and admitted the only reason he ever went to the movies as a kid was to see the cowboy heroes. He felt that people looked down on him because he was from a cowboy family. The movies, however, restored his self-image, because—in the movies, at least—the cowboys were heroes.

Sons and daughters of working cowboys, enrolled in classes at Sul Ross State University in Alpine, said that they often wondered how the heroes in the movies always managed to fight and roll down hillsides without tearing their clothes, losing their hats, or getting dirty. They knew the movies were not real, but liked them anyway, although they were amused at the way the language was cleaner and more sophisticated than the language often heard around the cow lots. When going to the movies with their working cowboy fathers, the sons and daughters had to endure Dad criticizing the hero for having the wrong kind of horse's bit and for wearing dress boots for roundups. Dad made fun of the movie cowboy who rode a fancy saddle with glitter for working cattle. The trick mountings on a horse always brought chuckles from Dad. "No one," said Dad, "can jump from a building and land on a horse without hurting himself."

The truth is, working cowboys never had time to be a hero of the B movie type. There were few bad guys, if any, and never would Dad have time to ride off looking for wrongs to be made right. Singing? The sons and daughters said Dad tried to sing, but never to the cows.

Jack Yarbo was an old-time cowboy of the Big Bend, admired by just about everyone. When local cowboys gathered to watch videotapes of the B Westerns, Jack always gave running commentaries on the way Roy mounted Trigger, the gear used, and the kind of gun he shot. Jack loved to watch those Westerns and greatly admired the heroes of the silver screen. He said that it was good for cowboys to see themselves as heroic and as role models for the kids.

Practically all of the working cowboys expressed frustration at being able to vote for only three of the movie cowboys in the survey. One commented that the survey was "splitting hairs way too hard." Reluctantly, he settled on "Sunset" Carson, William S. Hart, and John Wayne.

Some of the working cowboys submitted write-in votes. Clayton Moore (a.k.a. The Lone Ranger) received six write-in votes. He had not been included on the list because he was mainly a television cowboy, not one of the silver-screen cowboys, although *The Lone Ranger* did have a large following as a radio show during the 1940s.

Consider these comments made by the working cowboys as they voted for their favorite movie cowboys:

Hopalong Cassidy and Topper enjoyed a rebirth of popularity when the films were run on television in the early 1950s. Topper was buried in a nine-foot casket, but reports persist that his ghost still haunts the Los Angeles Pet Memorial Park. Courtesy of Albert B. Tucker.

"This is a hard choice to make. Too bad the kids of today can never know the thrills of the silver screen and to know before the movie started that the good guys always won!"

"Gene Autry was the toughest movie cowboy because he went into the bar and ordered Buttermilk. That's tough!" [Note: Johnny Mack Brown is the one who ordered buttermilk. Hopalong Cassidy liked to order milk. Gene Autry usually drank sarsaparilla (root beer).]

"I thought Topper was a great horse, and Hoppy was white-haired like my granddaddy."

"I liked Charles Starrett (Durango Kid) because he could jump on his horse from behind and from the top of a building."

"My favorite cowboys are the ones I met personally, such as Sunset Carson."

"I prefer the cowboys who were cowboys before they became movie cowboys, such as Lash LaRue."

"I don't want to vote for the obvious ones, such as Gene Autry and Roy Rogers. I picked the more obscure ones, such as Buck Jones, Monte Hale, and William S. Hart."

"I liked Tom Mix because I got a picture taken with him one time."

"I liked the parts that recurred in every movie, such as the stampede scene. It was the same scene shown over and over again in every movie, and the cowboys always rode past the same rocks, over and over again."

"William S. Hart was without doubt the toughest of all the movie cowboys. He looked tough. I wouldn't want to cross him!"

"Tom Mix is my favorite because he did all his own stunts. Most of the other movie cowboys let Yakima Canutt do the stunts for them."

"I know everyone liked Lash LaRue, but another whip artist I liked was a Pecos, Texas, rodeo star named Whip Wilson. He made more movies than Lash."

Allen Chandler, one of Alpine's hard-working, hand-calloused cowboys, said, "We didn't get to go to the movies except when we went to town to get a haircut. I didn't get to see many movies, mind you, but I know this much, Gene Autry could beat the hell out of Roy Rogers."

After they completed the survey, the cowboy poets continued to talk about the B Western heroes. They drilled each other on the name of the movie cowboy's horse, his sidekick, and in which movie did Gene Autry actually kiss the girl. To be able to describe the movies and to spout trivia about the heroes is a test of intelligence. Some cowboys polarized on one side of the room or the other, with the Gene Autry fans against the Roy Roger fans. The John Wayne and Hopalong Cassidy fans just looked on with amusement.

Not all of the comments were based on knowledge. When one cowboy looked at the survey list, he asked if Sunset Carson was the same as Kit Carson. Another insisted that Warren Oates and Tommy Lee Jones should have been included on the list.

The Cowboy Heroes and Their Values

John Wayne (a.k.a. Marion Michael Morrison) received the second highest number of votes in the survey. In ways unequaled by the others, he probably deserves the title of "King of the Cowboys" more than the others combined. He, along with Randolph Scott and Gary Cooper, managed to bridge the gap between the low budget Westerns and the high budget, more adult Westerns. Wayne started his acting career in the Lone Star Production Westerns in the 1930s, but his acting ability and durability caught the eye of larger producers, who cast him in *Stagecoach, The Angel and the Badman,* and *Red River.* His popularity spans at least three generations, and at the rate his movies are still replayed on television, he will remain a popular star for another generation or two. John Wayne, Randolph Scott, Gary Cooper, Jimmy Stewart, Ben Johnson, and Ronald Reagan transcended the limitations of the B Western.

The other top vote getters were Gene Autry, Roy Rogers, and Hopalong Cassidy. Gene Autry always played himself in the movies, whether set in 1880 or 1950. He wrote most of his own songs, and

Allan "Rocky" Lane starred in twenty-one Red Ryder features from 1945 to 1948. Other Red Ryders included Monte Hale, Don "Red" Barry, and Jim Bannon. Courtesy of Albert B. Tucker.

started the practice of naming the movies after the lead song. Gene deliberately demonstrated in his movies an ethical system compatible with the message in his theme song, "Back in the Saddle," where he declared that "the only law is right." The values found in his movies were not put there by accident. While he admitted "no claims for improving their intellects," Gene knew that the front row kids would try to imitate what they saw in the movies. He set down a code of ethics for the genre and urged the other cowboy heroes to adopt the standard. Gene knew that he took some "ribbing" about the Cowboy Code, but he took his role seriously. "I never felt there was anything wrong with striving to be better than you are."[11] Consider these rules for cowboys written by Gene Autry and infused into his movies.

Gene Autry's Cowboy Ten Commandments

1. The cowboy must never shoot first, hit a smaller man, or take unfair advantage.
2. He must never go back on his word, or a trust confided in him.
3. He must always tell the truth.
4. He must be gentle with children, the elderly, and animals.
5. He must not advocate or possess racially or religiously intolerant ideas.
6. He must help people in distress.
7. He must be a good worker.
8. He must keep himself clean in thought, speech, action, and personal habits.
9. He must respect women, parents, and his nation's laws.
10. The cowboy is a patriot.

The Cowboy Ten Commandments were readily adopted by the motion picture industry, and approved by church groups, parents, and kids.[12] Pat Buttram, one of Gene's sidekicks, is fond of relating how Gene used to ride off into the sunset, and then he came to own the sunset. Buttram added that Gene Autry and the B Western message saved a generation of young people.[13] Gene Autry and Roy

Rogers demonstrated that clean living and honest labor have their own rewards. Consider Roy's Rules for membership in the Roy Rogers' Riders Club, founded in the 1940s.

Roy's Rules

1. Be neat and clean.
2. Be courteous and polite.
3. Always obey your parents.
4. Protect the weak and help them.
5. Be brave but never take chances.
6. Study hard and learn all you can.
7. Be kind to the animals and care for them.
8. Eat all your food and never waste any.
9. Love God and go to Sunday School regularly.
10. Always respect our flag and our country.

Hopalong Cassidy, played by William Boyd, grew out of a literary series created by Clarence E. Mulford. The Bar 20 foreman was a middle-aged man who had to hop a little as he walked due to a wound received during the Civil War. After writing twenty volumes, Mulford, who had never lived out West, subcontracted the last four volumes to a new author who wrote under the name of Tex Burns, a.k.a. Louis L'Amour. David Niven turned down the movie role, and Harry "Pop" Sherman of Paramount Studios rejected John Wayne because he was too young and inexperienced. Sherman settled on William Boyd, a has-been actor who had fallen into bad luck. Paramount released more than sixty Hopalong movies. In 1947 Boyd bought the screen rights from Sherman, and with the advent of television, William Boyd made a fortune. William Boyd turned down all subsequent screen roles, stating that his image had been cast and he would appear only as Hopalong Cassidy. The kids saw Hoppy as a father figure who could outsmart as well as outshoot the bad guys. He was not a "singing" cowboy, which made the non-singers among the kids feel a little more comfortable.

Gene and Roy largely dominated the B Western popularity race from 1930 to 1950, and Hoppy became popular following his reruns on television. John Wayne's popularity came later, during the 1960s and 1970s. Roy Rogers recognized the role that he, Gene, and Hoppy had in the lives of the youngsters who watched their movies. Roy released this song in the 1980s, and it was used in the popular *Tribute to Roy Rogers* album released in 1990.

Hoppy, Gene and Me

Hoppy, Gene and me,
We taught you how to shoot straight.
You had to be a cowboy,
That's how it had to be.
Stories from the silver screen,
Now most of them forgotten,
Double feature Saturday,
With Hoppy, Gene and me.

History and the B Western

The screenwriters based the B Westerns on history—sort of. They at least made the front row kids aware of the significance of the Civil War, the Pony Express, the gold rush, the cattle drives, the Indian wars, the wars between the sheepherders and the cattlemen, and the conflicts over barbed wire and water and grazing rights.[14] The concepts of mortgage payments, indebtedness, and ownership of cattle and land were common topics of conversation among the front row kids. The roles of the railroads, the Winchester, and the sheriff were well-known and respected. When the front row kids began to study American history in school, they already knew the vocabulary and some of the famous people of the West.

Although the content may have been foggy on historical accuracy, the event itself, the B Western, had historical impact on the viewers as if the events actually happened as the films portrayed. Sure, Red Ryder may not have rescued the Butterfield Stageline from premature demise, and Wild Bill Elliot may not have prevented General U. S. Grant from making a big mistake at Gettysburg (Grant was not there), but the important thing was that the kids talked about these events.

The Greek myths and the Indian legends never enjoyed the social incorporation into folklore that the B Westerns enjoy. "Westerns are America's unique contribution to that body of mythic folklore familiar to most of the human race."[15] People from other countries were never actually convinced of their myths as much as the kids—especially those from rural communities—were convinced of the cowboy myth as portrayed in the movies.[16] Even the adult working cowboys in the 1930s and 1940s hoped that if they worked a little harder, lived a little cleaner, and always told the truth, they would somehow find that B Western paradise. As the reporter commented at the end of *The Man Who Shot Liberty Valance*, "When the legend becomes fact, print the legend."[17] The cowboy hero may only be a fantasy in the working cowboy's mind, but because the image persists, the cowboy hero has achieved a certain measure of immortality.[18]

Contributions to Society

The B Western not only helped with understanding history and with distinguishing right from wrong, it became a part of the history of the West. Phil Hardy in his *Encyclopedia of Western Movies* commented that Southern California and Hollywood became the last real employer of cowboys, who in turn became the inspiration of a new generation of cowboys. Just as the B Western was a product of Hollywood, it can be argued that Hollywood, to a large extent, was the product of the B Western. It was the main moneymaker during the 1930s Depression when Hollywood was struggling to exist. Thanks to Gene Autry and others, the moviegoers filled the theaters during the Depression years.[19]

The "reel" cowboys contributed toward raising a generation of "straight shooters" who had an appreciation for and a conceptual idea of the historical West. They understood the meaning behind Roy and Dale's theme song. Wherever working cowboys gather, when the guy with a guitar starts to sing "Happy Trails," everyone joins in, and they all know the words and tune.

The "real" cowboys have a deep respect for the "reel" cowboys, but the demands of ranching keep them from living in the fantasy world of the B Western. Ann Sochat and Tony Cano attended the 1998 Cowboy Poetry Gathering in Alpine. Nestled among their poetry collection in *Cowhide 'n Calico* is the following poem, which sums up the respect.[20]

Roy 'n Gene

When I was young, the biggest treat
Was when we'd get to go
On Saturday afternoon to town
To see a picture show.
It was a real small theater.
They charged but a small price,
And they didn't mind if we stay around
And watched the movie twice.
Our world was centered round the action
On that silver screen,
As we watched our favorite heroes,
The cowboys Roy an' Gene.

As Pa would drive us home from town,
We couldn't wait to play,
And recreate the scenes we'd seen
Upon the screen that day.
But when we reached the ranch,
There were always chores to do.
Ma said, "No playin' cowboys
Until you all are through!"
So we helped her with the milking,
Put the livestock in their pens,
An' we had to put out evening feed
For the horses an' the hens.
But then we'd grab our cap guns
An' head for the ravine,
Where we'd take turns at bein'
Our heroes Roy an' Gene.

It took years of livin' at the ranch
To open up our eyes.
Diggin' fence posts, mendin',
Hard work made us realize,
That throughout all of the trying times
Our family had to weather,
It was the courage of my Ma an' Pa
That kept our ranch together.
An' cowboyin' ain't 'bout
Chasin' outlaws with a gun,
But it's working hard with no reward
Except for a job well done.
Those movie cowboys should have been more honest,
Like I figger,
They should have had Roy shovelin'
Behind that horse named Trigger!
An' though they shore ain't famous
An' they ain't quick on the draw,
The REAL cowboy heroes
Are just plain men like my Pa.
They ain't the dashin' figures
That you see up on the screen.
Some things you learn as you grow up!
Sorry, Roy an' Gene!

And so as both the "reel" and the "real" cowboys ride off into the sunset, may they go, as they typically did in the summer of 1947, with Roy's famous television blessing: "Goodbye, good luck, and may the good Lord take a likin' to you."

Notes

1. Jerry Mitchell, Associate Professor of Education, Sul Ross State University, Alpine, Texas. 1996. Used with permission.
2. Don Graham, *Cowboys and Cadillacs* (Austin: Texas Monthly Press, 1983), 26.

3. Jon Tuska, *The Filming of the West* (New York: Doubleday & Company, 1976).

4. Phil Hardy, *Encyclopedia of Western Movies* (New York: Woodbury Press, 1984), xiv, 20.

5. Ibid., 149. In "Song of Wyoming" (1945), Eddie Dean used his considerable singing ability to reform the leading outlaw, Alfred Lash LaRue. LaRue proved so popular to the front row kids, he was given his own series.

6. Kalton C. Lahue, *Riders of the Range* (New York: A.S. Barnes & Co., 1973), 159. "LaRue had a sulky downright mean look about him (about which he could do nothing) and even his broadest smile reminded you of a prohibition racketeer who had just heard that his crosstown rival had turned up in concrete shoes . . . . No villain could possibly prove a match for anyone with as mean a look in his eyes as good old Lash LaRue."

7. James Horwitz, *They Went Thataway* (New York: E. P. Dutton & Co., 1976), 5.

8. Lewis Atherton, *The Cattle Kings* (Bloomington: Indiana University Press, 1961), 75. Buck Jones and Tom Mix were participants in Joe Miller's Wild West Show, where they acquired the rudimentary skills necessary to portray cowboys on the movie screen.

9. Buck Rainey, *Saddle Aces of the Cinema* (A. S. Barnes, P.O. Box 3051, La Jolla, California 92038, 1980), 146-50. Buck Jones died trying to rescue others in the Coconut Grove fire in 1942, a hero in legend and in fact. He was probably the most respected of all screen cowboys.

10. Wild Bill Elliot will probably be most remembered for his role as Red Ryder, wearing his guns backwards for fast backhand draws. His best line was, "I'm a peaceable man," and after saying this he proceeded to knock the tar out of the troublesome bad guy.

11. Gene Autry with Mickey Herskowitz, *Back in the Saddle Again* (Garden City, New York: Doubleday & Company, 1978), 184.

12. George Fenin and William K. Everson, *The Western* (New York: Orion Press, 1962), 20.

13. Alex Gordon, *The Gene Autry Story*. A two-hour radio program on two audio cassettes. Gene Autry Western Heritage Museum, 4700 Zoo Drive, Los Angeles, California 90027, 1989.

14. Bill O'Neal in *Cattlemen vs. Sheepherders* (Austin, Texas: Eakin Press, 1989) cites Gene Autry's *Springtime in the Rockies* and Roy Rogers's *Roll on Texas Moon* as examples of how the sheep vs. cattle controversy affected the West, 161.

15. William T. Pilkington and Don Graham, *Western Movies* (Albuquerque: University of New Mexico Press, 1979), 1.

16. The B Western stars themselves were convinced of the myth. William Boyd came to believe that he was Hopalong Cassidy, and never made public appearances without his outfit on. Clayton Moore fought a court battle over maintaining his right to be the Lone Ranger. Once while traveling the Los Angeles freeway in his white Cadillac and wearing his Lone Ranger outfit, Moore stopped to help a stranded lady motorist. After getting the car started, he drove away, leaving the lady to ask the sheriff who had just driven up, "Who was that masked man?"

17. Hardy, *Encyclopedia of Western Movies,* 281. *The Man Who Shot Liberty Valance* starred John Wayne and Jimmy Stewart. It was filmed in 1962, after the B Western had largely declined.

18. Michael T. Marsden, "Savior in the Saddle: the Sagebrush Testament," *Focus on the Western,* Jack Nachbar, ed. (Englewood Cliffs, New Jersey: Prentice-Hall, 1974), 93.

19. Hardy, *Encyclopedia of Western Movies,* x, xv.

20. Ann Sochat and Tony Cano, *Cowhide 'n Calico: A Poet Looks at the West* (Canutillo, Texas: Reata Publishing, 1997), 17–19.

16 Today's Cowboy: Coping with a Myth

LAWRENCE CLAYTON

When Joe B. Frantz and Julian Ernest Choate Jr. published their fine study *The American Cowboy: The Myth and the Reality*[1] in 1955, many may have erroneously assumed that the final word had been said on the subject. If anything, however, the discussion of the cowboy image has continued unabated and is as controversial today as it has ever been. Of the various interpretations of the figure, there are two noticeable extremes and, it seems to me, a tenable middle ground reasonably close to actuality. Atop this triad, however, still sits the mythic cowboy, apparently as secure as ever.

One approach has been to discredit the figure. Even if people feel threatened by the image casting its shadow from the past on the present, one can but wonder why the demythologizers attack the image so vehemently as Jane Kramer, a New Yorker, does in *The Last Cowboy*.[2] Her protagonist often sits on a rotting chuck wagon dreaming of the good old days rather than facing contemporary life. This kind of hostile response comes typically only when people feel threatened by something they cannot accept or understand. Jack Shaefer's *Monte Walsh*[3] is one of the best novels downplaying the cowboy myth, but it deals with a character who clings to the cowboy role even when he grows old. He can adjust to no other life. I doubt this cowboy is the only person to cling to a career when he has outlived his usefulness.

The debunkers describe the cowboy as only a hired hand working for low wages. The man himself has been depicted as lazy, shiftless, celibate, free-spirited, unsophisticated, stubborn, and antisocial. He was just a laborer who happened to ride a horse to do his

Previously published in *West Texas Historical Association Year Book*, 60 (1984): 178-184.

work. No doubt, his work was dirty, demanding, smelly, and dangerous. OSHA certainly would not have approved, and the Wage and Hour Commission would have had a field day with his employers. There was no health insurance, retirement plan, disability insurance, or union dues. His life, seen in this light, certainly appears unattractive, or can be made to appear so.

One of the most significant events dealing negatively with the cowboy is the 1983 exhibit on the figure done by the Library of Congress.[4] The opening of the display was complete with live performers who sang, exhibited skills, and generally demonstrated the continuation of this kind of life in the present day and emphasized the realistic view. One of the obvious results of such an occasion is the demythologizing of the idea of the cowboy by presenting much of the reality of his life and depriving the public of the preferred conception. The furor raised over the negative kind of presentation in the exhibit—and the figure—has been significant, and resulted in the withdrawal of some of the items from the exhibit by the National Cowboy Hall of Fame in Oklahoma City.[5]

Although currently out of vogue, film parodies of the western hero—cowboy, gunfighter, etc.—have had their effect as well. *Blazing Saddles, Water Hole #3, Cat Ballou, The Apple Dumpling Gang,* and others have belittled the whole idea of serious western heroes— usually gunfighters instead of cowboys—but the effect is the same. A cowboy figure as hero is rare anyway. Most cowboy films had their climax in gunfights and other life-threatening situations, not in just handling cattle. The status of the hero is usually determined by his ability with a gun or with his fists, not with a rope at roundup time. *The Outlaw Josey Wales,*[6] starring Clint Eastwood, though not a cowboy movie, is the only really successful serious western film in recent years, and is actually a treatment of one rebelling against the injustices of society, a popular theme even today.

At the other extreme, the most prominent recent romanticized view of the cowboy followed in the wake of a box office smash film, *The Urban Cowboy,* stemming from an article on that phenomenon that Aaron Latham published in *Esquire Magazine* in 1978.[7] The film created a subculture rivaling that triggered by the Beatles and their rock music in the 1960s. The urban cowboy image, of course, has

Swenson Land and Cattle Company. Catching and saddling horses, ca. 1920. Courtesy of Southwest Collection, Texas Tech University. SWCPC File no. 123(A), Envelope 10.

little to do with life on a ranch, but instead highlights the nightlife in the huge club of country and western music star Mickey Gilley in a suburb of obviously non-rural Houston, one of the nation's largest metropolitan areas. Millions have watched actor John Travolta play out the fantasy of a generation of discontented city dwellers living a version of "rural" life sporting mechanical bulls and flowing with cold beer, preferably Lone Star, instead of with milk and honey. One wonders why the attraction was so strong, except that it offered through the costume an escape into the neon-lit cowboy life without the sweat, dust, and boredom found in actual ranch life. They took to heart a line in "The Cowboy's Lament," a folk song containing the statement "I see by your outfit that you are a cowboy."[8] For these, clothes make the man—or woman, as the case may be—as the wide appeal of western style clothing evidences. Dressing like cowboys are supposed to dress is enough for those who will never do any cowboy work anyway. Even though by 1982 Texas "chic" was dead

Most of these cowboys (ca. 1915) wear a mix of cowboy clothes and lace-up or button shoes; high-heeled boots did not serve well for ground work. Courtesy of Southwest Collection, Texas Tech University. SWCPC File no. FR 265.

as a national fashion trend, the boom to the clothing industry in the late 1970s and early 1980s was economically as well as culturally significant. It is likely that the feeling is only waning, waiting for another stimulus to wax again.

These barroom cowboys are not real cowboys, however. In a perceptive contemporary song, Billy "Crash" Craddock sings, "A real cowboy don't care how he fills out his jeans/He just cares how he feels in his heart."[9] Clothes do not make real cowboys, as old photographs tend to confirm. These old-timers looked pretty motley in their range getups. Real cowboys still exist, much to the surprise and chagrin of many skeptics, who would see them gone if they had their way. The proof is before us on such ranches as the Matthews, Nail, Caldwell, and Green in the area north of Albany, as well as on the Pitchfork, 6666, King, Swenson, Double U, and others in the state. Also, a comparison of the photographs Irwin Smith made in the early days of this century (*Life on the Texas Range*[10]) and Ray Rector in the 1930s (*Cowboy Life on the Texas Plains*[11]) with the contemporary ones made by Martin Schreiber (*Last of a Breed*[12]) conclusively proves that though some technology has evolved, much of ranching life and, at least as important, attitudes in it remain virtually unchanged.

The real cowboy continues to draw other printed comment as well. *Buckaroos in Paradise*[13] is a modern classic documentation of cowboy life in Nevada. Closer to home, John Erickson, a modern-day cowboy, author, and folklorist, has written two books on the subject. *Panhandle Cowboy*[14] reflects his experiences running a ranch in the Oklahoma Panhandle in the 1970s and *The Modern Cowboy*[15] can actually serve as a training manual for anyone who would like to develop cowboy skills but cannot find a teacher. Like those listening to army veterans talking about the "joys" of boot camp, Erickson's readers are regaled with humorous versions of sometimes dull but dangerous activities that can become exciting when judgment errs only slightly. Andy Adams in his *Log of a Cowboy*[16] and Teddy "Blue" Abbott in *We Pointed Them North*[17]—both early cowboy classics—reflect the same kind of experience.

A seemingly minor incident, with far-reaching implications, appears in Elmer Kelton's *The Good Old Boys*,[18] a fine novel dealing creatively with the end of the West. In it there is a clear recognition that the old days are over and a realization that the situation can be faced without altering the idea or, worse, degrading it. Roy Rasmussen is an aged drifter following the old ways in 1906, well past the time amenable to his ways. He is found dead beside his horse where he stops, ironically, to open a pasture gate in a barbed wire fence, a barrier that altered open range practices forever. A pauper by modern standards, Rasmussen faces burial at county expense. The young "cowboys" at the bar in town refuse to allow this injustice and pool their meager resources to pay the cost of the funeral. This truthful view is sobering, but not disgusting like that of the demythologizers.

Although we know that the daily life of this real American cowboy was filled with dust, sweat, toil, loneliness, and frequently death, the "popular" image of the figure—that held by many people today—is infused with an aura of romance that elevates him to a pure-hatred knight-errant. This cowboy, this figment of the imagination fed on B-Western movies and pulp magazine stories and novels and dude ranches, must be considered part of the myth of the West, for indeed he still looms larger than life. He is usually imagined trotting over the prairie on a cowpony, working the herd around a branding fire, or sitting around an evening campfire singing songs of lonely and dying cowhands. In reality, trotting on horseback can be painful to the inexperienced or clumsy, and working around the branding fire is difficult and dangerous work requiring skill, patience, and often daring. In addition, few sing well enough to garner accolades for their voices, even if they had the time and fortitude to sit around a campfire harassed by buzzing insects and campfire smoke and enduring the extremes of the weather.

One must ask, what influence does the truth have on the image of the mythic cowboy? The response is certainly a resounding "None!" Myths meet little resistance from verifiable truth.[19] Instead, they help us visualize the exotic, the supernatural, the unexplainable. Mythology deals with matters of cosmic importance, with gods and creation, and with the source of life itself. Disregarding actuality, we see these knights of the range in our mind's eye wandering the open prairies, free from the confines of the city, of marriage, of capitalism, and of the many other fetters and fetishes that plague our lives. People today are attracted by the very things cowboys did not depend upon, definitely a contrast as we try to get all that the "good life" allows and requires. We want bucolic, not urban, cowboys, though the urban version gets a fair share of the attention through imitation. I propose that emulation of the mythic cowboy results in the attire, the urban cowboy image. Actually the existence of the Library of Congress exhibit proves the lasting appeal of the figure, especially since it was put on display at the Institute of Texan Cultures in December of 1983. The mythic figure has an appeal we cannot deny. We care little that real cowboys may have been thin because of poor diet, celibate because their wages were too small to allow marriage, and broken and battered by the kind of life they lived and thus unsuited for other kinds of physical labor. We subconsciously, or perhaps even consciously, reject the negative type of treatment and perhaps feel melancholy about the realistically depicted figure. It is the mythic one, however, that we cling to, and we will continue to do so. The figure refuses to disappear because it appeals to a subconscious yearning that many of us steadfastly—

perhaps romantically—refuse to give up completely. It wells from so deeply within us that without it and our other myths we could not exist.

Notes

1. Joe B. Frantz and Julian Ernest Choate Jr., *The American Cowboy: The Myth and the Reality* (Norman: University of Oklahoma Press, 1955).

2. Jane Kramer, *The Last Cowboy* (New York: Harper and Row, 1977).

3. Jack Shaefer, *Monte Walsh* (1963; rpt. Lincoln: University of Nebraska Press, 1981).

4. A catalog of the exhibit is Lonn Taylor and Ingrid Marr, *The American Cowboy* (Washington, D.C.: The Library of Congress, 1983).

5. Much criticism of the exhibit has been heard. See, for example, the comments of Dean Krakel, Director of the National Cowboy Hall of Fame, in such sources as the *Abilene-Reporter News,* Friday Evening, October 28, 1983, 12D.

6. These films appeared as follows: *Blazing Saddles* by Warner Brothers, 1974; *Water Hole #3* by Paramount, 1967; *Cat Ballou* by Columbia, 1965; *The Apple Dumpling Gang* by Walt Disney Studio, 1975; and *The Outlaw Josey Wales* by Warner Brothers, 1976. Excellent recent studies of Western films are Jay Hyam, *The Life and Times of the Western Movie* (New York: Galley Books, 1983); and Brian Garfield, *Western Films: A Complete Guide* (New York: Ransom Associate, 1982).

7. "The Ballad of the Urban Cowboy: America's Search for True Grit," *Esquire* 90 (September 12, 1978), 21-30.

8. See, for example, John A. Lomax and Alan Lomax, *Cowboy Songs and Other Frontier Ballads* (New York: Macmillan, 1938), 417-22.

9. Capitol EMI Records.

10. Text by J. Evetts Haley (Austin: University of Texas Press, 1952).

11. Margaret Rector, ed. *Cowboy Life on the Texas Plains.* Introduction by John Graves (College Station: Texas A & M University Press, 1982).

12. Martin Schreiber, *Last of a Breed.* Introduction by Louis L'Amour (Austin: Texas Monthly Press, 1982).

13. Howard W. Marshall and Richard E. Alhborn, *Buckaroos in Paradise: Cowboy Life in Northern Nevada* (Lincoln: University of Nebraska Press, 1981).

14. John Erickson, *Panhandle Cowboy* (Lincoln: University of Nebraska Press, 1980).

15. John Erickson, *The Modern Cowboy* (Lincoln: University of Nebraska Press, 1981).

16. Andy Adams, *Log of a Cowboy* (1903; rpt. Lincoln: University of Nebraska Press, 1964).

17. Teddy "Blue" Abbott, *We Pointed Them North*. With Helena Huntington Smith (1939; rpt. Lincoln: University of Nebraska Press, 1978).

18. Elmer Kelton, *The Good Old Boys* (New York: Doubleday, 1978).

19. Definitions of *myth* abound, of course, but in the sense used here, see "Myth," *Handbook to Literature,* 3rd ed. C. Hugh Holman, ed. (Indianapolis: Odyssey Press, 1972), 333-34.

A Cowboy Bibliography

FREEDONIA PASCHALL

People from many parts of the world have long been interested in American cowboys and their counterparts in Latin America, western Asia, and elsewhere. In the United States the interest began in the late nineteenth century with Wild West shows, dime novels, and western ranch tourists, and it has proceeded to the present time.

Scholarly interest began in earnest in the early 1930s with the publication of Walter Prescott Webb's *The Great Plains* (1931). Webb wrote that the western cattle industry with its vaqueros and cowboys began in South Texas with the region's first Spanish ranchers and spread quickly through the West after the Civil War. In fact, he writes that for rapidity of expansion there is not a parallel movement in North America. From Webb's point of view, cowboys and the western cattle industry are Spanish in background—a view that prevailed for many years.

In the early 1950s Charles Towne and Edward Wentworth published *Cattle and Men* (1955). They offered a challenge to Webb's arguments. Towne and Wentworth took the position that herders (cowboys) and cattle breeds from England came to America and then moved westward along successive frontiers of a new, expanding nation. Their focus was primarily on the colonial Carolina backcountry and the westward push of cattle from there through the Old South of Georgia, Mississippi, and Louisiana to Texas. Such colonial cowboys and cattlemen brought to Texas methods of herding, livestock paraphernalia, and cattle breeds that were of English, rather than Spanish, background.

Then in the late 1960s Sandra Myres, in *The Ranch in Spanish Texas, 1691-1800* (1969), reaffirmed the Webb hypothesis for a Spanish background to cowboys and the western cattle industry. There the debate—if it can be called that—stood; most people who studied the matter accepted arguments for a strong Spanish connection.

Then came Terry G. Jordan. A proud revisionist with a cultural geography pedigree, Jordan in the early 1980s published *Trails to Texas: Southern Roots of Western Cattle Ranching* (1981). Although a short work, his book nails with precision the arguments for an American South background to cowboy work and western cattle raising. He revived the Towne and Wentworth position in the debate.

Historians, particularly Texas historians, countered. The best arguments were those in Jack Jackson's impressive and award-winning *Los Mesteños: Spanish Ranching in Texas, 1721–1821* (1986). In detail, but with clarity, Jackson describes Spanish ranching in Texas and shows again how the industry, with its herders, equipment, methods, and terms, came out of Mexico and spread through Texas.

Jordan responded with *North American Cattle-Ranching Frontiers: Origins, Diffusion, and Differentiation* (1993). While now acknowledging the strong Spanish influence, Jordan in this masterful work also reminds scholars that the western cattle industry and cowboy traditions have roots not only in the American South but also in the Midwest. He emphasizes that a final answer to questions about cowboy roots must be many-sided.

Four years later Richard Slatta, who has studied cowboys from Canada to Argentina, brought out *Comparing Cowboys & Frontiers* (1997). In a solidly written narrative, Slatta compares American cowboys with gauchos, llaneros, vaqueros, and charros of South America and Mexico, and in the process again affirms the Spanish and Mexican background for American cowboy and cattle industry origins.

The books in this "cowboy bibliography" do not all touch on the issue of cowboy and cattle industry origins. Rather, most treat various aspects of cowboy life and culture. They examine biography, work and recreation, myth and popular culture, and clothes, dress, and equipment of cowboys. They represent, one hopes, a good cross section of the best literature that is available on perhaps America's favorite icon, one that is recognized the world over and one that professional scholars continue to take seriously.

Abbott, Edward Charles "Teddy Blue." *We Pointed Them North: Recollections of a Cowpuncher*. New edition. Norman: University of Oklahoma Press, 1955, ©1939. This is a popular work that recounts the experience of cowboy life and the sometimes dull but dangerous work involved.

Adams, Andy. *The Log of a Cowboy: A Narrative of the Old Trail Days*. New York: Houghton, Mifflin, and Company, 1903. The classic tale of both humor and danger associated with driving cattle from Texas to Montana, this book continues to attract a large following.

Adams, Ramon F. *Come an' Get It: The Story of the Old Cowboy Cook*. Norman: University of Oklahoma Press, ©1952.

Adams, Ramon F. *The Old-Time Cowhand*. Reprint. Lincoln: University of Nebraska Press, 1989, ©1961.

Adams, Ramon F. *The Rampaging Herd: A Bibliography of Books and Pamphlets on Men and Events in the Cattle Industry*. Norman: University of Oklahoma Press, 1959.

Adams, Ramon F. *Western Words: A Dictionary of the Range, Cowcamp, and Trail*. Revised and enlarged edition. Norman: University of Oklahoma Press, 1968.

Allmendinger, Blake. *The Cowboy: Representations of Labor in an American Work Force*. New York: Oxford University Press, 1992. This is a postmodern deconstructionist account that is as laborious to read as it is unfriendly to ranch hands.

Beckstead, James H. *Cowboying: A Tough Job in a Hard Land*. Salt Lake City: University of Utah Press, 1994. A pictorial history of ranching and related activities in Utah, the book is presented in a large coffee table format.

Blasingame, Ike. *Dakota Cowboy: My Life in the Old Days*. New York: Putnam, 1958. Autobiography.

Branch, E. Douglas. *The Cowboy and His Interpreters*. Reprint. New York: D. Appleton, 1961.

Brooks, Connie. *The Last Cowboys: Closing the Open Range in Southeastern New Mexico, 1890s–1920s*. Albuquerque: University of New Mexico Press, 1993.

Burns, Mamie S. *This I Can Leave You: A Woman's Days on the Pitchfork Ranch*. College Station: Texas A & M University Press, 1986. This delightful book contains more than thirty wonderful tales of ranch life in West Texas during a thirty-year period after

1930. The author, the wife of the ranch manager, recorded her observations about amusing events, special sayings, and family and ranch memories.

Cannon, Hal, and Thomas West, eds. *Buckaroo: Visions and Voices of the American Cowboy.* New York: Simon and Schuster, 1993. A collection of poetry and music of the American cowboy, the book is a delight.

Clayton, Lawrence. *Clear Fork Cowboys: Ranch Life Along the Clear Fork of the Brazos River.* Austin: Eakin Press, 1997. Clayton, who owns a ranch in the Clear Fork country, has written several books about ranches and rural life in Texas. This one includes pictures and descriptions of modern working cowboys.

Clayton, Lawrence, and Kenneth W. Davis, eds. *Horsing Around: Contemporary Cowboy Humor.* Detroit: Wayne State University Press, 1991. Reissued 1999 by Texas Tech University Press, Lubbock, with a new afterword by Mary Evelyn Collins. An anthology of cowboy humor, the book contains cartoons, stories, folktales, and thoughtful essays.

Cleaveland, Agnes Morley. *No Life for a Lady.* Lincoln: University of Nebraska Press, 1941.

Clements Monday, Jane, and Betty Bailey Colley. *Voices from the Wild Horse Desert: The Vaquero Families of the King and Kenedy Ranches.* Austin: University of Texas Press, 1997.

Collins, Michael L. *That Damned Cowboy: Theodore Roosevelt and the American West, 1883–1898.* New York: Peter Lang, 1989.

Coolidge, Dane. *California Cowboys.* Tucson: University of Arizona Press, 1985.

Dale, Edward Everett. *Cow Country.* Norman: University of Oklahoma Press, 1942. Something of a personal history of cowboys and ranching on the Great Plains, the book covers ranching from the author's viewpoint. It is a great book and covers music, dance, work, and other aspects of cowboy life.

Dale, Edward Everett. *The Range Cattle Industry: Ranching on the Great Plains from 1865 to 1925.* New edition. Norman: University of Oklahoma Press, 1960.

Dary, David. *Cowboy Culture: A Saga of Five Centuries.* Lawrence, Kansas: University Press of Kansas, 1989.

Dearen, Patrick. *A Cowboy of the Pecos*. Plano: Republic of Texas Press, 1997. The book is a history of cowboys and the Pecos River through the 1920s.

Dobie, J. Frank. *A Vaquero of the Brush Country*. Reprint. Austin: University of Texas Press, 1981. A true story of cowboy life and cattle ranching in the Texas brush country below San Antonio, this book became a classic in western literature. It is the tale of John Young, a man of imagination, who was a sheriff, trail driver, cattle raiser, and range and ranch manager, among other things.

Durham, Philip, and Everell L. Jones. *The Negro Cowboy*. Lincoln: University of Nebraska Press, 1983. One of the first books to examine African American cowhands in the West, the book remains the standard treatment of the subject.

Dusard, Jay. *The North American Cowboy: A Portrait*. Prescott, Arizona: Consortium Press, 1983.

Ellison, Glenn R. *Cowboys Under the Mogollon Rim*. Tucson: University of Arizona Press, 1968.

Erickson, John R. *Panhandle Cowboy*. Lincoln: University of Nebraska Press, 1990. Highly readable account of modern cowboys in the Texas Panhandle, the book is a delight.

Frantz, Joe B., and Julian E. Choate Jr. *The American Cowboy: The Myth and the Reality*. Norman: University of Oklahoma Press, 1955. The story of the cowboy illustrated with photographs from Erwin E. Smith, who spent many years with a camera on the range.

Freedman, Russell. *Cowboys of the Wild West*. New York: Clarion Books, 1985. The book describes the daily life of the early cowboy.

Gipson, Fred. *Fabulous Empire: Colonel Zack Miller's Story*. Boston: Houghton Mifflin Co., 1946.

Grant, Bruce. *The Cowboy Encyclopedia: The Old and the New West from the Open Range to the Dude Ranch*. Chicago: Rand McNally, 1951.

Hough, Emerson. *The Story of the Cowboy*. 2nd ed. New York: D. Appleton, 1923, ©1897. An early history of the cowboy, the book is by an early author of western topics.

Iverson, Peter. *When Indians Became Cowboys: Native People and Cattle Ranching in the American West*. Norman: University of Oklahoma Press, 1994. A study of Native Americans as cowboys and cattlemen in the Southwest and on the Great Plains. Iverson shows how Indian men worked on ranches and cattle drives, and how many of them became successful ranchers.

Jackson, Jack. *Los Mesteños: Spanish Ranching in Texas, 1721–1821*. College Station: Texas A & M University Press, 1986. A thorough study of Spanish ranching activities in early Texas, the book argues for a Spanish background to the western livestock industry.

Jordan, Teresa. *Cowgirls: Women of the American West*. New York: Anchor Books, 1982.

Jordan, Terry G. *North American Cattle-Ranching Frontiers: Origins, Diffusion, and Differentiation*. Albuquerque: University of New Mexico Press, 1993. The book is a detailed and well-written history of cattle ranching in North America. It contains an enormous number of maps and charts.

Jordan, Terry G. *Trails to Texas: Southern Roots of Western Cattle Ranching*. Lincoln: University of Nebraska Press, 1981. A brief revisionist account of the origins of the cattle industry, the book argues for a southern background to the cattle industry.

Lanning, Jim, and Judy Lanning, eds. *Texas Cowboys: Memories of the Early Days*. College Station: Texas A & M University Press, 1984. It is an attractive and altogether wonderful collection of reminiscences from early Texas ranch lands; the pieces came from the WPA's Writer's Project in the 1930s. They clearly show how real cowboys were young.

Manns, William. *Cowboys & the Trappings of the Old West*. Santa Fe: Zon International Publishing Co., 1997.

Marrin, Albert. *Cowboys, Indians, and Gunfighters: The Story of the Cattle Kingdom*. New York: Atheneum, 1993.

Marshall, Howard W., and Richard E. Ahlborn. *Buckaroos in Paradise: Cowboy Life in Northern Nevada*. Lincoln: University of Nebraska Press, 1981.

McCallum, Henry D. *The Wire That Fenced the West*. Norman: University of Oklahoma Press, 1965. This book is a detailed study of barbed wire and related incidents in the West. It contains many of the traditional stories of barbed wire.

McCracken, Harold. *The American Cowboy*. Garden City, New York: Doubleday, 1973. A readable history of the cowboy.

McDowell, Bart. *The American Cowboy in Life and Legend*. Washington: National Geographic Society, 1972.

Mora, Joseph J. *Californios: The Saga of the Hard-Riding Vaqueros, America's First Cowboys.* Garden City, New York: Doubleday, 1949. This is a history of the vaquero of California.

Mora, Joseph J. *Trail Dust and Saddle Leather.* New York: Scribner's Sons, 1946. A first-person history of cowboy life, the book is an old standard.

Myres, Sandra L. *The Ranch in Spanish Texas, 1691–1800.* El Paso: Texas Western Press, 1969. Arguing for a Spanish background for the Western cattle industry, the book is short and well-written.

Rollins, Philip Ashton. *The Cowboy: An Unconventional History of Civilization on the Old-Time Cattle Range.* Revised and Enlarged Edition. Norman: University of Oklahoma Press, 1997 (1932, 1936). An entertaining and valuable early study of cowboys and the cattle industry. Detailed and thorough, the book covers clothes, saddles, ropes, brands, and other items associated with cowboy life and culture.

Savage, William W. Jr. *The Cowboy Hero: His Image in American History and Culture.* Norman: University of Oklahoma Press, 1979. An intelligent study of the cowboy as a popular icon. This is a superb study of how modern folks see the cowboy of the past.

Savage, William W. Jr. *Cowboy Life: Reconstructing an American Myth.* Norman: University of Oklahoma Press, 1975. This book makes use of essays and photographs to tell the story of the American cowboy.

Seidman, Laurence. *Once in the Saddle: The Cowboy's Frontier, 1866–1896.* New York: Knopf, 1973.

Siringo, Charles A. *A Texas Cowboy: or, Fifteen Years on the Hurricane Deck of a Spanish Pony, Taken from Real Life.* Bison edition. Lincoln: University of Nebraska Press, 1979.

Skaggs, Jimmy M. *The Cattle Trailing Industry: Between Supply and Demand, 1866–1890.* Lawrence: University of Kansas Press, 1973. Written in a clear, narrative style, the book details the commercial movement of cattle from Texas northward in the post–Civil War era.

Slatta, Richard W. *Comparing Cowboys and Frontiers.* Norman: University of Oklahoma Press, 1997. Slatta compares cowboys of the American West with charros, vaqueros, llaneros, gauchos, and other herders in the Americas.

Slatta, Richard W. *The Cowboy Encyclopedia*. Santa Barbara, California: ABC-CLIO, 1994.

Slatta, Richard W. *Cowboys of the Americas*. New Haven, Connecticut: Yale University Press, 1990. This book is a study of the cowboys of North and South America. It covers their origin, character, equipment, lifeways, and mythic culture. It is a serious and award-winning study.

Starrs, Paul F. *Let the Cowboy Ride: Cattle Ranching in the American West*. Baltimore: Johns Hopkins University Press, 1998.

Sullivan, Tom R. *Cowboys and Caudillos: Frontier Ideology of the Americas*. Bowling Green: Bowling Green State University Popular Press, 1990.

Truettner, William H., ed. *The West as America: Reinterpreting Images of the Frontier, 1820-1920*. Washington: Smithsonian Institution Press, 1991.

Vernan, Glenn R. *Man on Horseback: The Story of the Mounted Man from the Scythians to the American Cowboy*. Lincoln: University of Nebraska Press, 1964.

Wagoner, Junior Jean. *History of the Cattle Industry in Southern Arizona, 1540-1940*. Tucson: University of Arizona Press, 1952.

Walker, Don. *Clio's Cowboys: Studies in the Historiography of the Cattle Trade*. Lincoln: University of Nebraska Press, 1981.

Ward, Fay E. *The Cowboy at Work: All About His Job and How He Does It*. New York: Hastings House, 1958. One of the best books ever written about cowboys, this work is illustrated with six hundred drawings by the author. The book examines in detail cowboy equipment, dress, and other paraphernalia.

Webb, Walter Prescott. *The Great Plains*. Boston: Ginn and Company, 1931. Contains an important and very readable chapter on what Webb called the "Cattle Kingdom," where he argues for Spanish and South Texas origins for cowboys and cattle raising.

Westermeier, Clifford P., ed. *Trailing the Cowboy: His Life and Lore as told by Frontier Journalists*. Caldwell, Idaho: Caxton Printers, 1955.

Worcester, Don. *The Texas Longhorn: Relic of the Past, Asset for the Future*. College Station: Texas A & M University Press, 1987. This book provides a brief history of the development of Longhorn cattle.

About the Authors

Thomas A. Britten is associated with BrittenMedia in Traverse City, Michigan. He has published several articles on Native American history and two books, including the highly acclaimed *American Indians in World War I: At Home and at War*.

Paul H. Carlson is professor of history at Texas Tech University. He is the author of a number of articles and six previous books, including *Empire Builder in the Texas Panhandle: William Henry Bush* and *The Plains Indians*.

Lawrence Clayton is Dean of the College of Arts and Sciences at Hardin-Simmons University. He is the author of numerous articles on western topics and has written several books, including *Historic Ranches of Texas* and *Watkins Reynolds Matthews: Biography of a Texas Rancher*.

Kenneth Davis is emeritus professor of English and American Folklore at Texas Tech University. He has published several books, the most recent of which, *Horsing Around: Contemporary Cowboy Humor*, he co-edited with Lawrence Clayton and Mary Evelyn Collins.

Susan Karina Dickey is Historian of the Catholic Diocese of Springfield, Illinois. She has written on architectural history, clothing, and furniture, and her published work on the history and development of the Barbie Doll toys has attracted wide attention.

Jim Fenton is book review editor for the *Permian Historical Annual* and lives in Lubbock, Texas. He has published numerous articles on frontier history and Texas history, and he is completing a book-length study of Lieutenant John L. Bullis, a leader of black-Seminole army scouts in the West.

Douglas Hales is a graduate student at Texas Tech University completing a dissertation on black political leader Norris Wright Cuney. He was a contributor to the new *Handbook of Texas* and has presented papers at several conferences, including the Texas State Historical Association and the Dallas African American Museum.

Jorge Iber is assistant professor of history at Texas Tech University. His primary research areas are Mexican American and Hispanic social history and minority business history.

Freedonia Paschall is associate archivist at the Southwest Collection/Special Collections Library at Texas Tech University. She also serves as an editor for the *West Texas Historical Association Year Book*.

J'Nell L. Pate is professor of history at Tarrant County College. She has written more than a dozen articles on western history and four books, including the award-winning *Livestock Legacy: The Fort Worth Stockyards, 1887-1987* and *Document Sets for Texas and the Southwest in U.S. History*.

Judy Greaves Rainger teaches French in Lubbock, Texas. She has researched the culture of the French *gardians* and Buffalo Bill's travels in France with the help of several grants from the National Endowment of the Humanities. In 1995-1996 she was named as a Teacher-Scholar by the NEH.

J. Boyd Trolinger received a master's degree in history from Texas Tech University. He is currently a graduate student in Computer Science at California State University at Chico. In addition to his work on cowboys he has published two previous articles.

Albert B. Tucker, who died just before this book went to press, was Director of Institutional Research at Howard Payne University. He wrote, spoke, and sang about the so-called B Westerns for several years.

James R. Wagner is a freelance writer and historian who lives in Ridgefield, Washington. In addition to his work on cowboys, he has written on African American recipients of the Congressional Medal of Honor.

Robert G. Weiner is reference librarian at the Mahon Library in Lubbock, Texas. He has published a number of articles on popular culture, library studies, and music. He is co-author of *The Grateful Dead and the Deadheads: An Annotated Bibliography*.

Robert E. Zeigler is Executive Vice President of San Antonio College. He has taught American history, coordinated his college's extensive program for teaching through television, and served as Interim President.

Index

Hoffer, Lt. John, 83
"Home on the Range" (song), 147
honky-tonks, 1, 2, 3
horsebreakers, 25
horsemen, 3: and equestrian culture, 22
homesteaders, 47
horses, 6, 135; of Camargue, 168; and cowboys, 6, 115; and cowboy music, 145; and Indians, 46; stinks, 3, 9; and "80 John" Wallace, 34
Houston, Texas, 203
Howard County (Texas), 63, 65
huaso, 22, 23
humor: characteristics of, 131; of cowboys, 131-40
hunting, 67-68
Hyer, Charles Henry: as bootmaker, 104, 105

I

"I'd Like to Be in Texas" (song), 150
"I'm Bound to Follow the Longhorn Cow" (song), 149
"I'm Going to Leave Old Texas Now" (song), 151
Indians. See American Indians and Native Americans
Ingraham, Prentis, 5
Institute of Texan Culture, 206
Iron Tail (Sioux), 173
irrigation: and cowboy songs, 146
itinerant workers: as cowboys, 6
Iverson, Peter, 45, 58

J

jacales, 26
Jackson, Jack, 24
Jackson, Joaquin, 187
jeans: as clothing, 179
jobs: for cowboys, 17-18, 78, 84, 113, 138; and cowboy strike, 80; and sheepherders, 114; and winter, 113
Johnson, Bart, 104
Johnson, Ben, 185, 191

Johnson, Hugh "Huckleberry Slim," 160
"Jolly Cowboy" (song), 149
Jones, Buck, 185; as favorite cowboy, 190; and Gene Autry, 183
Jones, Tommy Lee, 191
Jordan, Terry G. 13-14, 210
Jungle, The, 121
Justin, Herman Joseph (bootmaker), 104, 105

K

Kansas, 149; bootmakers in, 105; cattle drives to, 150; railheads in, 119
Kelton, Elmer, 137, 205
Kenedy family, 29
Kenedy ranch, 26
Kennedy, R. F., 66
Kent County (Texas), 38
Kerrville, Texas, 113
Kilgore, Martin H., 112, 113
King Cattle Company, 51
"King of the Cowboys," 191: Gene Autry as, 183; Buck Taylor as, 5; Roy Rogers as, 183; John Wayne as, 191
King Edward VII, 72
King family, 29
King ranch, 26, 27, 113; and vaqueros, 29; and working cowboys, 204
King, Richard, 26, 113
Kiowas, 54
Kramer, Jane, 201
Kuenster, John, 187
Kupper, Winifred, 8; and sheepherders, 109, 115

L

laborers, 6-7, 106; cowboys as, 17-18; at stockyards, 126
labor force, 77, 85
labor unions: and cowboy strike, 7
Lakotas, 46
L'Amour, Louis, 45, 194
Lampasas County (Texas), 34

On The Trail

– Malibu to Santa Barbara –

By Cathy Philipp

Cathy Philipp Publishing
Thousand Oaks, California

Printed and Bound by
Trade Service Publications
San Diego, CA, USA
Photographs by Kathy Gillman
as well as the author and her family.
Cover Photo: Lynn Road area of Thousands Oaks
at foothills of Santa Monica Mountains.
Photo by John Blanchard/Concepts Photography.
Cover inset: Cycling in Cheeseboro Canyon, Agoura.
Cover inset: Viewing Tidepools at Goleta Point,
Santa Barbara.
Maps and illustrations by Shawnn Jelletich,
John Johnson & CPS Communications, Inc.

This is the first edition of this book.
Library of Congress Catalog Card Number:
97-91534

ISBN 0-9655848-0-1

Please note:
The trail information in this book is as accurate and true as
possible at the time of printing. However, natural phenomena
such as fire and floods sometimes change man-made pathways
through wilderness parks. The author does not assume any
responsibility for loss, damage or injury caused through the
use of this book. Each trail user must assume responsibility
for their actions while using the public trails described in
On the Trail.

Printed on recycled paper.

On the Trail is dedicated to future generations...
May there always be wild places to explore.

Table of Contents

The Hikes

Chapter Two

Conejo Valley **79**

Chapter Three

Oak Park to Malibu Canyon **112**

Chapter Four

Chapter Five

Chapter Six

Chapter Seven

Welcome to *On the Trail!* This book is designed to help families find suitable outdoor adventures in Southern California's coastal mountains between Malibu and Santa Barbara. The outings are based on my "On the Trail" column in the Ventura County Star Newspapers and consist mainly of easy nature hikes. Along with the hikes, I have included several of my family's favorite, easy mountain bike rides.

When our children were just babies, my husband and I lived in West Los Angeles. We were very interested in sharing our love of nature with the girls, but having grown up in rural, upstate New York we were at a loss as to how to accomplish our goal in Southern California. Local hiking guides were of little help as they were all written with adults in mind. On several outings, trails that were described as easy reduced the kids to tears within 100 yards of the trailhead.

Frustrated after striking out on our own, I joined Nursery Nature Walks, which is a docent group made up of families leading other families on nature walks throughout the Los Angeles area. Hiking with NNW helped me to appreciate Southern California's Mediterranean climate. As my knowledge of the area grew, hiking sites that at first seemed inhospitable, if not downright threatening, began to pique my curiosity as well as my family's. Since that time I have worked as a naturalist for a number of outdoor education organizations from the Las Virgenes area north to the City of Ventura. Coming into daily contact with school children and their parents, I met many people who were interested in hiking with their children but simply didn't know where to go.

This bit of knowledge was the impetus for my weekly hiking column in the then Thousand Oaks News Chronicle. Since that time the Ventura County Newspapers have been consolidated under the Scripps Howard banner, taking my column into households throughout Ventura County. For a while, I honored requests for copies of those early articles, but that soon became physically impossible to do. This book has come about as an effort to make these hikes more readily available to the general public. I hope that you and your family enjoy these trails as much as my family and I do.

– Cathy

Acknowledgements

I would like to thank my family and friends who took the time
to go exploring open spaces with me as well as those faithful
Ventura County Star newspaper readers who encouraged me over
the years. But most of all, I would like to thank my husband Tom,
without whose guidance and understanding _"On the Trail"_
would not be possible.

Nojoqui Falls trailhead with the Philipps

How to Use This Book

This book is intended to get you and your family out on the trail, whatever your level of hiking experience may be. Many people are experienced in the outdoors but unfamiliar with Southern California's unique climate, while other people may be very gung-ho themselves and very familiar with the area but not know how to "tone-down" their experiences to share them with less experienced members of their family. (An example of this is the adult hiker that walks at the rate of 3 or 4 mph and can't understand why the rest of the family doesn't like to go along.)

The majority of outings in this guide are meant to be enjoyed by the whole family. Some of the hikes are so easy that they can best be described as walks and others are included to give those that want it, a more physical challenge. Many of the sites lend themselves to this quite easily as the first part of the hike might be easy enough for almost anyone, yet the farther you go on the trail, the more challenging it becomes. I also have included trails that are accessible to strollers and wheelchairs. Most of the hikes are one or two hours long.

The outings have been arranged geographically. After writing my column for a couple of years, I found that a number of trails and parks have similar names. Hopefully, this method of grouping will help avoid confusion about trail names as well as help the reader decide where to go hiking or biking.

While this book is not intended to be a field guide about the area's fauna and flora, I have included a glossary of terms. A number of excellent guides already have been written about the area between Los Angeles and Santa Barbara. Please check the bibliography of this book for a list of field references as well as a list of books that are intended to help parents share their love of nature with their children. I also have listed park agencies and local interest groups that present outdoor recreation activities as well as provide public service information about open space topics.

Kid Appeal, or what do those little stars mean? I have ranked each outing as to the level of appeal it will have for children. Five stars - * * * * * - is the highest rating and signifies the outings with the most Kid Appeal. You may notice that two stars -* * - is the lowest rating and these are generally what I consider to be adult hikes. In other words, these are outings that are more strenuous or more esoteric in nature.

A word about the maps. The road maps in this book are not to scale but they will get you to the trailhead. Trail maps and the step by step directions are intended to keep you on the designated trail without taking away that sense of personal discovery that brings people out into open space in the first place. Also, Mother Nature has a way of changing trails over the years. Fires, floods and earthquakes can make detailed accounts of pathways confusing as time goes by.

Notes:

Helpful Hints to Hiking with Children

Very often I am asked what to bring along on the trail. Happily enough, hiking is an activity that requires very little in the way of specialized equipment. Besides a love of nature and enjoying the outdoors with your loved ones, I would have to say the most important thing to bring with you is, of course, your own common sense. I have found that many people will offer many different suggestions and trail techniques but unless you are comfortable with them, other people's ideas often don't work. That said, I would like to take this opportunity to make some suggestions that have worked for me or other hiking families I know. Please feel free to do your own thing within the framework of general and specific park rules that everyone must follow. Let's start with some of the more basic guidelines for enjoying the out-of-doors in Southern California's unique Mediterranean climate.

1. What to bring:

Besides common sense, water is a must. Bring at least one quart of drinking water per hiker. I also usually bring a small spray bottle filled with water; misting the kids throughout the hike is one of the most effective ways to cool-off quickly. A light snack is very helpful for small children who tend to burn off large amounts of energy, usually before the adults are ready to leave the parking lot. Maps and field guides are nice, as are cameras, binoculars, bird callers and hand lenses; they are all wonderful extras that enhance the experience. Something to pack trash out is recommended for all outings.

2. What to wear:

The best advice is to wear comfortable, layered clothing. Long-sleeved shirts and long pants are probably best in all but the hottest weather. This type of covering keeps poison oak and biting insects from direct contact with your skin as well as offering some protection in the event of a fall. I also suggest hats, visors and sunscreen. Bandannas are easy to wear or carry and have a number of uses on the trail, from blowing your nose to becoming an impromptu sling.

As far as footwear goes, I suggest lightweight hiking boots for most trails, especially rocky ones. Very often you will be able to get away with sturdy shoes or sneakers but if you hike often, boots are a good investment. Boots offer better protection for your feet and it may be cheaper in the long run to avoid ruining your everyday shoes.

Of course, helmets are necessary protection for mountain bike rides. And for specialty hikes such as tide pools, old sneakers or surf-sox are a good idea.

3. Bring a friend:

Share your hiking experience by bringing friends along and carpooling. Not only is this more fun, it is ecological and safer too. That way if the unexpected does occur, there is always someone to go for help. When hiking alone, be sure to tell someone where you are going and when you plan to be back. If you have a change of plans once you get to the trailhead, simply leave a note visible in your car with an updated itinerary.

4. If your friend is a furry one:

Many families enjoy bringing the dog along. If you like hiking with your dog, be sure to check the site first. Different park agencies have different rules about dogs on the trail. The National Park Service and most local agencies will allow leashed dogs on the trail, while some places state "dogs under their owners control are welcome on the trail". Usually sites that are designated wilderness or wildlife areas will not allow dogs at all. Some places allow pets in campground areas but not on trails, such as California State Parks, which categorically does not allow dogs on any trails. The reason that dogs are restricted on the trail is twofold: 1.) Dogs scare wildlife and 2.) Dogs scare people.

Many wildlife experts agree that dogs' scent on the trail frightens wild animals from using the pathway. Also unleashed animals often chase wildlife if given the chance, scaring them off the trail and deep into the park lands. This is particularly upsetting at the ocean where shore birds depend on the changing tide for their dinner, or during the breeding season when animals already feel stress. While most owners see their pets as friendly and unintimidating, many people are actually afraid of dogs, especially small children. Considering that most park users want to see wildlife when they visit the park and use the trails in peace, it does seem to make sense to keep dogs off the trail.

Of course, most dog owners feel that they should be able to walk their pets in the open space parks when they go hiking. Not only does the dog get lots of exercise, but many people feel that dogs provide them with company as well as protection. A good compromise has been to allow leashed dogs on the trail. While most people want to give their pet the opportunity to run free, the leash can be very useful to the dog-owner. For one thing, it keeps the dog out of the poison oak; dogs can actually bring the toxins home with them on their fur and make the whole family itch! If you bring your dog hiking with you, remember to bring extra water along as dogs can get dehydrated on the trail just as people do. Also, at the end of the hike be sure to check your pet for ticks and other tag-alongs such as foxtail seeds.

5. *Impact on the land:*

Hiking with your family is a terrific opportunity to foster a "hands-on" feeling of stewardship for the land. This is a great time to teach the kids about packing out whatever you pack in. Be sure to bring a trash bag along; we usually make a point of spending 5 minutes picking up trash in the parking lot as well as along the trail. Most of the parks have signs at the trailheads reminding us to "leave only footprints and take only memories." It is much easier to convince a toddler to leave that pretty rock or flower in the park for others to see if you have remembered to bring a camera along and allow her to take a picture of her treasure.

6. *Go often and be flexible:*

Visiting the same site several times throughout the year helps everyone to understand our unique climate a little better. It is especially good for very young hikers to witness seasonal changes in a familiar setting. Going hiking at different times of the day can put a whole new face on an old trail while at the same time have the comfort of being in a familiar setting. Being flexible enough to cut the hike short if participants are having a bad day or conditions make it uncomfortable can be a real benefit. If the weather is terribly hot or it is extremely windy, don't feel that you have to make it to the end of the trail on every hike. Once again, use your common sense and cut your hike short if you must or make frequent rest stops in the shade. The idea is for every one to have a good time and not burn-out on the idea of hiking.

Extremely hot, dry and/or windy weather in the late summer marks what Southern Californians know as "fire season". If temperature is too high and humidity too low during this time of year the parks will probably be closed because of fire danger. Be sure to call the particular park you are planning to visit on the day of your hike to make sure it is open. Weather conditions can change radically from day-to-day.

7. Kid appeal or Kids don't hike for the view:

It's not that kids don't enjoy the view, it's just that they prefer crossing streams on stepping stones or natural log bridges, climbing trees, exploring caves, finding animal tracks and scat. When our girls were little, stream-crossings were their favorite part of many hikes. Often times we would spend more time in the water than on the trail! When I worked with inner-city school groups, I was surprised to learn that many of the children had never had the opportunity to even climb a tree. Be sure to take the time to let the kids explore their natural environment. Let them get their feet wet in the stream while checking out those tadpoles and crawdaddies, but be sure to explain why they can not drink the water no matter how clean it looks.

Other kid turn-ons include: waterfalls, swimming holes, single-track trails, tide pools, fossils and animal tracks. Animal migrations, from monarch butterflies to gray whales occur all along California's diverse coastline. Timing your visit to coincide with these events will add another dimension to your outing. Field-guides about wildlife, birds, wildflowers, geology and the like are popular as long as they are geared to your child's level of interest. Some of the newer guides even include "hands-on" experiments that intrigue the kids while at the same time emphasizing a respect for nature.

8. Learn more about the area

It is possible to learn more about our coastal mountains through educational programs presented by local agencies. The Santa Monica Mountains National Recreation Area rangers put on a number of free hikes and lectures throughout the year to teach the general public about the open space around us and the wildlife that lives there. Most of these scheduled events are free and can be experienced with your children. Many local agencies and volunteer interest groups sponsor field days and outings both for school-age children and families. Group activities such as these provide a lot of outdoor fun in a safe setting and are excellent ways to meet other families interested in the out-of-doors. CPR and

Community First-Aid Classes are very helpful as far as a source of knowledge and supplies. If you spend a lot of time at outdoor activities, a formal wilderness first-aid and/or survival class might appeal to you. Kids can take these classes too: encourage your child's scout leader to arrange a class for the whole troop. Not only is the information helpful, but they can earn merit badges at the same time.

9. Safety first

Now that we have everyone ready to hit the trail, let's talk a little about what to watch out for to make your trip into open space as hassle-free as possible. Foremost in this discussion will, of course, be common sense. Number one: Be prepared. First-aid supplies and extra water are always a good idea for any outdoor excursion and especially so in the backcountry. It's also a good idea to have a working knowledge of the dangers particular to this area. When asked what to watch out for on local trails, Conejo Recreation and Park District Ranger Doug Tait says, "STP -- snakes, ticks and poison oak."

While any snake may bite you if disturbed, there is only one poisonous snake that resides in our local mountains and that is the Western Pacific Rattlesnake. These are not aggressive animals. They would much rather shake their rattles to scare you off than bite you, but accidents do happen and people do end up getting bit. The best course of action is to avoid getting bit in the first place! Do not try to pick up wild snakes or put your hands, or anything else, down their holes.

If someone does get bit, first aid as recommended by the American Red Cross includes: moving the victim to prevent more bites, washing the wound and transporting the victim to a medical facility as quickly as possible. You may also want to splint the extremity, have the victim stay motionless and treat for shock. *Do Not* cut the wound and try to suck the venom out. Besides the danger of contracting blood-borne diseases such as AIDS, you could get the venom into your own bloodstream as well from small abrasions inside your mouth. *Do Not* use ice or a tourniquet. You have several hours to get the victim to the hospital before serious damage results from the bite. If at all possible have the victim remain motionless and send someone to the ranger station and call 911 for help. The idea is to slow the movement of the venom through the body. Once at the hospital an anti-venom can be administered. **_Do Not_** attempt to capture, kill or restrain the snake, you could get bit.

Ticks are most generally a seasonal problem and are usually seen in the early spring. These tiny spider-like creatures live in the grass and get on you or your dog as you brush past them. Your best defense against ticks is to stay on the trail and out of the grass. Check yourself and others in your party shortly after each outing. If you are lucky, you can usually find the critters before they become attached and it's simply a matter of picking them off and disposing of them. If the ticks do become attached to someone's skin, it's best to remove them with a special tick removing tool that is like a tweezers. These inexpensive items can be purchased at backpacking, gardening and hardware stores. Old war stories tell of removing ticks with lit cigarettes or matches, however I think that more prudent action might be to go to a doctor or emergency room if the tick can't be removed easily with the tool. While no cases have been reported in the Santa Monica Mountains, ticks have been known to carry Lyme disease. Other common biting or stinging things to watch out for include: black widow spiders, scorpions and bees.

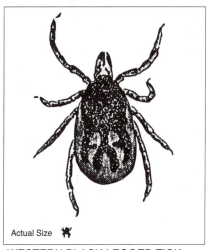

Actual Size

WESTERN BLACK-LEGGED TICK

<u>VENOMOUS</u>
WESTERN PACIFIC RATTLESNAKE

<u>NON-VENOMOUS</u>
GOPHER SNAKE/CAL. KING SNAKE

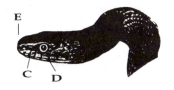

A- TRIANGULAR-SHAPED HEAD
IS LARGER THAN NECK

A- NARROW HEAD IS ABOUT
SAME SIZE AS NECK

B- THICK, HEAVY BODY WITH
BLUNT TAIL & RATTLE

B- SLENDER BODY WITH
LONG, TAPERED TAIL

C- LOREAL PIT BETWEEN
NOSTRIL AND EYE

C- ABSENCE OF PIT

D- ELIPTICAL (OR CAT'S EYE)
IRIS OPENINGS

D- CIRCULAR IRIS OPENING

E- TURNED UP NOSE

E- SMOOTH, ROUND NOSE

21

Poison oak is a plant that grows throughout Southern California's open space areas. It is easily recognizable from the shape of its leaf. The old scout adage "leaves of three...let it be" is a good rule of thumb when it comes to identifying this noxious weed. Should you encounter poison oak on the trail, remember that washing the affected area with soap and water is very effective. Even rinsing the skin with your drinking water will help reduce the irritation. Very often oils and resins of the poison oak plant will end up on your clothes, or your dog's fur, and sometimes irritate your skin hours after returning home. A good way to avoid this is to change your clothes, or brush your dog, as soon as you get home.

10. *Beyond STP*

It is very rare to see large mammals such as bobcats, coyotes, mountain lions and bears. Usually all we are lucky enough to see of these magnificent creatures is their tracks. While any wild animal may attack if annoyed, bobcats and coyotes are generally frightened of people and will run away if given the opportunity. Mountain lions and bears have no natural predators and are usually less afraid. If you should encounter one of these animals, remember to remain still, as running only enhances their predator instincts and they WILL give chase. Try to avoid looking like carry-out to these creatures by staying away from them at feeding time, which is early morning or around dusk.

Breakdown of a typical First-Aid kit:

Band-Aids	*alcohol wipes*
gauze	*tape*
tweezers	*scissors*
pen & paper	*gloves*
mouth shield	*cold pack*
cellular phone	*change for phone*
water	*soap*
whistle	*mirror*
matches	*tick remover*

Poisonous Plants

Castor Bean

Poison Hemlock

Poison Oak

<u>Poisonous Plants</u>

Elderberry

Datura

Indian Tobacco

Chumash Indians

According to many theories, the first people to settle in Southern California were nomadic hunting parties who found the area by following the game that roamed the continent between 7,000 and 10,000 years ago.

As the climate changed and became drier and the large animals disappeared, these nomadic people came to rely heavily on gathering plant materials for food. The ocean to the west and the rugged mountain ranges inland acted as a barrier that kept the tribes isolated from other Native American cultures. This isolation along with a fantastic source of natural resources allowed for the development of an artistic lifestyle free of famine and hostilities. As the climate changed, the Chumash learned to use the plants that grew in the mountains as well as those of the inland valleys. When these people discovered that acorns were edible once the tannins were leached out, there was a dramatic increase in their population. As a response to this situation, the Chumash pushed inland. Seasonal hunting camps became thriving villages that developed a peaceful, stratified society of hunter-gatherers. Isolated by mountains and sea, the Chumash people had virtually no enemies to contend with for thousands of years.

The Chumash were the only true sea-faring Indians along the California coast. While many tribes constructed rafts for use on lakes and streams, the Chumash built large, redwood canoes called "tomols". The tomols were very seaworthy and enabled the Chumash to colonize the Channel Islands. An intricate system of trade developed and a standard form of money was used as currency between the various village communities. The sea also provided an important source of food for the Chumash.

Very little is known of these magnificent people whose territory stretched along the coast from Malibu to Morro Bay and the neighboring inland valleys. Once in contact with Europeans eager to settle in new lands and convert the natives to Christianity, the Chumash culture declined rapidly. In an attempt to bring Christian ways to the indigenous heathens, many Native Americans were enticed, by hook or by crook, into the Catholic Missions. Unfortunately, mission life was a harsh experience, especially so to people who had been used to a great deal of personal freedom and very comfortable lifestyle. Overcrowding and poor sanitation nearly devastated the Chumash population with disease in less than 50 years time. Also, once a large part of the population was moved to the missions, the Chumash villages became ghost-towns unable to sustain their traditional way of life. As the struggle for control of California continued to escalate, the plight of the Chumash worsened almost to the point of annihilation. Once a peaceable people with a rich and diverse lifestyle, the Chumash were left homeless in their homeland.

What little we do know about the Chumash comes from the work of one man: John Peabody Harrington. By means of interviews with local Indians, Harrington recorded volumes of first hand experiences. Unfortunately, Harrington recorded his information using a private shorthand that has proven difficult to translate. Also Harrington paid for the interviews by the hour: the more a client talked the more he/she was paid. This leads some scholars to believe that some of the information may have been somewhat embellished. Another factor in all of this is that none of those interviewed were alive when the Chumash culture was in full flourish; so most of the information was about the time when the culture was already in decline.

Many of the original Chumash village names remain a part of our 20th century culture:

Modern Name	Chumash Name	Meaning of Name
Ojai	'Awhay	moon or nest
Malibu	Humaliwo	surf sounds loudly
Lompoc	Lompo'	stagnant water
Point Mugu	Muwu	beach
Nojoqui Falls	Nakhuwi	meadow
Nipomo	Nipumu'	village
Pismo Beach	Pismu'	tar
Saticoy	Sa'aqtiko'y	sheltered from wind
Sespe	S'eqp'e	kneecap money
Simi	Shimyi	windy
Anacapa Island	'Anyapakh	mirage
Santa Cruz Island	Michumash	makers of shell money

Chapter One

Santa Monica Mountains National Recreation Area

Rest stop at Big Sycamore

HIKES AT A GLANCE 1
Santa Monica Mountains National Recreation Area

	DIFFICULTY	FACILITIES	FEATURES	FEES	OPEN TO
RANCHO SIERRA VISTA/SATWIWA		located at Ranch Center			
1. Satwiwa Loop Trail	easy	w, t, p, rs	Chumash Cultural Center, views	no	H, C, E, D, S
2. Windmill Trail	easy	none	wildflowers, views	no	H, C, E, D
3. Old Cabin Site Trail	moderate	w, t, p, rs	memorial, waterfalls, views	no	H, E
4. Big Sycamore to the Sea	easy/moderate	w, t, p, rs, c	sycamore savanna, Danielson Ranch Center, beach, wildflowers	no	H, C, E, S
POINT MUGU STATE PARK		located at Ranch Center & campgrounds			
5. Wood Canyon Bike Ride	easy/moderate	w, t, p, rs, c	shady canyon, stream	yes	H, C, E
6. Scenic Overlook Trail	easy	w, t, p, rs, c	ocean view, wildflowers	yes	H
7. La Jolla Trail	easy/moderate	w, t, p	waterfall, wildflowers, hike-in camp	yes	H, E
8. Chumash Trail	strenuous	none	ocean view, wildflowers	no	H
BACKBONE TRAIL					
9. Castro Crest Trail	moderate	none	views, fossils	no	H, C, E, D
10. Newton Canyon Trail	moderate	none	stream, shady canyon	no	H, C, E, D
11. Upper Zuma Canyon Trail	easy	none	waterfall, grotto	no	H, C, E, D
12. Rocky Oaks Walkabout	easy	w, t, p	pond, oak woodlands	no	H, C, E, D

FACILITIES: w - water, t -toilet, p - picnic table, rs - ranger station, c - camping, m - marina, s - store
OPEN TO: H - hikers, C - cyclists, E - equestrians, D - dogs on leash, S - stroller/handicap access

Overview of Hiking Sites in the
Santa Monica Mountains National Recreation Area

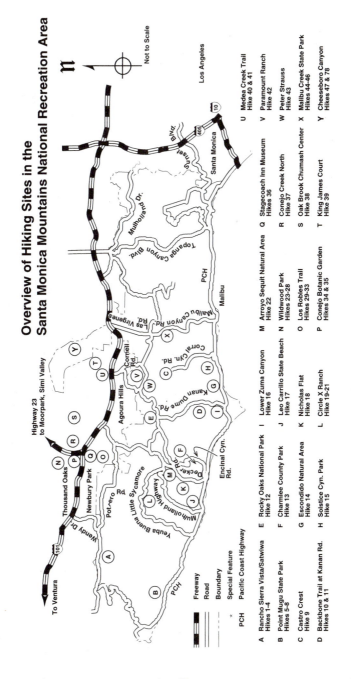

A Rancho Sierra Vista/Satwiwa
Hikes 1-4

B Point Mugu State Park
Hikes 5-8

C Castro Crest
Hike 9

D Backbone Trail at Kanan Rd.
Hikes 10 & 11

E Rocky Oaks National Park
Hike 12

F Charmlee County Park
Hike 13

G Escondido Natural Area
Hike 14

H Solstice Cyn. Park
Hike 15

I Lower Zuma Canyon
Hike 16

J Leo Carrillo State Beach
Hike 17

K Nicholas Flat
Hike 18

L Circle X Ranch
Hikes 19-21

M Arroyo Sequit Natural Area
Hike 22

N Wildwood Park
Hikes 23-28

O Los Robles Trail
Hikes 29-33

P Conejo Botanic Garden
Hikes 34 & 35

Q Stagecoach Inn Museum
Hikes 36

R Conejo Creek North
Hike 37

S Oak Brook Chumash Center
Hike 38

T King James Court
Hike 39

U Medea Creek Trail
Hike 40 & 41

V Paramount Ranch
Hike 42

W Peter Strauss
Hike 43

X Malibu Creek State Park
Hikes 44-46

Y Cheeseboro Canyon
Hikes 47 & 78

Freeway
Road
Boundary
* Special Feature
PCH Pacific Coast Highway

Rancho Sierra Vista/Satwiwa National Park and Point Mugu State Park

Starting at Potrero Road in Newbury Park, these state and national parks link up with the local Conejo parks, effectively stretching open space all the way to the ocean. There are a number of easy hikes that loop around the upper Rancho Sierra Vista/Satwiwa National Park site off Potrero. The two parts that make up this national park's name remind us of this area's early history. Satwiwa is a Chumash Indian word meaning "the bluffs" and refers to the rolling grasslands. At least one Chumash village is known to have occupied these bluffs and many feel that this was an ancient tribal gathering and ceremonial site. When Europeans settled The Conejo the area was divided up into large ranchos, or homesteads, for retired soldiers loyal to the Spanish government. At this point the area became known as Rancho Sierra Vista for the dramatic view of Boney Mountain that looms over the grassland. Celebrating both cultures, the names have been combined.

Beyond the rim of Big Sycamore Canyon, which cuts through the heart of the parkland, is Point Mugu State Park. Big Sycamore Canyon Road is an old ranch road that follows the canyon a little more than 8 miles to the sea. There is a substantial loss in elevation over the first mile, as the road switchbacks its way down the back wall of the canyon. Once at the canyon bottom there is a more gradual loss in elevation the rest of the way to the ocean. About halfway through the park there is a pleasant group campsite in a shady oak grove near the old Danielson Ranch.

As the road follows the creekbed, stately sycamore trees line the nearly year-round creek making for a shady outing through the canyon's rolling oak savannas. Steep, chaparral covered slopes rise up on either side of this river of green along the bottom land.

1. Satwiwa Loop Trail

Difficulty: Easy
Length: 1 1/2 mile loop
Elevation: 150 foot gain/loss
Type of trail: mostly fire road
Kid appeal: * * * * *

In the summertime, the longer days make hiking in the evening hours a cool and refreshing idea. Not only is this a pleasant way to work off dinner, but evening hikes also put you in the wild when the animals are beginning to stir. Walking the Satwiwa Loop Trail at Rancho Sierra Vista National Park is an excellent way to spend a summer evening. Be sure to bring a picnic supper; there are picnic tables and restrooms located at the ranger station at the ranch compound, northeast of the parking lot. For thousands of years the Chumash celebrated the summer and winter solstices here in view of Boney Mountain. As you walk this trail, try to imagine the view through the eyes of someone from another "high tech" society long ago.

Directions: *Rancho Sierra Vista is in Newbury Park at the corner of Pinehill Avenue and Potrero Road. From Highway 101 take Wendy Road south toward the mountains until it deadends at Potrero. Turn right and take Potrero to Pinehill.*

Step by Step:

1. From the parking lot, follow the service road into the park toward the Native American Indian Center. The Satwiwa Loop Trail begins on the dirt trail leading past the center. Before you go too far you'll reach the old ranch pond. Stop here for a moment and see how many animals call this oasis home.

2. Leaving the pond, take the trail south through the grassland. As the trail nears the chaparral, it crosses a seasonal stream, then climbs uphill slightly using rough wooden steps and a few switchbacks. After 1/2 mile, at the rim of Big Sycamore Canyon, the Satwiwa Trail joins with the Old Boney Trail. Take the path northwest along the edge of the canyon toward the water tank at the service road.

3. At the road, turn right and follow it 1/2 mile back to the center. From here, bear to the left and follow the service road back to the parking lot.

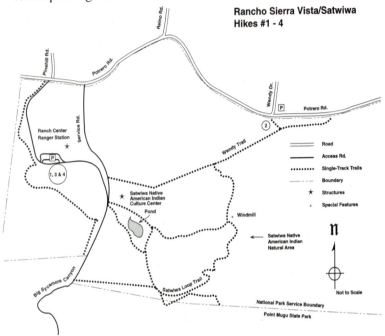

Rancho Sierra Vista/Satwiwa Hikes #1 - 4

2. Windmill Hike Trail

Difficulty:	Easy
Length:	1 mile round-trip
Elevation:	150 foot gain/loss
Type of trail:	single-track trail
Kid appeal:	* * *

Hiking over the rolling hills from Potrero Road to the windmill takes one by a number of vistas as it crosses three plant communities. This large grassland made the area attractive to the Chumash Indians. The grassland served them as a giant playing field for the intertribal games, similar to soccer, that inevitably took place during religious gatherings.

Animal tracks and scat in the trail are strong evidence that the trails are used daily by wildlife living in or passing through the park. Close examination of the trail through the grassland yields many clues as to the kind of prey that is hunted. Look for tiny star bursts in the soft dirt at the edge of the trail; these are the tracks of field mice, kangaroo rats and rabbits.

Directions: *Today's hike on the Windmill Trail begins at the west Potrero Road trailhead. From Highway 101 take Wendy Road south toward the mountains until it deadends at Potrero Road. Park in the wide dirt area on the west side of Potrero Road.*

Step by Step:

1. The trail begins near the information kiosk just inside the park boundary. Follow the dirt path straight ahead down a slight grade to where it crosses a seasonal stream under some large oak and sycamore trees. The trail then rises up to follow the contour of the hillside in the transition zone between chaparral and grassland.

2. Before long, the trail ascends a short but steep grade to a major trails junction. Continue straight up the hill, on the signed hiking trail for Boney Mountain. (The multi-use trail to the right is signed for the Big Sycamore Trail.) At the top of the grade there is another trail junction; take the trail to the left, signed for the Windmill. At this point the service road shrinks down to a footpath through the chaparral forest.

3. Just as you are wondering where the windmill is, you are there. At the windmill the trail joins the Satwiwa Loop Trail. Take the trail to the right, heading downhill through an oak-canopied ravine. The trail soon leads back to the grassland again. Stay to the right at the next trails junction. Keep the windmill to your right as you're walking .

4. In several hundred yards you will be back at the top of the hill mentioned in Step #2. At the trail junction turn left, go down the hill and you will be heading back to the Potrero Road Trailhead.

3. Old Cabin Site Trail

Difficulty: Moderate
Length: 5 1/2 mile round-trip
Elevation: 1,200 foot gain/loss
Type of trail: fire road
Kid appeal: * * *

    Today's hike begins in the grasslands of Rancho Sierra Vista/Satwiwa, then winds its way up the fire road through the chaparral of Point Mugu State Park to Boney Mountain. Along the way the road passes through the Upper Sycamore Canyon watershed, an area that was severely burned in the Green Meadow Wildfire of 1993. This is an exciting hike because of the variety of scenery along the way. A short side trip to the waterfall is a real gem. Nestled in an open oak woodland, the Old Cabin site with the rock chimney, flagstone memorial and quaint fresh water spring makes a quiet picnic retreat.
Directions: *Same as Hike #1 - Satwiwa Loop Trail.*

Step by Step:
1. From the parking lot, follow the service road into the park past the Native American Cultural Center all the way to the rim of Big Sycamore Canyon. Turn left and follow Old Boney Road east along the canyon's edge. The trail heads uphill for a ways before dropping gradually down through the chaparral to the canyon bottom.
2. In about 1/2 mile, the road crosses the stream under a canopy of bay, sycamore and oak trees. Proceed up the road to the first switchback to the right. Here an unmarked spur trail leads to a surprisingly picturesque waterfall. After exploring the falls, continue uphill along the road.
3. In 1/3 mile, after a little rise, Boney Road forks off to the right; however we will go straight ahead. Check to the north for views of the Channel Islands. As the road meanders through the oak woodland, look for a rock chimney a little uphill from the stream. This is the Old Cabin site.
4. From the Old Cabin area there is a trail that leads to Boney Mountain and then on to Sandstone Peak at the Circle X Ranch

Site via the Backbone Trail. Once you've finished exploring the mountainside, return to your car the way you came.

4. Big Sycamore Canyon to the Sea

Difficulty: Easy going into canyon, difficult coming back
Length: 8 miles one-way
Elevation: 1,000 foot loss/gain
Type of trail: fire road
Kid appeal: * * * * *

This ride through Big Sycamore Canyon is one of the most popular bike rides in the Conejo Valley. Basically, it's downhill all the way from Rancho Sierra Vista to the ocean on this route. A car shuttle waiting at the beach for the return trip home makes this an easy ride for beginners, while the entire round-trip can even be a challenge for more experienced riders. Winter and springtime outings give riders the opportunity to literally get their feet wet, as the trail crisscrosses the seasonal stream that runs the length of the canyon bottom.

Directions: Same as Hike #1 - Satwiwa Loop Trail.

Step by Step:
1. From the parking lot follow the paved service road into the park, passing the Native American Culture Center on your way to Point Mugu State Park which begins 1/2 mile away at the edge of Big Sycamore Canyon.
2. Controlling your downhill speed, follow the service road as it winds down the canyon's steep back wall. Two miles of easy, slightly downhill riding on the paved road will bring you to the Danielson Ranch, which is about mid-way through the canyon. Near the ranch the service road changes to a dirt surface as it continues another 4 1/2 miles to the sea.
3. Along the way to the ocean you will pass the Ranch Center Road, the Wood Canyon Road and the Scenic Overlook Road; all of which are open to bicycles and make good riding loops that come back to the main Big Sycamore Canyon Road.
4. The trail ends in the Big Sycamore Campgrounds at the mouth of the canyon.

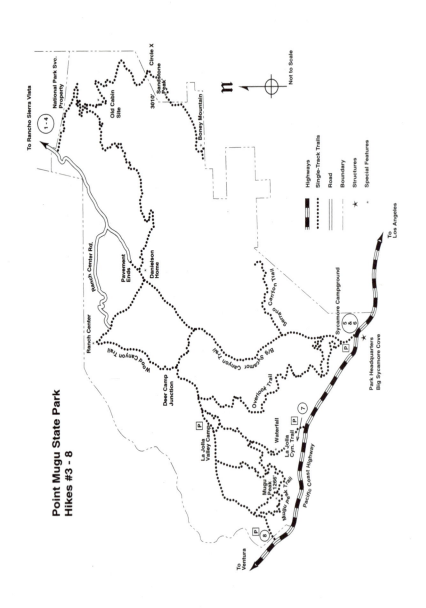

Point Mugu State Park
Hikes #3 - 8

To Rancho Sierra Vista

1 - 4

National Park Svc.
Property

Circle X

Old Cabin
Site

Sandstone
Peak

3010'

Boney Mountain

Not to Scale

Highways
Single-Track Trails
Road
Boundary
Structures
Special Features

To
Los Angeles

Ranch Center Rd.

Pavement
Ends

Danielson
Home

Ranch Center

Wood Canyon Trail

Serrano Canyon Trail

Deer Camp
Junction

Big Sycamore Canyon Trail

Overlook Trail

Sycamore Campground

5 & 6

P

Park Headquarters
Big Sycamore Cove

La Jolla
Valley Camp

P

Waterfall

La Jolla
Cyn. Trail

P

7

Mugu
Peak
1266'

Mugu Peak Trail

Pacific Coast Highway

P

8

To Ventura

5. Wood Canyon Bike Ride

Difficulty:	Easy/moderate
Length:	13 1/2 mile loop
Elevation:	800 foot gain/loss
Type of trail:	fire road
Kid appeal:	* * * * *

The Wood Canyon bike loop is one of my family's favorite rides. Beginning at the ocean end of Point Mugu State Park, today's mountain bike loop ride encounters a number of plant communities as well as a variety of riding surfaces. Much of the ride is quite easy through the oak savanna along the bottom lands of Big Sycamore Canyon on a well-maintained service road. The paved surface of the Ranch Center Road makes the hill up to the high meadow and chaparral somewhat easier. Speaking of hills to climb, using the trailhead off Pacific Coast Highway avoids the intense ride (or walk) up the back of the canyon to Newbury Park.

Oak-canopied Wood Canyon provides one of the most enjoyable "off-road" bike rides in the county. Early morning and evening rides often give rise to sightings of mule deer and coyotes. Springtime wildflower displays in the meadows and chaparral covered hillsides also are a visual treat.

Directions: *Point Mugu State Park is on Pacific Coast Highway about 4 miles west of the Ventura/Los Angeles county line. From Highway 101 in Newbury Park, exit on Wendy Drive and head south toward the mountains. Take Wendy to Lynn Road, turn right and take the Lynn Road extension to Potrero Road and turn right. Take Potrero Road 5 miles through Long Canyon and past Camarillo State Hospital to Lewis Road and turn left. This puts you on Los Posas Road, which leads to Pacific Coast Highway. Turn left (south) and continue to Big Sycamore Canyon in Point Mugu State Park. There is a $6 parking fee at state parks.*

Step by step:

1. From the day-use parking lot, proceed through the campground to campsite #10. Today's bike ride begins here on the

37

nearly level Big Sycamore Canyon fire road. Up the main canyon about 4 miles is the signed junction for Wood Canyon; this trail will be your return route from the Ranch Center.

2. Farther up Big Sycamore Canyon you will pass the Danielson Ranch multi-use area; about 1/2 mile beyond the group campsite is the turn-off for the Ranch Center Road. Take this paved road to the left as it heads uphill toward a high meadow.

3. A few miles of moderate riding will bring you to the Ranch Center in Wood Canyon, which is now a ranger's residence. Take the signed Wood Canyon Trail to the left as it heads back to Big Sycamore Canyon. This is the best part of the ride as the trail is slightly downhill through a beautiful riparian area for the next 2 miles. Once past the signed Deer Camp picnic area the main canyon road is about 1/2 mile away. Turn right on the fire road and return to the trailhead the way you came.

6. Scenic Overlook Trail

Difficulty: Easy/moderate
Length: 2 1/2 mile loop
Elevation: 900 foot gain/loss
Type of trail: single-track trail
Kid appeal: * * * *

The Scenic Overlook Trail offers the casual hiker a good way to view the seasonal changes in the park. The sycamore savanna which follows the streambed through this coastal canyon has a golden hue during the Fall. After the rains in the winter and spring, the chaparral and grassland that this trail pass through are a showcase for native wildflowers. On summer mornings, the trail is often shrouded in fog, giving a ghostly look to the white-barked sycamore trees.

Directions: Follow the directions given for the Big Sycamore Canyon Trailhead for #5 - Wood Canyon Bike Ride..

Step by step:

1. From the day-use parking lot proceed through the campground to campsite #10 and start walking up the Big Sycamore Canyon fire road. The Scenic Overlook Trail begins

on the left side of the road about 50 yards past the gate. Take the indistinct trail through the grass to the edge of the creekbed, where several wooden steps lead down and up the other side of this small ravine. Look for the trail sign on the far side ·of the creek. (Following a rainstorm this crossing may be flooded, in which case a little rock-hopping is called for.)

2. After crossing the creek, the trail winds through a small forest of sumac bushes before climbing the hill to a vista point of the campground. Continuing uphill another 1/2 mile the trail changes direction and then traverses an open grassland. Before long you are at the top of the hill. A spur trail leads straight ahead to an ocean overlook. Many use this as a turn around for their hike, returning to the trailhead the way they came.

3. To continue on the Overlook Trail, head north and slightly uphill from the spur trail and in a few hundred yards the trail will join with a park service road. Take the road to the right as it heads back to the canyon bottom. In about 1 mile the Scenic Overlook Trail comes to an end at the Big Sycamore Canyon fire road. Turn right and follow it about 1 mile back to the campground.

7. La Jolla Canyon Trail

Difficulty: Easy/moderate
Length: 4 miles round-trip
Elevation: 700 foot gain/loss
Type of trail: fire road, single-track trail
Kid appeal: * * * * *

La Jolla Canyon Trail makes for an outstanding family adventure. The trail passes by a rocky waterfall, through a narrow gorge and finally ends up at a shady walk-in camp near a grassy meadow. Because of this beautiful assortment of terrain, La Jolla Canyon is a popular weekend hiking spot. Spring is an excellent time of year to make this hike. The waterfall is at its best following a rain and the wildflowers are simply stunning. This is one of the best local areas to view the giant coreopsis, one of the rarest plant specimens found in the Santa Monica Mountains.

Directions: *La Jolla Canyon Trail uses the Ray Miller Trailhead in Point Mugu State Park which is located about 1 mile west of the Big Sycamore Canyon entrance to the park. (see Hike #5). Look for the signed Ray Miller Trailhead on the north side of Pacific Coast Highway.*

Step by step:

1. The signed La Jolla Canyon Trail begins at the north end of the parking lot on an old dirt service road. Take the level service road about 1/2 mile to the waterfall. Here some steep wooden steps lead up to the falls where the trail continues up the gorge as a narrow, but passable, single-track trail. Be sure to look for marine fossils in the rocky cliffs next to the trail.

2. At the top of the gorge, there is a trails junction. Stay to the right on the signed La Jolla Valley Loop Trail as it heads through a ceanothus forest along the edge of the valley. After another 1/4 mile of easy walking there is another trails junction. This time, take the trail slightly downhill and to the left.

3. In a few hundred yards the trail skirts the edge of a small pond. To get to the walk-in camp, turn right at the next trails junction on the far side of the pond. The campsites are in an oakwoodland a few hundred yards east of this junction. There are picnic tables, water, restrooms and shade available at the camp, making this a good lunch spot. Following a leisurely picnic, return to your car the way you came.

Optional: The La Jolla Trailhead is the western terminus of the Backbone Trail. The signed, single-track trail begins on the far side of the creek near the gate at the fire road into the canyon. The hillside overlooking the group campsite is an excellent spot for an early spring wildflower walk. From here the trail heads aggressively to the top of the ridge before dropping down into Big Sycamore Canyon on its way east toward Boney Mountain and Circle X Ranch.

8. Chumash Trail

Difficulty: Strenuous
Length: 2 mile loop
Elevation: 1,300 foot gain/loss
Type of trail: single-track trail
Kid appeal: * *

If you are looking for a hike that presents a physical challenge as well as a spectacular vista point, the Chumash Trail is just the ticket. Starting at sea level, this trail follows an ancient Chumash Indian foot path the back way into La Jolla Canyon. Believed to be thousands of years old, this trail was used by the Indians as they transported shellfish harvested from the ocean back to their inland village.

A number of trails crisscross through this beautiful and remote end of Point Mugu State Park. If you want to make this a longer hike, pick up a detailed map of the park's 35 miles of trails from the ranger kiosk at the main Big Sycamore Canyon entrance to the park.

Directions: This trail uses an unsigned parking area on the north side of Pacific Coast Highway as its trailhead. The parking area is about 1 mile west of the Ray Miller Trailhead, across the Highway from the Seabee firing range near county road marker 8.07. This is a primitive area; no dogs, bicycles or horses are allowed on this trail.

Step by step:

1. The unsigned Chumash Trail begins near the middle of the roadside parking area by a large California State Park sign. The indistinct trail more or less assaults the steep hillside, gaining 900 feet in elevation over the first 1/4 mile of trail.

2. At the crest of the hill there is a trails junction. Go to the right on the unsigned Mugu Peak Trail as it contours the ocean-facing side of Point Mugu. After an easy 1/2 mile of walking there is another trails junction with the spur trail that leads to the summit of Mugu Peak. After enjoying the view, continue west over the hilltop to join once more with the Chumash Trail. Stay to the left on this trail as it descends the hill toward the ocean and your car parked by PCH.

Big Sycamore offers many challenging cycling trails.

The Backbone Trail

With the Ray Miller Trailhead at Point Mugu State Park as its western terminus, the Backbone Trail is an east-west, ridgeline route through the Santa Monica Mountains. It was originally planned so that when completed, it would have comprised more than 65 miles of continuous trail connecting this remote open space park in Ventura County with the urban wilderness surrounding Griffith Park in Los Angeles. Unfortunately, spiraling real-estate prices have stopped work for the time being east of Will Rogers State Park in Pacific Palisades.

At its western end, the Backbone trail takes you through some of the highest as well as the most ruggedly beautiful parts of the Santa Monica Mountains. The dramatic land forms, spectacular vista points and outstanding array of plant communities along the trail make the Backbone Trail very popular with the local hiking crowd.

As I have arranged the hiking venues in this guide by location, most of the Backbone Trail hikes I mention will be presented "along the way", so to speak. Except, of course, for the three in this chapter which are located close to one another. Rocky Oaks isn't actually part of the Backbone Trail, but because you pass it on the way to these trailheads, I have included it in this chapter.

9. Castro Crest Trail

Difficulty: Moderate
Length: 3 1/2 mile round-trip
Elevation: 850 foot gain/loss
Type of trail: fire road
Kid appeal: * * * *

Castro Peak, at 2,824 feet above sea level, seems to be the top of the world. Today's walk offers the casual hiker some spectacular views of the Santa Monica Mountains as well as Catalina Island. Part of the Backbone Trail, this ridgeline hike along the old Castro Motorway is also a favorite of rock hounds looking for fossils of mollusks millions of years old.

Castro Motorway intersects several other mountain fire roads, providing a good access point to the multi-use trail system of the central part of the Santa Monica Mountain range. Making full use of modern day utility and fire roads as well as ancient Chumash footpaths, the trails from Malibu Creek State Park and Solstice Canyon are linked at this point by the Backbone Trail.

Numerous mini-trails explore the huge rock outcroppings along the motorway. Many animals favor these rocky sites for their nests and dens. This is also a popular site for fossil hunts. Be sure to explore this area carefully, but remember that on National Park Service property all "found" material must be left for others to enjoy.

Directions: *The Castro Crest site is in the Santa Monica Mountains National Recreation Area at the end of Corral Canyon Road in Malibu. From Highway 101 in Agoura, take Kanan Road 12 miles to Pacific Coast Highway and turn south toward Los Angeles. Take Pacific Coast Highway approximately 3 miles to Corral Canyon Road and turn inland. Follow Corral Canyon 5 miles to the parking lot at the end of the road. From Los Angeles simply follow Pacific Coast Highway directly to Corral Canyon Road; it is opposite the highway from Dan Blocker State Beach.*

Step by step:

1. Castro Peak Motorway begins at the north end of the parking lot. Walk around the gate and follow the dirt road as it climbs to the peak about 1 3/4 miles away. After about 3/4 mile of steady uphill walking, on the right is the Bulldog Motorway. This service road is the Backbone Trail into Malibu Creek State Park.

2. Continue west on the Castro Crest Motorway and before too long the road runs along the top of the ridge near some very large rock outcroppings. Another mile of easy walking brings you to the junction with the Newton Motorway and the western continuation of the Backbone Trail.

3. About 1/8 mile beyond this junction is the spur trail to Castro Peak. After enjoying the views from the summit, return to your car the way you came.

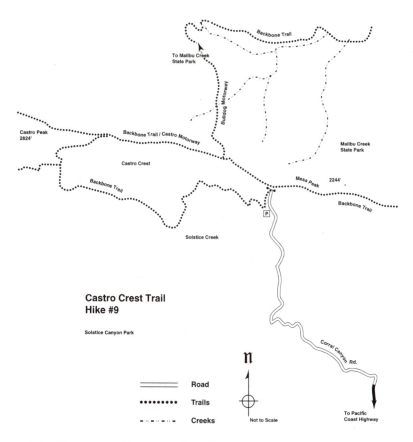

Castro Peak
2824'

Backbone Trail / Castro Motorway

Bulldog Motorway

To Malibu Creek
State Park

Backbone Trail

Malibu Creek
State Park

Castro Crest

Mesa Peak   2244'

Backbone Trail

Backbone Trail

P

Solstice Creek

**Castro Crest Trail
Hike #9**

Solstice Canyon Park

Corral Canyon Rd.

𝔫

Road

Trails

Creeks

Not to Scale

To Pacific
Coast Highway

10. Newton Canyon Trail

Difficulty:	Moderate
Length:	5 mile round trip
Elevation:	200 foot gain/loss
Type of trail:	single-track trail
Kid appeal:	* * * *

    One of my favorite sections of the Backbone Trail is the 2 1/2 mile stretch through Newton Canyon. This beautiful portion of the trail connects Kanan Road to Latigo Canyon Road and with large parking lots at either end of the trail, this makes a good "shuttle-hike". As the Backbone Trail follows the contour of the canyon's north facing slope, much of the trail is through a woodland and under a canopy of oaks. However, near the end of the canyon the trail climbs through the chapar-

45

ral to provide a dramatic contrast of the two most prominent plant communities: the oakwoodland and the chaparral. Not only does the plant life of the two areas contrast each other but the soils also are very different. Chaparral soil is sandy and well drained whereas the woodland soil is made of clay and has lots of organic material in it to absorb water. See how many differences you can identify on your hike.

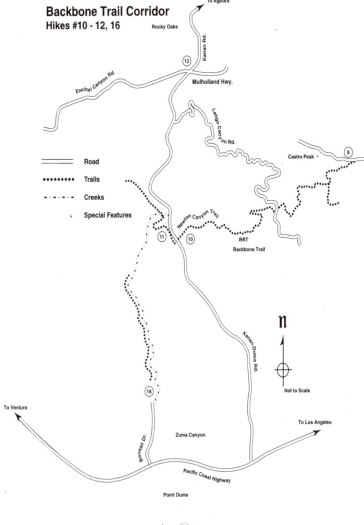

Backbone Trail Corridor
Hikes #10 - 12, 16

To Agoura

Rocky Oaks

Kanan Rd.

Encinal Canyon Rd.

12

Mulholland Hwy.

Latigo Canyon Rd.

Castro Peak ·

9

————	Road
••••••••	Trails
- · - · - · -	Creeks
·	Special Features

Newton Canyon Trail

11

10

BBT

Backbone Trail

Kanan-Dume Rd.

n

Not to Scale

To Ventura

16

To Los Angeles

Bonsall Dr.

Zuma Canyon

Pacific Coast Highway

Point Dume

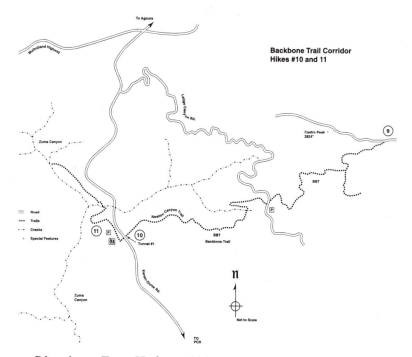

**Backbone Trail Corridor
Hikes #10 and 11**

Directions: From Highway 101 in Agoura, exit on Kanan Road heading south toward the ocean. Take Kanan through the mountains and about 2 miles south of the Mulholland Highway intersection look for a large parking area on the west side of the third tunnel (T-1, the one closest to the ocean) on Kanan Road. From Los Angeles take Pacific Coast Highway north of Malibu to Kanan Road and turn right. Head inland on Kanan and the parking area for the Backbone Trail is on the left side of the road on the inland side of the first tunnel.

Step by step:

1. The unsigned Backbone Trail begins along the fire road located at the south end of the parking lot. It starts by heading aggressively uphill from the locked gate at the foot of the fire road to climb to the top of the tunnel, T-1, on Kanan Road. At the top of the grade there is a trails junction as well as a terrific view of Catalina Island. Take the trail to the left as it heads east over the tunnel.

2. A little past the tunnel there is another trails junction. Once

again stay to the left and take the single-track trail that heads gradually downward into the canyon through an oak-covered hillside. Before too long the path crosses a paved private road. Be sure to stay on the trail through this area as the Backbone Trail crosses private property via an owner's granted easement.

3. The trail continues to wind down along the shady north-facing slope of the canyon for another mile or so. Once past the seasonal stream crossing however, the path starts to climb up through the chaparral to meet Latigo Canyon Road about 1/2 mile away. Once at the road turn around and return to your car the way you came.

Optional: The Backbone Trail continues on the other side of Latigo Road all the way to Castro Crest, about 4 miles away.

11. Upper Zuma Canyon Trail

Difficulty: Easy
Length: 1 1/2 mile round-trip
Elevation: 200 foot loss/gain
Type of trail: single-track trail
Kid appeal: * * * * *

Even though it is right off Kanan Road, the National Park Service property in Upper Zuma Canyon is one of the most remote areas in the central part of the Santa Monica Mountains. There are no roads through this part of the canyon and there are no designated trails all of the way through either. Today's hike is a short one to a grotto waterfall along this section of the Backbone Trail. The National Park Service has created a new multi-use trail that goes from the parking area off Kanan Road down a rather steep, north-facing slope to the creek. The tree-canopied single-track trail is in excellent condition making the hike through a dense woodland to the creek an ideal family adventure.

While the final spur trail to the creek is rather steep, exploration along the stream will provide hours of fun. Not only are the waterfalls exciting, but there is the added attraction of numerous fossils embedded in the rocks near the trail. The only trouble with exploring this stream is how extremely dark it can be under all of those trees and the abundance of poison oak growing on the creekbank.

The first half of the trail, going down the north-facing slope to the creek, is through a ceanothus forest that soon gives way to larger trees such as oak and sycamore. Ferns and tiger lilies growing along the trail here in the shade of the larger plants indicate a softer climate than just a few yards away in the exposed chaparral. It is these extremes of growing conditions that make the Santa Monica Mountains one of the most diverse botanic regions in the world.

Directions: From Highway 101 in Agoura, exit on Kanan Road heading south toward the ocean. Take Kanan through the mountains and about 2 miles south of the Mulholland Highway intersection look for a large parking area on the west side of the third tunnel (T-1, the one closest to the ocean) on Kanan Road. From Los Angeles take Pacific Coast Highway north of Malibu to Kanan Road and turn right. Head inland on Kanan and the parking area for the Backbone Trail is on the left side of the road on the inland side of the first tunnel.

Step by step:

1. The unsigned Backbone Trail into Upper Zuma Canyon begins at the north end of the parking area. The path leads slightly uphill at the first National park Service trail sign. Once past the trail sign the path begins its descent down into the canyon. At the first major switchback there is a spectacular vista of the junction of Upper Zuma and Newton Canyons.

2. As the woodland gives way to chaparral, look for a spur trail to the left that leads under a very large oak tree. This trail leads to a vista point at the top of the grotto waterfall. Return to the main trail and continue a couple of hundred yards through the chaparral. Look for another spur trail to the left. This is the path that leads down to the creekbed. It is quite steep, so use caution.

3. Just before the creek the trail comes to a "T" junction. Follow the trail to the left about 50 yards and you will be at the bottom of the waterfall. The trail to the right enters the creekbed and follows the creek downstream. This part of the trail is indistinct but if you don't mind rock-hopping and bat-

tling poison oak, you can hike to another, larger waterfall about 1/8 mile away.

4. After exploring the creek area you can use the spur trail to return to the main Backbone Trail. If you are ready to head back to your car, turn right and take the same path back to the parking area. If you want to hike farther, the Backbone Trail continues uphill through the chaparral for another 1/2 mile or so. The trail traverses the canyon ridgeline as it parallels Kanan Road. After awhile the trail becomes indistinct, crosses a meadow and then enters the chaparral once more. This makes a good turnaround spot as it very close to the NPS boundary line.

12. Rocky Oaks Walkabout

Difficulty: Easy
Length: 2 mile loop hike
Elevation: 200 foot gain/loss
Type of trail: fire road, single-track trail
Kid appeal: * * * * *

The Rocky Oaks Site within the Santa Monica Mountains National Recreation Area features several shady oakwoodlands with picnic tables and even a rustic amphitheater. Look for the landmark "Turtle Rock" that looms over the area like a huge snapper coming out of its shell. There is also a pond that supports a variety of wildlife that usually is found only in deep canyon recesses. The five signed hiking trails that crisscross and loop around each other throughout this 198 acre park highlight various natural features making for an excellent "walk-about" hike. Rocky Oaks' convenient location, diverse terrain and plant communities make it a popular site for cyclists, equestrians and hikers alike. In the spring, the Glade Connector trail through the meadow is a riot of color thanks to the wildflowers that grace the park after the rains. Several are rare/endangered species endemic (found nowhere else) to the area. The wooden fence across the meadow protects this fragile environment from foot traffic. Be sure to stay on the trail and leave the flowers for others to enjoy. The oak-canopied woodlands make the park a good summertime spot as well, especially for evening hikes.

Directions: *The Rocky Oaks Site is in Agoura at the intersection of Kanan Road and Mulholland Highway. From Highway 101, take Kanan Road south 6 miles to Mulholland Highway and turn west, right. The park entrance is immediately on the right.*

Step by step:

The Rocky Oaks Loop Trail defines the perimeter of this site and provides access to the park's other trails: Creek Trail, Overlook Trail, Pond Loop Trail and the Glade Connector Trail.

1. From the parking lot, cross the small wooden bridge near the bathroom to the park's information kiosk. Start your walk here. Up the trail 25 yards there is a "T" junction. The signed Rocky Oaks Loop Trail goes in both directions. To the right it goes toward the pond and to the left it goes toward the signed Creek Trail.

2. For the 2 mile walk-about hike, take the Rocky Oaks Trail to the left along with the Creek Trail. In a few hundred yards, after an open area, is a trails junction; the Creek Trail goes off to the left through an oakwoodland about 1/4 mile to a seasonal creek near the park boundary. Straight ahead, starting with a couple of railroad tie steps, is the Overlook Trail. This trail climbs to a summit vista in about 1/2 mile, then joins the Rocky Oaks Loop Trail once again as it goes off to the right through the chaparral.

3. For the 1-mile hike, take the Rocky Oaks Loop Trail to the right from the "T" in step #1. As you walk along you can see the picnic area and amphitheater to your right in a large oakwoodland. In 25 yards the Rocky Oaks Loop Trail intersects the signed Pond Loop Trail to the left and the Glade Connector Trail to the right.

4. The Pond Loop Trail, as one might expect, loops around the pond. Stay to the right on the single-track trail that follows the eastern side of the pond along the meadow. To the left, the trail is a little wider and follows the western side of the pond along the chaparral at the foot of the hill.

5. In a shady area at the north end of the pond, the signed Rocky Oaks Loop Trail crosses the Pond Loop Trail once again. To see more of the park, take the Rocky Oaks Loop Trail to the right toward the highway. In a little bit there is another fork in the road. The Rocky Oaks Loop Trail continues to the left as it heads toward an oakwoodland near the park's perimeter. To the right, going across the open meadow is the signed Glade Connector trail. Both trails return to the major trail junction where we started the Pond Loop Trail.

6. To return to the parking lot, you can retrace your steps on the Rocky Oaks Loop Trail or take the Glade Connector trail through the woodland to the picnic area and on to the parking lot.

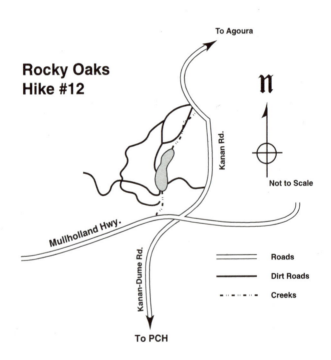

Coastal Canyons

Coastal woodlands

HIKES AT A GLANCE 2
Santa Monica Mountains National Recreation Area

	DIFFICULTY	FACILITIES	FEATURES	FEES	OPEN TO
COASTAL CANYONS					
13. Charmlee County Park Walkabout	easy	w, t, p	nature center, ocean views woodlands, wildflowers	weekends & holidays	H, E, D, S
14. Escondido Natural Area	easy	w, t	waterfall, shady canyon	no	H, E, D, S
15. Solstice Canyon Park Walkabout	easy/moderate	w, t, p	shady canyon, historic buildings, stream, wildflowers, waterfall	yes	H, E, C, D, S
16. Lower Zuma Canyon Trail	easy	w	shady canyon, stream	no	H, C, E, D
17. Leo Carrillo State Beach Walkabout	easy/moderate	w, t, p, c, s	tidepools, pirate cave, views, nature center, wildflowers	yes	H, C, E, S
18. Nicholas Flat Trail	easy/moderate	none	views, pond, wildflowers	no	H, C, E, S
CIRCLE X RANCH SITE					
19. Grotto Trail	moderate	w, t, p, rs, c	grotto, waterfall, shady trail	no	H, C, E, D
20. Sandstone Peak/via Backbone Trail	moderate	none	views, wildflowers,	no	H, C, E, D
21. Mishe Mokwa Trail	easy/moderate	none	views, shady trail	no	H, E, D
22. Arroyo Sequit Natural Area Walkabout	easy	w, t, p	views, wildflowers, canyon stream	no	H, D

FACILITIES: w - water, t -toilet, p - picnic table, rs - ranger station, c - camping, m - marina, s - store
OPEN TO: H - hikers, C - cyclists, E - equestrians, D - dogs on leash, S - stroller/handicap access

13. Charmlee County Park Walkabout

Difficulty: Easy
Length: 2 mile loop
Elevation: 100 foot gain/loss
Type of trail: fire road
Kid appeal: * * * * *

High on a hill overlooking the Malibu coastline, Charmlee County Park makes for a terrific family outing. The gentle, wide dirt trails through the park offer a variety of ocean views as well as several plant communities to explore. This is an excellent site for that long overdue Sunday picnic in the park.

Acquired by the Los Angeles County Department of Parks and recreation in 1968, the park derives its fanciful name from the combination of the first names of its previous owners: Charmain and Leonard Swartz. One of the side trails on today's walkabout of the park takes you by the burnt out ruins of their home on a small rocky knoll overlooking the ocean.

Today's figure-8 walk is but one of the many trail combinations through the park's gently rolling 460-acre hillside expanse of open space. The oakwoodlands surrounding the grassy meadow that makes up the heart of the park offer many shady and secluded picnic spots among large rocky outcroppings and boulders hidden from a distant view.

Open on weekends, the park's Nature Center serves also as the park office. As you pay your parking fee, be sure to check the list of upcoming park activities and nature programs. The beautiful mural on the center's inside wall provides an excellent pictorial representation of both the animals that live in the park as well as a glimpse of how the park may have looked in pre-Columbian times. The mural also depicts the Chumash Indian history and legends of the area. One of our favorite scenes is the old storyteller re-telling the tale of the "Rainbow Bridge" to a rapt group of youngsters.

Charmlee Park is not in a coastal canyon. Why is it in this section then? I placed it here because it, like Nicholas Flat, makes up the headlands of the coastal canyons below.

Unfortunately, the trails from Charmlee Park run into the obstacle of private property before they reach the ocean. Maybe one day an easement may be granted through this area and connect this inland park to the ocean.

Directions: Charmlee County Park is on Encinal Canyon Road, 4 miles north of Pacific Coast Highway in Malibu. From the Conejo Valley take Highway 101 to Westlake Boulevard and go south several miles through the Santa Monica Mountains. Stay to the left, on Highway 23 south, at the second intersection with Mulholland Highway. Proceed on Highway 23 for another mile and turn left at Lechusa Road. Stay to the right at the intersection of Lechusa Road and Encinal Canyon Road. Take Encinal Canyon Road another 2 miles to the park. Watch for mileage and entry signs for Charmlee Park, turn right and follow the paved service road into the park.

Step by step:

1. From the parking lot the trail leads slightly uphill past the oak-grove picnic area as it makes its way south toward the ocean. Near the top of the hill there is a trail that goes off to the left which will be used for the return portion of today's walkabout hike. Continue straight ahead on the main trail as it leads to another, larger trails junction about 100 yards ahead.

2. At this junction the trail splits off three ways: to the right is a fire road that leads through the chaparral toward the ocean and out of the park; straight ahead a short spur trail leads to a shady knoll with large rock outcroppings as well as the foundation of Charmain and Leo's home; and the main trail turns left and leads to the rest of the park. To continue the loop take the trail to the left as it heads downhill toward the large meadow that makes up the heart of the park.

3. Once at the meadow there is a 4-way trails junction; the trails on the left will be part of the return route. Take either the main trail straight through the meadow or the trail to the right as it skirts the western edge of the grassland; both trails meet again on the other side.

4. At the ridgeline overlooking the ocean, take the trail slightly uphill to an abandoned reservoir. At this point the trail heads east and then downhill to an outstanding ocean vista point

about 1/2 mile away. From the vista the trail turns around and heads north, back uphill toward the meadow. On the way back the main trail skirts the eastern edge of the grassland as it passes several small oakwoodlands.

5. At the north end of the meadow, stay to the left as the trail brings you back to the major trail junction in step #3. This time take the trail to the right as it heads uphill and to the east around a small hill. At the top of the hill this trail joins the main trail as mentioned in step #2. Turn right and follow the main trail back the way you came, downhill toward the picnic area and parking lot.

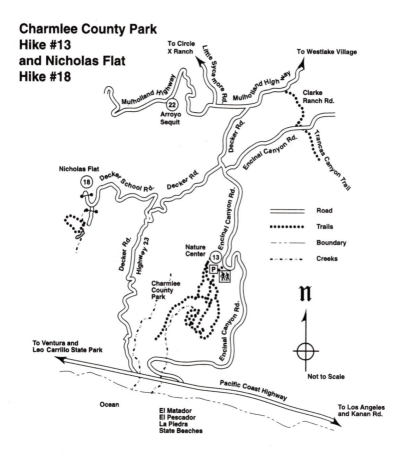

Charmlee County Park
Hike #13
and Nicholas Flat
Hike #18

14. Escondido Canyon Natural Area

Difficulty: Easy
Length: 4 1/2 mile round-trip
Elevation: 300 foot gain/loss
Type of trail: fire road, foot path along road
Kid appeal: * * * *

Escondido Canyon is another of those shady coastal canyons blessed with a year-round stream that just begs to be explored. This narrow box-canyon is tucked away from the coast by a small hillside and ends dramatically at the sheer rock face of Escondido Falls.

The word escondido means "hidden" in Spanish: what a perfect name for this picturesque little canyon tucked away behind some small hills. The falls are, of course, the major draw to the area. Collectively the upper and lower falls drop several hundred feet. The lower falls are about 50 feet high and have a sheer face encrusted with a unique form of calcium deposit.

Public access to the Escondido Natural Area has been in the news over the years as relations between parks users and residents have become strained. Hopefully, this age-old issue will be resolved in the near future to the satisfaction of both parties as plans are in the works to establish alternative access into the canyon. Please respect private property when hiking here. Also, please resist the urge to climb the falls. The falls themselves and the surrounding land are under private ownership and are closed to the public. Most weekends you will see guards at the end of the trail to ensure that no one attempts to climb the cliffs.

Directions: *Escondido Natural Area is off Winding Way in Malibu. From Highway 101 in Agoura Hills, take Kanan Road south 12 miles to Pacific Coast Highway in Malibu. Turn left, south, toward Los Angeles on PCH and take it approximately 1 1/2 miles to Winding Way and turn left. Park in the signed parking lot provided at the trailhead.*

Step by step:

1. The signed Winding Way Public Access Trail to the Escondido Canyon Natural Area begins just north of the parking lot. The trail leads you into the canyon as it follows along the left, then the right side, of a private road for about a mile. As you approach the tree-canopied canyon, the trail leaves the roadside to cut across an open meadow.

2. At the canyon bottom, the trail crosses the stream just before coming to a trails junction. As we are looking for the falls, turn left and follow the unsigned trail deeper into the canyon. Follow the creek upstream for about a mile, while the trail crosses it again via several stepping stones.

3. The trail comes to an open area where the park boundary is noted on a small sign post. This is the end of public property. The current owner has sanctioned trail use up to the falls, so you may continue on the trail as it heads up from the canyon bottom to traverse some chaparral.

4. The trail gains slightly in elevation as it passes by a large level area. There are several small paths that crisscross each other and then head out into the chaparral-covered hills. Stay on the main trail as it continues deeper into the canyon following the creek upstream.

5. Before too long you will catch a glimpse of the falls between the trees. Then in a few hundred yards the trail dead-ends at the base of the waterfall. After enjoying the view, return the way you came back to the first stream crossing.

6. This time instead of going over the creek, continue straight ahead on the trail and follow it along the creekbed to the lower part of the National Park Service road. Take the paved road up the hill to get a view of the bird life that fills the canyon's oak canopy. Return on the access trail along Winding Way back to your car parked in the lot off Pacific Coast Highway.

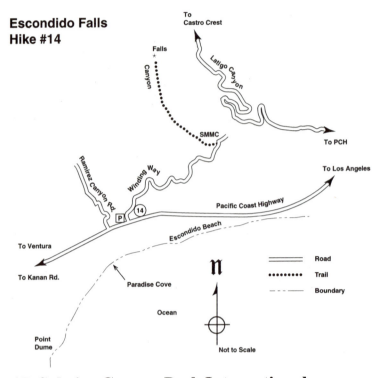

Escondido Falls
Hike #14

To Castro Crest

Falls

Canyon

Latigo Canyon

SMMC

To PCH

To Los Angeles

Ramirez Canyon Rd.

Winding Way

P 14

Pacific Coast Highway

To Ventura

Escondido Beach

To Kanan Rd.

Paradise Cove

Ocean

Point Dume

n

Road
Trail
Boundary

Not to Scale

15. Solstice Canyon Park International and Sostomo Trails

Difficulty:	Easy to moderate
Length:	3 1/2 mile round-trip
Elevation:	250 foot gain/loss
Type of trail:	fire road, single-track trail
Kid appeal:	* * * * *

Solstice Canyon is the quintessential coastal canyon. The year-round stream provides ample water to support this 550-acre park's numerous plant communities and a diverse wildlife population. Watch for movement in the tree canopy above the streambed and you will see squirrels using the huge branches as we would use the freeway to drive around town. Also look for birds busy building nests and attracting mates.

Solstice Canyon was part of the large Spanish land grant known as the Rancho Topanga Malibu Sequit holding.

Because the gentle slopes at the mouth of the canyon were ideal for grazing sheep and cattle, many of the native plants have been displaced by weeds and other non-natives. To help remedy this, California Native Plant Society volunteers have been working to revegetate the park with native plants. Several of the wildflower gardens in the park owe their existence to these dedicated individuals.

The Roberts family purchased Solstice Canyon for a retreat in the 1930's, originally staying in the stone Keller House located about half-way up the canyon. The Tropical Terrace House later was designed for the family by renowned architect Paul L. Williams. On-site features such as trees, the creek and waterfalls were integral to the design of this unique home. Thanks to the 1982 fire and subsequent flooding, all that remains of this once beautiful dwelling is the foundation, the fireplaces and a few outside gardens.

After the loss of their home to fire and flood, the Roberts sold the lower portion of Solstice Canyon to the Santa Monica Mountains Conservancy in 1986. The silo-like building overlooking the Caballero Picnic area is the Conservancy office headquarters and was built originally as a remote test site for satellite instruments made by Space Technology Laboratories, part of the TRW Corporation, in the 1960's. The park has been open to the public since 1988.

Directions: Solstice Canyon Park is off Corral Canyon Road in Malibu. From Highway 101 in Agoura, take Kanan Road 12 miles to Pacific Coast Highway and turn left, south, toward Los Angeles. Take PCH approximately 3 miles to Corral Canyon Road and turn inland. Drive 50 yards to a large white gate fence on the left side of the road. Enter the park and pay the parking fee to the "iron ranger", then proceed up the service road to the parking area for the Caballero picnic area by the Santa Monica Mountains Conservancy office.

Step by step:

1. The Caballero Picnic area is the trailhead for today's hike on the International Trail through Solstice Canyon. The hike follows Solstice Creek about 1 1/2 miles upcanyon. From the parking lot you can follow the paved service road or take the signed

TRW Trail, a single-track path on the far side of the creek, through the oakwoodland.

2. About a half mile from the trailhead, the two trails meet at the El Alisar picnic area. Continue up the road, or along the gentle path, to the wooden bridge over the stream. The road follows the creekbed for another 1/2 mile or so to the Tropical Terrace. Just before the bridge the signed TRW Trail heads straight through the woodland a short way before climbing up a chaparral covered slope as it loops back to the trailhead.

3. Continue on the International Trail by following the road. Along the way you will pass the Sostomo Trail junction on your left. You also will see the Keller House, which is believed to be the oldest structure in Malibu, on the far side of the creek. Beyond the Keller House there are two Texas-style creek crossings and the lovely Fern Grotto picnic area before reaching the back of the canyon.

4. Only the foundation and chimneys of the once-extraordinary Roberts House remain in today's Topical Terrace. After exploring the ruins and paying homage to the shrine, you can return the way you came or take the Rising Sun Trail back to the trailhead.

5. The single-track Rising Sun Trail crosses the creek as it leads up to, then along, the canyon's east wall. This path heads aggressively uphill a short way before leveling out along the rim. Near the trail's end it passes by the Conservancy offices as it makes its way back down to the parking lot following the route of the TRW Trail.

Roberts' Family Shrine

Sostomo Trail

For a more strenuous workout, try the Sostomo Trail. It starts by following the main park service road upcanyon along the International Trail to the Tropical Terrace. Once out of the sheltered canyon bottom, the path takes you through the open chaparral, a dense ceanothus forest and two creek-crossings before reaching the loop to the vista point. This multi-use, single-track trail leads to the Deer Canyon Overlook as it explores the north side of the canyon. It is a moderate round-trip hike that is about 4 miles long and has an 800 foot gain/loss in elevation. Beginning near the ruins of the Tropical Terrace House at the back of the canyon, the trail heads aggressively up the canyon wall. At first following Solstice Creek, the path then climbs to the vista point before looping back along the canyon's steep western ridge.

16. Lower Zuma Canyon Trail

Difficulty: Easy
Length: 2 mile loop
Elevation: nearly level
Type of trail: mostly fire road
Kid appeal: * * * * *

Lower Zuma Canyon offers a terrific family outing any time of year. This wide, nearly level coastal canyon is a bit primitive, but it is a beautifully lush riparian habitat filled with wonderful places to explore. Many large trees such as sycamores, live-oaks and poplars make this a pleasant, shady hike for any time of day or season of the year. Old growth on some of the full-covering trees like sumac and elderberry hang over the trail in several areas creating a tunnel effect.

Farther from the creek and on the hillsides that define the canyon, the plants make a transition from streamside to chaparral conditions. Notice that the leaves of the chaparral plants are generally small in size and usually fragrant from the resin inside the leaves. These two adaptations help many of the plants through the long dry summers here in Southern California by reducing transpiration, or plant sweat.

The numerous creek crossings make the Zuma Canyon trail a big hit with children. When asked what they like best about the hike most kids will tell you "crossing the creek". Zuma Creek can become dangerously swollen just after a good rain storm, so some parental guidance is required, but generally the creek is shallow enough to cross on rocks without getting too wet. Some youngsters may prefer to just wade across, thoroughly enjoying the walk in the water.

The creekbed often reveals evidence of the wildlife that lives in the canyon. Be sure to check both the soft dirt on the trail and the mud next to the creek for animal tracks. See how many different ones that you can identify. Also, encourage everyone to be quiet for a few moments and to listen carefully. How many different bird songs do you hear near the stream? How many in the chaparral? See if you can tell where the most birds are just by listening.

Directions: *Lower Zuma Canyon is in Malibu off Bonsall Drive. From Highway 101 in Agoura Hills, take Kanan Road 12 miles to Pacific Coast Highway. Turn right on PCH and take it north 3/4 mile to Bonsall Drive and turn right. Go north on Bonsall about 1 1/2 miles until the road deadends and park in the lot by the trailhead.*

There are several trails through Lower Zuma Canyon. Described below is the main Zuma Canyon Trail that follows the creek through the canyon bottom lands. Most of these lateral trails explore the chaparral a short way before connecting back with the main trail through the canyon.

Step by step:
1. From the parking lot proceed north along the signed Zuma Canyon Trail as it heads into the canyon. After 1/2 mile of easy walking, first through an open meadow and then through a forest of sumac and elderberry trees, you come to the junction with the signed Zuma Loop Trail on the left. Continue straight ahead on the Zuma Canyon Trail.
2. A few hundred yards deeper into the canyon brings you to the first creek crossing. This can be either an exercise in rock-hopping or a fun wade in the cool creek. Farther up the trail about another 1/8 mile is the junction with the signed Canyon

View Trail on the right. Continue on the Zuma Canyon Trail as it goes to the left, close to the creek and then parallel to it.

3. Before too long there is another trail junction, this time with the signed Scenic Trail, also going off to the right. Again, stay on the main trail along the creek. At the next stream crossing the two paths will meet again; on the far side of the creek the Zuma Loop trail also returns to the main trail.

4. If you have had enough hiking for one day, your return to the trailhead could be on either of these two trails. The Zuma Loop trail is the longer of the two and it gains a little elevation as it leaves the canyon floor for the chaparral covered hillside. The Scenic trail explores more of the canyon bottom. If you are still in hiking mode, Zuma Canyon Trail continues deeper into the canyon, crisscrossing the creek every few hundred yards or so. The canyon becomes narrower and the cliffs get steeper, requiring major boulder-hopping to proceed upstream. At this point you must return to your car the way you came. *Trail map on page 46; Backbone Trail.*

One of many interesting trailside characters

17. Leo Carrillo State Beach

Difficulty: Easy/moderate
Distance from kiosk:

          1/3 mile to pirate cave
          1 1/2 miles to vista point
Elevation: 35 foot loss/gain
          620 foot gain/loss
Type of trail: single-track trail
Kid appeal: * * * * *

Whether it's for a weekend campout or a day trip to observe tidepool life, Leo Carrillo State Beach is one of my family's favorite getaways. The park's 2,000 acres of back-country and more than a mile of beachfront make it a model Southern California playground. Indeed, Leo Carrillo State Beach is a haven for outdoor enthusiasts of all types.

Besides being physically beautiful, this state beach has a rich archeological history. The Chumash Indians inhabited this lush coastal canyon centuries before being "discovered" by Juan Rodriguez Cabrillo in 1542. In fact, the main campground is on an ancient village site.

It's hard to imagine a more complete outdoor playground than Leo Carrillo State Beach. The park's three campgrounds provide easy access to a number of ocean sports as well as miles of backcountry hiking trails that reveal some prime ocean vistas. Great for close-to-camp sunset hikes and winter whale watching, these trails also provide access to inland hills and valleys via the Backbone Trail to Nicholas Flat. The cliff trail leading along rocky Sequit Point is an exciting way to get to the beach; it goes to a secret, pirate cave beneath the point.

Directions: Leo Carrillo State Beach is on Pacific Coast Highway in Malibu. It is at the intersection of PCH and Mulholland Highway at the Ventura / Los Angeles county line. From Highway 101 in Agoura Hills, take Kanan Road south 12 miles through the mountains to the coast. Turn right on PCH and take it north 8 miles. The entry to the park is on the canyon side of the highway.

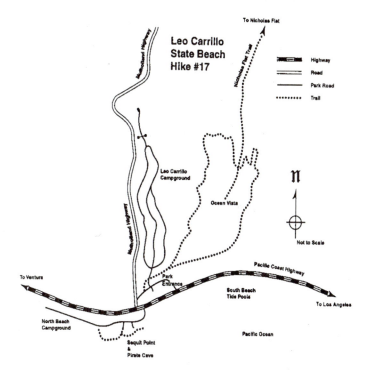

Leo Carrillo State Beach Hike #17

To Nicholas Flat

Mulholland Highway

Nicholas Flat Trail

Highway
Road
Park Road
Trail

Leo Carrillo Campground

Ocean Vista

𝔫

Not to Scale

Mulholland Highway

Pacific Coast Highway

To Ventura

Park Entrance

South Beach Tide Pools

To Los Angeles

North Beach Campground

Pacific Ocean

Sequit Point & Pirate Cave

Cliff Trail to Pirate Cave

Step by step:

1. Park in the canyonside day-use lot near the entry kiosk. Follow the paved service road under the highway overpass to the beach. As the road climbs a slight grade, take the wooden steps on the left toward Lifeguard Tower #2. (At the top of the grade, a dirt service road also leads to the cliff trail and guard towers.)

2. Take the footpath to the right along the cliff and go north toward Lifeguard Tower #3. Take the wooden steps down to the beach. Continue north on the beach and, depending on the tide, either go around or over the rock outcropping to reach the pirate cave situated in the cliff under the Lifeguard tower.

3. After exploring the cave area, return south along the beach if the tide is low enough; otherwise, return the way you came along the cliff trail.

Ocean Vista Point Trail

Step by step:

1. Park in the canyonside day-use lot near the entry kiosk. Follow the paved service road toward the canyon campground sites. The backcountry trails begin near the large green trails sign on the right-hand side of the service road. Follow the footpath up out of the campground area for about 100 yards until the trail comes to a fork by another trail sign.

2. Stay to the right on the signed Willow Creek Trail that parallels the ocean as it climbs up through the coastal sage scrub plant community. The trail to the left is the Nicholas Flat Trail and will be our return route.

3. As the trail turns inland, it overlooks the arroyo formed by seasonal Willow Creek. A mile or so of steady uphill walking and a couple of switchbacks bring you to a saddle (a low spot between two hilltops), where there is a trails junction. This is where the Willow Creek Trail meets the Nicholas Flat Trail that came up and around the hill from the campground side. (Nicholas Flat Trail continues north through the chaparral for another 3 1/2 miles before arriving at Nicholas Flat.)

4. Take the trail to the left toward the ocean to the Ocean Vista Point. After enjoying the views, make this a loop hike by returning to the campground via the Nicholas Flat Trail. As you descend the hill the path winds down through some stands of native bunch grass overlooking the canyon campsites. After a mile or so, the trail returns to the first trails junction and you can return to your car the way you came.

18. Nicholas Flat Trail

Difficulty: Easy
Length: 1 mile round-trip
Elevation: nearly level
Type of trail: mostly fire road
Kid appeal: * * * * *

Nicholas Flat, a highland meadow dotted with ancient oak trees, provides an excellent outdoor adventure for families of all abilities. The hike is along a dirt road that leads under a tree canopy to a pond. This road can easily accommodate strollers or wheelchairs. For hikers seeking a more challenging workout, the trail continues all the way to Leo Carrillo State Beach 3 1/2 miles away.

This beautiful mountaintop is close enough to the seashore to benefit from cool ocean breezes and the summer fog banks, making Nicholas Flat a comfortable hike site in any season. Early spring hikes offer the promise of ceanothus blooming on the hillsides, wildflowers in the meadow and tadpoles in the pond. There are many different animal tracks by the stream and at the edge of the pond. See if you can tell which tracks belong to the predators and which belong to the prey. This hike has something for everyone.

The view of San Nicholas Canyon from the boulder ridge at the end of the trail is nothing short of spectacular. The ridge is at an elevation of approximately 1,500 feet above sea level while the canyon immediately below it is about 300 feet above sea level. Looking across the divide you will see a number of caves: ideal nesting and den sites for local wildlife. Looking south you will see the ocean where eventually the canyon meets sea level. While enjoying the views, take extra care to avoid the poison oak growing in this and the other oakwoodlands at this site.

Directions: Nicholas Flat is at the end of Decker School Road in Malibu. For the scenic route from Highway 101 in Westlake Village, take Highway 23 (Westlake Boulevard) south through the mountains about 11 1/2 miles. Turn right onto Decker School Road and follow it 1 1/2 miles to a deadend. Park your car along the side of the road. To reach Nicholas Flat from Pacific Coast Highway in Malibu, take Decker Road (Highway 23) inland, or north, several miles and then turn left on Decker School Road and follow the directions given above.

Step by step:

1. From the end of Decker School Road, proceed around the old gate on the signed Nicholas Flat Trail. The first part of the hike goes through an oakwoodland paralleling the stream on its way to the pond. After a few hundred yards, the road rises above the woodland and enters a forest of ceanothus that soon becomes an oak savanna.

2. A little further on there is a trails junction. Turn right and follow the trail that crosses the stream. About 50 yards past the stream, the trail comes to another junction. Stay on the left fork of the signed Nicholas Flat Trail as it follows the edge of the meadow. Before too long you will be at the pond.

3. The trail goes along the pond's west bank toward an open oak-woodland at the south end of the pond. (Along the way you will pass by another fork in the trail that heads north up a small grade. This is the way to Leo Carrillo State Beach via the Backbone Trail.) From the woodland the trail narrows to a footpath that leads up to a wall of gigantic boulders that effectively keeps the pond from spilling into San Nicholas Canyon on the other side of the ridge.

4. After enjoying the dramatic views of the canyon and the shade of the oaks around the pond, you can return the way you came.

Circle X Ranch Site

Circle X Ranch site contains the highest parts of the Santa Monica Mountains - Sandstone Peak at 3,111 feet - as well as some of the wildest. Chosen for national recognition area status, primarily for its diverse and outstanding examples of Mediterranean plant communities, Circle X is like a jewel in the crown of the park lands that make up the Santa Monica Mountains National Recreation Area.

The drive through the backcountry to get to Circle X Ranch is by far the most splendid of the scenic routes through the Santa Monicas. At these remote and high elevations of the mountains, there are a number of plants that are endemic to the area. These plants once flourished throughout the range but now are found only in these areas. One of the most fascinating is the red shanks, a large shrub with unique peeling bark that is found here, in Palm Springs and in northern Baja. This ruggedly beautiful area has a rich history of habitation by Native Americans as evidenced by the midden (trash heap) area off a spur trail near the Happy Hollow Camp ground.

Circle X Ranch vista.

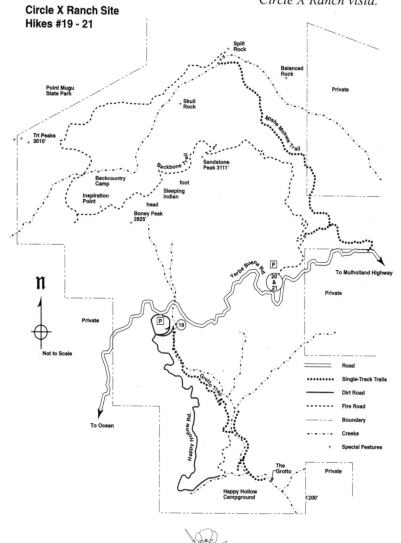

Circle X Ranch Site
Hikes #19 - 21

Split
Rock

Balanced
Rock

Point Mugu
State Park

Private

Skull
Rock

Mishe Mokwa Trail

Tri Peaks
3010'

Backbone Trail

Sandstone
Peak 3111'

Backcountry
Camp

foot

Sleeping
Indian

Inspiration
Point

head

Boney Peak
2825'

Yerba Buena Rd

P

20
&
21

To Mulholland Highway

Private

𝔫

P

19

Private

Not to Scale

Grotto Trail

Happy Hollow Rd.

To Ocean

The
Grotto

Private

1200'

Happy Hollow
Campground

	Road
••••••	Single-Track Trails
—	Dirt Road
- - -	Fire Road
-·-·-	Boundary
-··-··-	Creeks
*	Special Features

19. Grotto Trail

Difficulty: Moderate
Length: 3 1/2 mile round-trip
Elevation: 800 foot loss/gain
Type of trail: single-track trail
Kid appeal: * * * * *

The Grotto Trail crosses several plant communities as well as the west fork of the Arroyo Sequit as it leads down to Happy Hollow Campground. As you boulder hop down into the grotto area you can explore hidden caves, formed by massive landslides eons ago, that are found at the park's lower elevations.

Directions: Circle X Ranch is 5 1/2 miles from Pacific Coast Highway on Yerba Buena Road in Malibu. For the scenic route from Highway 101 in Westlake Village, take Westlake Boulevard (Highway 23) south; take it through the mountains several miles and it joins Mulholland Highway at the county line. They continue together for a short way; stay to the right, on Mulholland Highway at the unmarked intersection and take it to Little Sycamore Canyon Road. Turn right onto Little Sycamore and it will become Yerba Buena Road as you cross the county line once more. Proceed on Yerba Buena about 5 miles to the signed Circle X Ranger Station on the left side of the road.

Step by step:

1. From the ranger station follow Happy Hollow Road down to the group campground. The signed Grotto Trail begins by the outhouse on the far side of the parking area. As the trail starts down the canyon it passes through a woodland grove mixed with pine trees planted when Circle X was a Boy Scout Camp. Access to the park's other trails is marked by a wooden sign just before the Grotto Trail crosses the west fork of the Arroyo Sequit. From here, the trail crosses a meadow before leading downhill in earnest through a rare forest of red shanks chaparral.

2. At the bottom of the canyon the Grotto Trail comes to the service road for Happy Hollow Campground. It is your choice to follow the path straight ahead over the wooden bridge or to follow the path past the outhouse: both will lead you over the creek to the continuation of the Grotto Trail.

By following the signed paths through this primitive camp-ground you will soon come to an old wooden pump house.

3. The Grotto itself is located just past the pump house. The climb down into the Grotto is a rugged one consisting of major rock scrambling along a steep creekbed. You can explore this dramatic area until you come to the park boundary in about a half mile, then retrace your steps back through the grotto and campground.

Optional: It is possible to drive into Happy Hollow on a rough park service road. However, due to extreme mud, the road is often closed following a rainstorm. Check with the rangers.

20. Sandstone Peak via Backbone Trail

Difficulty:	Moderate
Length:	2 mile round-trip
Elevation:	950 feet gain/loss
Type of trail:	fire road
Kid appeal:	* * * *

Sandstone Peak, the eastern summit of Boney Mountain, offers hardy trail users one of the most far-reaching 360 degree vistas in Southern California. At 3,111 feet above sea level, it is the highest spot in the Santa Monica Mountains. On clear days, the Pacific Ocean and the Channel Islands dominate the view to the west, the Oxnard Plain is to the immediate north and the Topa Topa Mountains are visible beyond that. Closer at hand are Westlake Village and the Conejo Valley as well as the rest of the Santa Monicas with the eastern terminus, Griffith Park, also visible.

Severely ravaged in the firestorm of 1993, this area has been a showcase of fire-following wildflowers. Mountain peaks that have been covered with dense chaparral for decades and then denuded by these fires were turned into fields of blooms. The beautiful, rugged mountaintop known as Boney Mountain has long been a source of inspiration to the human eye. Archeological studies indicate that native peoples occupied this site. The service organization that set up the original camp here was known as the Exchange Club, hence the name Circle X. Later the camp was leased to the Boy Scouts of

America and its cause was championed by W. Herbert Allen. Through Allen's hard work, and even personal funds, the camp thrived until the late 1970s.

To commemorate his heroic efforts in this matter, a grassroots movement tried to rename Sandstone Peak in his honor. That's why the bronze plaque at the top of the peak reads Mount Allen. Interestingly enough, this rocky promontory is not made out of sandstone as its name implies, but formed by volcanic processes.

Directions: Follow the general directions given for Circle X Ranch site - Hike #9, except watch for and use the large signed parking lot for the Backbone Trail located about one mile east of the ranger station. If you are traveling from Westlake Village, you will come to the parking area before the ranger station. If you are traveling from the ocean, the parking area is a mile past the ranger station.

Step by step:

1. The signed Backbone Trail begins at the north end of the parking lot, a little beyond the information kiosk. Walk around the locked gate that closes the fire road to vehicle traffic. The trail follows the road uphill all the way to the summit.

2. After 1/3 of a mile of steady uphill walking the single-track Mishe Mokwa Trail separates from the Backbone Trail to head down into Carlisle Canyon. For today's hike continue up the fire road.

3. About a mile from the trailhead, the fire road widens as it overlooks the flat top of Boney Mountain to the immediate north. Watch for the signed "Hiking Only" trail to Sandstone Peak. The trail starts as a set of wooden steps that lead a few hundred yards from the trail junction up a steep, narrow indistinct path to the summit. There is a plaque honoring W. Herbert Allen as well as a sign-in registry. Looking over the registry we know that a number of families with children in tow have made it safely to this tiny space at the top of the mountain. After enjoying the tremendous views all around, return to your car the way you came.

21. Mishe Mokwa Trail

Difficulty: Easy/moderate
Length: 3 1/2 mile round-trip
Elevation: 250 foot loss/gain
Type of trail: single-track trail
Kid appeal: * * * *

The Mishe Mokwa Trail is one of the most beautiful in the Santa Monica Mountains high country. Part of the Backbone Trail system that leads to Sandstone Peak, this single-track hiking and equestrian trail explores the eastern side of Boney Mountain. This trail makes an exciting family adventure as it goes through two unique ecosystems and has beautiful views of the headlands of Carlisle Canyon.

There are several notable geologic features that can be viewed from this trail. One is Echo Cliffs that make up the north rim of the canyon. These remarkable formations are riddled with crevices and caves that provide homes for many wildlife species from mice to mountain lions. It is truly moving to hear the roars of a lonely lion echoing from the cliffs. Precariously perched on the western end of the cliffs is monolithic Balanced Rock. And Split Rock at trail's end, under a dense canopy of oak and bay trees, is a wonderful place to explore as the trail passes between the "splits".

Directions: *Same as for Sandstone Peak/Backbone Trail - Hike #20.*

Step by step:

1. The signed Backbone Trail begins at the north end of the parking lot, a little beyond the information kiosk. Walk around the locked gate that closes the fire road to vehicle traffic. The trail follows the road uphill all the way to the summit.

2. After 1/3 of a mile of steady, uphill walking, the single-track Mishe Mokwa Trail separates from the Backbone Trail to head down into Carlisle Canyon. Take the single-track trail as it heads east, down into the canyon. The trail also is marked by a large boulder painted yellow.

3. There is another trails junction in about 200 yards. Stay to the left as the Mishe Mokwa Trail heads north, deeper into the canyon and continue to watch for yellow paint marks to show

the way. (The trail joining from the right is another segment of the Mishe Mokwa Trail originating from a small parking area off Yerba Buena Road about 1 mile east of the signed Backbone Trail parking lot where you left your car.)

4. The trail continues along the canyon rim for another mile or so as it enters into upper Carlisle Canyon. It dips down aggressively into the canyon a few hundred yards before the shady Split Rock picnic area. The Mishe Mokwa Trail ends at the junction with an old fire road that is part of the Backbone Trail about 100 yards upstream from Split Rock. This is the turn around point for today's hike. After enjoying the scenery in the oak grove, return the way you came.

22. Arroyo Sequit Natural Area

Difficulty: Easy
Length: 2 mile loop
Elevation: 250 foot loss/gain
Type of trail: mostly single-track trail
Kid appeal: * * * *

Arroyo Sequit Natural Area isn't really part of Circle X Ranch. However, the plan is to connect it to the National Park via the Backbone Trail, once it is completed through this remote part of the Santa Monica Mountains. Starting in a high grassland meadow, this comfortable single-track trail makes an excellent family adventure as it leads down into the canyon to explore the headwaters of the east fork of the Arroyo Sequit.

The loop hike through this 155-acre park showcases several native plant communities, as well as providing some stunning vista points of Boney Mountain. Limited to weekend hiking use, the plant communities in this natural area are relatively undisturbed. Likewise, the trail itself seems more primitive than most, yet it is easy to follow as well as walk.

***Directions:** Arroyo Sequit is about 6 miles from Pacific Coast Highway at 34138 Mulholland Highway, Malibu. Follow the directions to Circle X Ranch from Westlake Village (see Hike #19) to the intersection with Mulholland Highway and Little Sycamore Road. This time go straight on Mulholland and the*

park entrance is on the left side of the road just before a large bend in the road. Look for the mailbox with the number 34138 and follow the driveway into the park.

Step by step:

1. From the parking lot, follow the paved road about 1/4 mile to the ranch house. Take the signed trail that begins at the south end of the driveway leading through the meadow. The grassland gives way to chaparral at the rim of the canyon. This is a natural vista point with great views of Boney Mountain juxtaposed with the huge government satellite dishes in the next canyon. There is a trails junction here, but continue straight ahead on the path leading down into the canyon.

2. This is the steep part of today's hike, with several switchbacks making the descent a relatively easy one. At the canyon bottom is another trails junction; stay on the left fork as it heads toward the seasonal creek. In about 100 yards, the trail crosses the creekbed and heads upstream along the bottom edge of the canyon's far wall.

3. There are a few more seasonal stream crossings before coming to the back of this box canyon. At this end, the climb out is not as steep as the way in. And after crossing the stream one last time, the footpath gives way to a graded roadbed making for some easier hiking.

4. Once out of the canyon there is another meadow and the footpath through it leads back toward the vista point in Step 3. For our loop tour, proceed straight ahead and slightly downhill along the dirt road toward the oak-shaded picnic area. From here follow the trail back to the paved road and then return to your car the way you came.

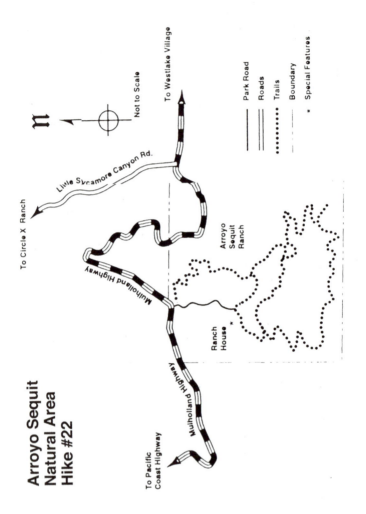

Arroyo Sequit
Natural Area
Hike #22

N

Not to Scale

To Westlake Village

Little Sycamore Canyon Rd.

To Circle X Ranch

Mulholland Highway

Arroyo
Sequit
Ranch

Ranch
House
*

Mulholland Highway

To Pacific
Coast Highway

Park Road
Roads
Trails
Boundary
* Special Features

Chapter Two

The Conejo Valley

Paradise Falls

HIKES AT A GLANCE 3
The Conejo Valley

	DIFFICULTY	FACILITIES	FEATURES	FEES	OPEN TO
WILDWOOD PARK					
Big Sky Trailhead		at Meadows Cave			
23. Indian Creek Trail	easy	w, t, p	shady stream, nature center, caves	no	H, C, E, D
24. Moonridge Trail	moderate	w, p	waterfall, wildflowers	no	H, C, E, D
25. Mesa Trail	easy	none	wildflowers, Lizard Rock	no	H, C, E, D, S
26. Santa Rosa Loop Trail	moderate	none	views from Mountclef Ridge	no	H, C, E, D
27. Mountain Bike Ride	easy/strenuous	w, t, p	overview of park, Paradise Falls	no	H, C, E, D
Fort Wildwood Trailhead		at trailhead			
28. Wildwood Nature Trail	easy/moderate	w, t, p	Fort Wildwood, playground	no	H, C, E, D
LOS ROBLES TRAIL		at Arts Council Center & Community Parks			
29. Oak Creek Canyon Whole Access Trail	easy	w, t, p	oak woodland, whole access trail	no	H, C, E, D, S
30. Los Padres Trail	easy/moderate	none	views, woodlands, wildflowers	no	H, C, E, D
31. Triunfo Canyon Trail	moderate	w, t, p	views, playground at park	no	H, C, E, D
32. White Horse Canyon Trail	easy	none	views of lake, wildflowers	no	H, C, E, D
33. LRT Mountain Bike Ride	strenuous	none	ridgeline views, rigorous workout	no	H, C, E, D

FACILITIES: w - water, t -toilet, p - picnic table, rs - ranger station, c - camping, m - marina, s - store
OPEN TO: H - hikers, C - cyclists, E - equestrians, D - dogs on leash, S - stroller/handicap access

Driving Map of Conejo Valley
Hikes #23 - 30 and #33 - 38

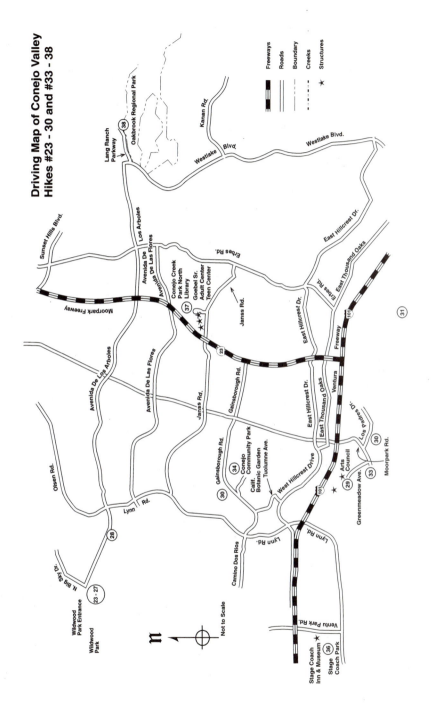

Freeways

Roads

Boundary

Creeks

Structures

Oakbrook Regional Park

Lang Ranch Parkway

38

Kanan Rd.

Westlake Blvd.

Westlake Blvd.

Sunset Hills Blvd.

Los Arboles

Avenida De Los Arboles

Avenida De Las Flores

Erbes Rd.

East Hillcrest Dr.

East Thousand Oaks

Moorpark Freeway

Conejo Creek Park North

Library

Goebel Sr. Adult Center Teen Center

37

Janss Rd.

East Thousand Oaks

Erbes Rd.

101

31

23

Avenida De Los Arboles

Avenida De Las Flores

Janss Rd.

Gainsborough Rd.

East Hillcrest Dr.

Ventura Freeway

Olsen Rd.

Gainsborough Rd.

Conejo Community Park

Calif. Botanic Garden

Tuolume Ave.

34

East Thousand Oaks

East Hillcrest Dr.

West Hillcrest Drive

Los Padres Dr.

30

Arts Council

29

33

30

Moorpark Rd.

Greenmeadow Ave.

Lynn Rd.

28

N. Big Sky Dr.

23 - 27

Wildwood Park Entrance

Wildwood Park

Camino Dos Rios

Lynn Rd.

101

Lynn Rd.

Ventu Park Rd.

Stage Coach Inn & Museum

Stage Coach Park

36

n

Not to Scale

81

Wildwood Park

Wildwood Park is like a treasure chest filled with outdoor recreational opportunities. The park's extensive trail system was designed to be used daily by hikers, cyclists and equestrians giving Conejo residents a variety of ways to enjoy nature right in their own backyard. Wildwood also acts as a buffer zone between development and wildlife corridors throughout the area.

To outdoor enthusiasts familiar with the area, Wildwood Park conjures up many images: the falls deep within the canyon, spring wildflowers dotting the mesa and the rugged form of Mountclef Ridge. Wildwood Park is an outstanding example of an "urban wilderness" area. Located within Thousand Oaks' city limits, Wildwood's multi-use trail system provides a safe, traffic-free venue for nearly every level of expertise.

Directions: *Wildwood Park is at the western end of Avenida de Los Arboles in Thousand Oaks. From Highway 101, take Lynn Road north to Arboles and turn left. To get to the park's main entrance, stay on Arboles until it ends at Big Sky Drive. Make a U-turn around the center divider and the park entrance is on the right side of the road. This is the Big Sky Drive Trailhead at parking lot #1.*

Facilities: There are picnic facilities with tables, drinking water and portable toilets in several locations throughout the park. Wildwood Park is open daily from 7am to sunset, weather permitting. The parking lots are open from 8am to 5pm.

What you'll see: The park's 1,700 acres of open space is a natural sanctuary for many species of wildlife. Wildwood's unique land forms and diverse plant communities provide excellent habitat: food, water, shelter and space for these creatures. Some species are seen much more readily than others. Mule deer are frequently spotted in Box Canyon and other more remote areas of the park. Some nocturnal animals such as coyotes and bobcats are rarely seen, but signs of their presence can be found throughout the park in the form of scat and tracks along the trail. The park is also home to more than 40 species of birds including red-tailed hawks, scrub jays, roadrunners and quail, to mention a few. See how many you can spot.

Vegetation and land forms within the park are quite diverse. Rugged Mountclef Ridge is a stunning example of the massive volcanic eruptions that occurred in the region about 30 million years ago. In contrast, sedimentary rock was left behind in the eastern end of the park some 15,000 years ago during the last Ice Age when a large lake covered most of what is now the Conejo Valley. Deep within the canyon, Paradise Falls acted much like the drain in a bathtub as the lake eventually drained into Santa Rosa Valley and then onto the ocean. The park's biodiversity and dramatic land forms made it an excellent "on location" spot for Hollywood filmmakers from the 1930's through the 1960's. Television westerns such as "The Rifleman", "Wagon Train" and "Gunsmoke" also were filmed here.

Rugged beauty in our backyard

Overview of
Wildwood Park Trail Guide
Hikes #23 - 28

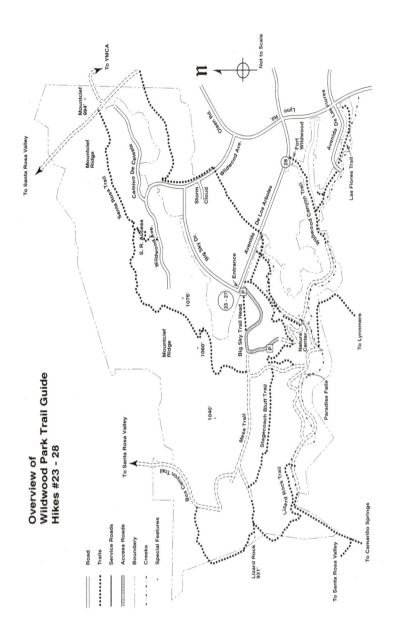

Road
Trails
Service Roads
Access Roads
Boundary
Creeks
Special Features

N

Not to Scale

To Santa Rosa Valley

To YMCA

Mountclef 994'

Mountclef Ridge

Santa Rosa Trail

Camino De Celeste

S. R. Access

Wildwood Ave.

Olsen Rd.

Wildwood Ave.

Storm Cloud

Big Sky Dr.

Avenida De Los Arboles

Lynn Rd.

Fort Wildwood

Avenida De Las Flores

28

Maywood Canyon Trail

Las Flores Trail

Entrance

23 - 27

P

1076'

Mountclef Ridge

1060'

Big Sky Trail Head

P

Nature Center

Paradise Falls

To Lynmmere

1040'

Mesa Trail

Stagecoach Bluff Trail

Box Canyon Trail

To Santa Rosa Valley

Lizard Rock Trail

Lizard Rock 931'

To Santa Rosa Valley

To Camarillo Springs

To Santa Rosa Valley

84

23. Indian Creek Trail

Difficulty: Easy
Length: 1/2 mile to Meadows Cave Nature Center
Elevation: 150 foot loss/gain
Type of trail: single-track trail
Kid appeal: *****

The easy Indian Creek Trail is an excellent way to explore several primary plant communities in the park: grassland, chaparral and riparian. It is also the shadiest and shortest way to the Meadows Cave Nature Center and picnic area that is the hub of the park's trail system off the main Wildwood Canyon Trail. The two creek crossings add a real sense of adventure to this hike into Wildwood Canyon.

Directions: See Wildwood Park introduction on page 82.

Step by Step:
1. From the Big Sky Trailhead, walk east along the footpath next to Avenida de Los Arboles toward Lynn Road. Follow the signed Moonridge Trail down several wooden steps into the park. In about 25 yards, take the signed Indian Creek Trail as it splits off to the left to follow Indian Creek to the canyon floor. The descent begins under a canopy of oaks and sycamores that soon give way to the chaparral along the canyon's rim.
2. Some crude wooden steps and a couple of switchbacks bring you to the creek where the trail skirts the water by hugging the canyon's steep, rocky walls. The trail widens as you stroll along the nearly level, densely wooded bottom lands. Before long, the trail crosses the creek on some man-made stepping stones. Then it's up a few steps as the trail continues through the woodland.
3. Along the way, Indian Creek picnic area is a secluded pocket off the trail under willow trees overlooking the stream. Follow the trail to the Meadows Cave Nature Center and picnic area. There are several well-shaded picnic tables near the creek. Be sure to check out the man-made nature center and the natural "Indian cave" on the opposite side of the creek.
4. Paradise Falls is another 1/2 mile down Wildwood Canyon Trail: watch for signposts to point the way. After exploring, return to your car the way you came on the Indian Creek Trail.

24. Moonridge Trail

Difficulty: Moderate
Length: 1 1/2 miles to waterfalls
Elevation: 300 foot loss/gain
Type of trail: mostly single-track trail
Kid appeal: **

The Moonridge Trail is a moderate hike through the chaparral from the Big Sky Trailhead to Paradise Falls. Following the contour of the canyon wall, this single track trail reveals a unique plant community with wonders all its own. The fragrant chaparral plants become a springtime mosaic as the subtle colored blooms of the sages contrast with the riotous monkey flowers and stately yucca plants. Be sure to look for California peonies growing in the shade of the larger shrubs along the trail. And once near the falls, Conejo buckwheat and Conejo dudleya grace the gorge's sheer volcanic cliffs.

Directions: *See Wildwood Park introduction on page 82.*

Step by Step:
1. The Moonridge Trail begins at the south end of the Big Sky Trailhead where a short flight of wooden steps brings you to the single track trail. Take the path to the right as it heads west into the park paralleling the route of the service road. (To the left the path leads around a small gully and through the grassland to join the Indian Creek Trail.)
2. Overlooking the ravine cut by Indian Creek, the trail trends downward into Wildwood Canyon. From parking lot #2, a half mile of easy walking through the chaparral brings you to the park service road.
3. The signed Moonridge Trail crosses the road and continues on the other side where the path follows the up and down contour of the ridge for another half mile before joining with another park service road, the Teepee Overlook Trail.
4. Take the road to the left as it heads toward the Teepee. (To the right the road leads to the Stagecoach Bluff Trail on its way back to the mesa.) From the Teepee Overlook, continue into the park by heading west on the roadway down a long slope toward Paradise Falls.

5. About a hundred yards down the hill, the single track Moonridge Trail leaves the road on the left to create a shortcut through the chaparral. In a short time, our trail joins the Wildwood Canyon Trail and several sets of wooden steps lead to the bottom of Paradise Falls. The service road joins with Wildwood Canyon Trail about a hundred yards beyond the falls and continues along the bottom of the canyon following the stream to the park's group picnic facilities in Oak Grove and Skunk Hollow.

6. After enjoying your view of the falls you can return to the Big Sky Trailhead the way you came or by a number of other routes. Take Wildwood Canyon Trail to the Meadows Cave Nature Center and from there either take the park service road across the mesa or take the Indian Creek Trail to get back to the trailhead.

25. Mesa Trail

Difficulty: Easy
Length: 1 1/2 miles to Lizard Rock
Elevation: 100 foot gain/loss
Type of trail: mostly fire road
Kid appeal: ****

At the base of rugged Mountclef Ridge, the Mesa Trail is perhaps Wildwood's easiest trail. There is very little change in elevation as you make your way across this treeless plain to Lizard Rock where there is a breathtaking panoramic view of the Conejo and Santa Rosa Valleys, the Channel Islands, the Santa Monica and Topa Topa Mountains as well as the Simi Hills.
***Directions:** See Wildwood Park introduction on page 82.*

Step by Step:
1. The Mesa Trail begins at the north end of the Big Sky Trailhead, near the information kiosk. This wide, single track trail heads north over a small grade as it leads into the park. In a short way the trail joins with the service road as they go into the park together and continue straight across the grasslands toward the Lizard Rock formation. (The service roads cuts away toward the lip of the canyon and the park's inner parking lot: this will be part of the return route.)

2. At the end of the mesa there is a fork in the road. Take the signed Lizard Rock Trail to the left toward the promontory. (The road to the right leads out of the park through Box Canyon toward Santa Rosa Valley.) The trail becomes somewhat narrow and indistinct as you approach the vista point and caution is advised. Several paths lead from the trail to the top of the rock.

3. After enjoying the terrific view, retrace your steps back to the Mesa Trail. The single-track Stagecoach Bluff Trail leaves the main Mesa Trail to follow the rocky lip of Wildwood Canyon back toward the trailhead. This trail is noted for its outstanding showcase of spring wildflowers: shooting stars, mariposa and chocolate lilies, ground pink, blue dicks, goldfields and tidytips are a few to watch for as you make your way along this rock - strewn ridge.

4. When the trail reaches the service road, turn left and follow it back to the Mesa Trail. From here, turn right and head east back along the roadway and it is about 1/2 mile back to the Big Sky trailhead.

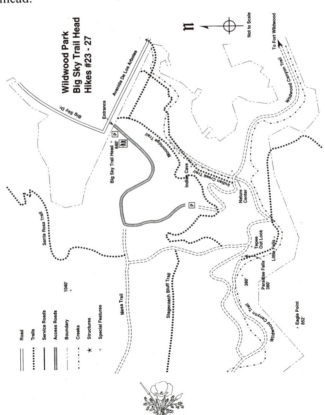

26. Santa Rosa Loop Trail

Difficulty: Moderate
Length: 4 1/2 mile loop trip
Elevation: 300 foot gain/loss
Type of trail: mostly single-track trail
Kid appeal: ***

The Santa Rosa Trail explores the rugged volcanic out-cropping known as Mountclef Ridge. In a number of places the trail passes around or over huge boulders made up of welded conglomerates formed millions of years ago. This hike along the ridgeline provides some tremendous views of both the Conejo and Santa Rosa Valleys. Long distance views of the Santa Monica Mountains to the south and the Santa Susanna and Topa Topa Mountains to the north make this hike well worth the effort.

Directions: *See Wildwood Park introduction on page 82.*

Step by Step:

1. The single track Santa Rosa Trail begins near the information kiosk at the Big Sky Trailhead. Take the trail as it heads west over a small grade on its way into the park. Before long the foot path joins with the service road to cross the mesa. Watch for the signed Santa Rosa Trail to the right of the road just past the locked gate that closes the mesa to vehicular traffic. The trail then turns north toward the summit of Mountclef Ridge. After a large switchback, the trail heads east toward a saddle defined by the two highest peaks.

2. At the top of the ridge, the signed Santa Rosa Trail continues east as it drops down to follow the contour of the north facing slope. This route gives you a bird's-eye view of the developing Santa Rosa Valley. A half mile of easy walking brings you to the spur trail that provides trail access from Wildwood Avenue. Continue straight ahead on the main trail.

3. Following the perimeter of the ridge-top community for another half mile, the trail brings you to a service road. Turn right on the road, continuing to skirt the houses; there is another trail access point at the end of the Camino de Celeste cul-de-sac. Check the trail sign and continue south on the

Lower Buttes Trail. As you approach the now defunct California Lutheran University Equestrian Center, stay to the right and take the single track trail that heads west between the buttes.

4. The Lower Buttes Trail returns to Wildwood Avenue; turn left and follow along the sidewalk about 1/4 mile. Cross the street at Sundance Street and return to Avenida de Los Arboles via the decomposed granite jogging path that runs along the arroyo. Once back at Arboles, turn right and the park is about 1/8 mile away.

27. Wildwood Mountain Bike Loop

Difficulty: Moderate with some easy & some steep spots
Length: 8 mile loop ride
Elevation: 400 foot loss/gain
Type of trail: mostly fire road
Kid appeal: ****

Besides hiking, Wildwood Park offers a traffic-free venue for mountain bike riding. While the Mesa Trail is a short, easy bike ride, today's 8-mile loop through the park is a more challenging off-road bike tour for riders with some previous experience. There are some strenuous uphill portions of the trail and some technical riding is required. (Technical mountain bike riding is when the trail is too steep, either going up or coming down, or rocky and Cathy has to push her bike.) This bike route takes in most of the more remote areas of the park as well as some of Santa Rosa Valley. Early morning and late afternoon rides often provide sightings of animals such as coyotes and mule deer foraging in and around the park.

Directions: *See Wildwood Park introduction on page 82.*

Step by Step:
1. From the Big Sky Trailhead, follow the service road into the park and across the mesa. As you approach Lizard Rock, watch for the signed Box Canyon trail on the right. Take this wide, washboard service road north as it leads to the valley farmland along Santa Rosa Road.

2. Once you reach the road, take it to the left about a half mile to the service road for the City of Thousand Oaks' Hill Canyon Sewage Facility. Turn left again and follow the service road about a mile to the plant. When you get there, look for the signed, single track trail along the canyon's north side that leads back into the park. This is part of the Lizard Rock Trail.

3. Stay on the single track trail as it goes around the back of the plant and crosses the creek. Past the creek there is a trails junction. Take the Wildwood Canyon Trail along the dirt service road to Paradise Falls. The trail crosses the stream several more times as it winds its way through the group picnic areas along the canyon bottom before coming to the falls.

4. At this point, there are some wooden steps up to where the service road shrinks down to a single lane trail again as Wildwood Canyon Trail hugs the cliff above the falls. Around the cliff the trail picks up on the service road once again.

5. Follow the service road next to the creek as it heads toward the Meadows Cave Nature Center. Take a few moments to explore this area and rest for the climb out of the canyon. To get back to your car, take the signed service road up to parking lot #2. Then follow the service road back to the park entrance at Big Sky Drive.

28. Fort Wildwood Nature Trail

Difficulty: Easy with some steep wooden steps
Length of trail from Fort:

               1 mile to Nature Center
               1 1/2 miles to Paradise Falls
               3 miles to Skunk Hollow

Elevation: 150 foot loss/gain
Type of trail: fire road, single-track trail
Kid appeal: *****

      There are a number of community trailheads located around Wildwood Park that provide easy access to the excellent system of trails through the park. One of my favorites is the Fort Wildwood Trailhead that heads up the Wildwood Canyon Trail beginning at Fort Wildwood and going to the Meadows Cave Nature Center about a mile away.

Today's outing at Fort Wildwood is a great one for families. Fort Wildwood Community Park, with its playgrounds and grassy lawns, is one of the more inviting trailheads into Wildwood Canyon. A few minutes of play time at the fort adds a real sense of adventure to the hike. The stream crossings also enhance the kid appeal.

Directions: This trailhead is behind Wildwood Elementary School on Avenida de Los Arboles. Once you turn onto Arboles from Lynn Road, look for the school on the left side of the street approximately two blocks from the intersection. If the school's parking lot is full there is additional parking available near the softball diamonds at Wildflower Playfields. Use the sidewalk leading through the underpass to cross Arboles and get to the park.

Step by Step:

1. The unmarked, but clearly visible Wildwood Canyon Trail begins next to a dense hedgerow on the northwest side of the fort. Walk along the hedgerow for 200 yards or so until you come to a signed trailhead. From here the trail starts to head down into the canyon via a rough wooden stairway. There are a few more sets of steps as well as many vista points of the stream below.

2. Near the bottom of the canyon the trail levels out and crosses a small bridge over the North Fork of Arroyo Conejo. A little further down the trail there is another stream crossing. This time you must do a little rock hopping to avoid getting your feet wet. On the opposite side of the second ford our single track trail joins up with the park service road and continues west on to Meadows Cave Nature Center about 1/2 mile away.

3. The Meadows Cave and picnic area marks the end of today's hike. After enjoying the shady picnic area, you can return the way you came.

Palefaces rest at Ft. Wildwood.

Los Robles Trail

The Los Robles Trail System is an extensive network of multi-use trails that explores the open space ridgeline separating the Conejo Valley from Hidden Valley. Following the footpaths of the Chumash Indians native to this area, this ancient transportation route connects the communities of Newbury Park, Thousand Oaks and Westlake Village by means of a non-motorized trail. The Los Robles Trail System also is a vital lynch-pin to the overall countywide trail system that links the inland valleys to the sea via Point Mugu State Park and the Backbone Trail through the Santa Monica Mountains. Numerous trailheads along the route make for easy residential access.

Consisting mainly of open chaparral, this east-west running ridge also features several outstanding oakwoodland and savanna areas. Springtime outings in the fragrant chaparral covered hillsides offer beautiful displays of wildflowers as well as some of the best bird's-eye views of the two valleys short of a helicopter ride. The oakwoodlands that fill the canyon bottoms along the various trails extend the invitation to picnic in a cool spot under a dense canopy of leaves.

Directions: *There are several trailheads spread out along the length of the Los Robles Trail System. The 10 mile trail system begins in Westlake Village and ends in Newbury Park as it traverses the ridgeline between the Conejo and Hidden Valleys. Directions to each trailhead are listed with its trail specifications.*

What you'll see: The trail was severely damaged in the Greenmeadow Fire of 1993. Starting in a single flash point behind the Conejo Recreation and Park District's Arts Council Center off Greenmeadow Drive in Thousand Oaks, this wildfire went on to ravage the ridgeline open space and continue on into the surrounding Santa Monica Mountains. Due to the firefighters' heroic efforts the eastern end of the system, around Westlake and Lake Sherwood, was saved and provided much needed habitat for the surviving wildlife. While devastating to human development in Southern California, fire is actually a rejuvenating agent in the chaparral. This is a natural condition in the chaparral and is known as the fire cycle. In the years immediately after a fire, new spring growth will often consist of many wildflowers whose seeds have been dormant for decades awaiting the next round of the cycle. The open space around the Los Robles Trail is one of the main wildlife corridors between the Simi Hills and the Santa Monica Mountains. Animal tracks abound in the dirt along the trail from their nightly forages; see how many you can identify as you walk along. Coyote and deer tracks are common in the chaparral, while raccoons seem to prefer the moister woodlands. Both chaparral and woodland offer good birding possibilities: be sure to watch for hawks, scrub jays, woodpeckers and roadrunners.

The Conejo Valley's namesake

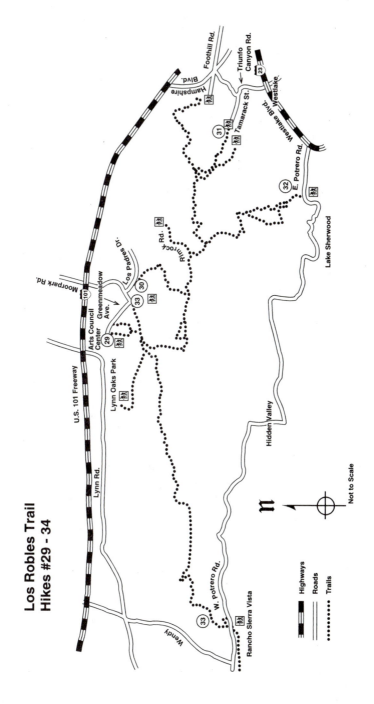

Los Robles Trail
Hikes #29 - 34

Legend:
- Highways
- Roads
- Trails

Not to Scale

95

29. Oak Creek Canyon Whole Access & Loop Trail

Difficulty:	Easy
Length:	1/2 mile round-trip on Whole Access Trail or 3/4 mile loop hike
Elevation:	50 foot gain/loss
Type of trail:	separate use trail single-track trail
Kid appeal:	* * * * *

The Oak Creek Canyon Loop Trail is without a doubt one of my family's favorite hike sites. Starting in one of the best preserved oakwoodlands, this trail serves as one of the main trailheads for the Los Robles Trail System. Under a canopy of oaks, the first quarter mile of the trail starts as a whole-use trail, which is a trail that has been adapted for use by disabled people. The original path has been widened to allow for a split-rail fence to be put down the center of the path. The fence plays a double role in that it separates the foot traffic from the equestrians and cyclists as well as supporting a guide line for sight-impaired trail users.

As always, this short and easy trail provides ample opportunity for nature exploration. Try walking the beginning portion of the trail with your eyes closed using just the guide line to lead you along the path. What do your other senses tell you? Feel the hot sun on your face, listen to the sound that your feet make on the path and smell the aroma of the sage around you.

Directions: This trailhead is next to the CRPD's Arts Council Center, 482 Greenmeadow Avenue, Thousand Oaks. From Highway 101 exit on Moorpark Road south to Green Meadow Avenue and turn right. The road deadends at the center in about 1/2 mile.

Step by step:

1. The trailhead is located at the south end of the parking lot near the restroom and information kiosk. Follow the signs and take the gentle, whole-use trail through the oakwoodland. On the far side of the woodland, the whole-use trail ends but the loop trail continues through the chaparral. At the top of a small grade a few hundred yards down the trail there is a trails junction to the right

that joins the Oak Canyon Loop Trail to the rest of the Los Robles Trail.

2. To continue on the Loop Trail, go straight at the junction as the trail rounds a small hill on its way back to the Arts Council Center. After a quarter mile of easy walking, the trail drops down into another small oakwoodland before returning to Greenmeadow Drive. Once at the street, turn left and follow it back uphill to the parking lot.

30. Los Padres Trail

Difficulty: Easy/moderate
Length: 3 mile round-trip
Elevation: 400 foot gain/loss
Type of trail: single-track trail, fire road
Kid appeal: * * * *

Los Padres Trail begins by meandering through a large oakwoodland. It then joins the main Los Robles Trail to traverse a chaparral covered slope, crossing a highland meadow to the Scenic Vista Loop Trail. The panoramic views of the Conejo Valley and the rugged peaks that rim Hidden Valley from this vantage point are magnificent.

The oak-canopied Los Padres Trail is a wonderful area to explore with small children. Like the nearby Oak Creek Canyon Trailhead, this is the site of many Nursery Nature Walks for young families. A local daycamp has developed an amusing legend of an escaped gorilla from the former Jungle Land. The aptly named Uncle Gor lives in a cave off the trail and the site is the destination of numerous camp nature hikes.

Directions: The Los Padres Trail is in Thousand Oaks. From Highway 101 exit on Moorpark Road south to Los Padres Drive and turn left. The trailhead is on the right side of the street across from Woodlet Way.

Step by Step:
1. The first 1/2 mile of trail is a single-track path that criss-crosses a seasonal stream several times as it goes through an oakwoodland. At the end of the canyon, several switchbacks lead sharply up to the fire road and signed Los Robles Trail.

Another 1/2 mile of steady uphill walking through the chaparral brings you to a meadow overlooking Hidden Valley.

2. From the wooden signpost turn right and follow the trail west across the nearly level grassland. On the far side of the meadow the signed Scenic Vista Loop Trail holds some beautiful views of the mountains. Take the trail to the right the last 1/4 mile along the northern ridgeline; it will soon bring you back to the main Los Robles Trail. You can then retrace your steps back down to Los Padres Trail and continue back to your car.

31. Triunfo Canyon Trail

Difficulty: Moderate
Length: 2 1/2 mile loop
Elevation: 400 foot gain/loss
Type of trail: single-track trail, fire road, sidewalk
Kid appeal: * * *

Beginning in beautiful Triunfo Canyon Park, today's hike on the Los Robles Trail explores the open space area surrounding Westlake Village. The spectacular views of Westlake and the mountain peaks behind Hidden Valley are highlighted by this loop hike through the chaparral. As the trail leads to the ridge, it winds through a forest of ceanothus (California lilac) that is very showy in the spring with blue and white flowers everywhere. There also is abundant evidence of area wildlife in the form of scat, tracks and animal trails crossing the hiking path. See how many of these items you can identify as you walk along the Triunfo Canyon Trail.

Directions: *Triunfo Canyon Park in Westlake Village is the trailhead for today's hike. From Highway 101 in Westlake, exit on Hampshire Road and turn south. Take Hampshire Road to Triunfo Canyon Road and turn right. After a couple of four-way stops, turn right again onto Tamarack Street and use the far entrance at the north end of the park.*

Step by step:

1. The signed Triunfo Canyon Trail begins near the map kiosk at the north end of the of the park. The trail starts by winding

through the open oak savanna, following the contour of the hillside. Before long the path enters the chaparral as it climbs toward the ridgeline.

2. Three switchbacks lead you up the steepest part of the grade to an outstanding vista point. The trails junction at the first ridge marks the halfway point of this loop hike. Proceed uphill to the left on the main Los Robles Trail and as you crest the next ridgeline you will come to another trails junction. Stay to the left once again to continue the loop back to Triunfo Park about one mile away. From here the trail is downhill all the rest of the way back to the park. At Brookview Avenue you must leave the trail and take the sidewalk a short ways through the neighborhood. At Stonegate Street turn left and take it downhill to Aranmoor Avenue. Once at Aranmoor continue through the park on the perimeter jogging path.

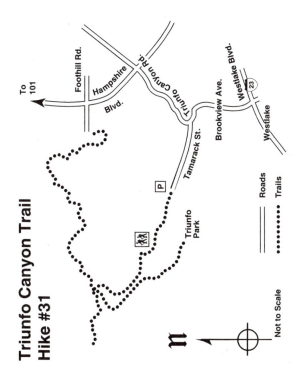

32. White Horse Canyon Trail

Difficulty: Easy
Length: 2 1/2 mile loop hike
Elevation: 450 foot gain/loss
Type of trail: mostly fire road
Kid appeal: * * *

White Horse Canyon Trail is the main trail through the rolling foothills surrounding Westlake Village and Lake Sherwood. This excellent trail is in good condition and provides easy, close-in views of many area wildflowers against the dramatic backdrop of cliffs rimming Hidden Valley.

Walk this trail on sunny spring days and you will notice how brushing against the new growth on the sage plants along the path releases a strong aroma. These fresh smells come from the plants' oils and resins that help them withstand dry chaparral conditions. Be sure to look for star and soap lilies, peonies and California poppies. As the chaparral becomes a mosaic of spring colors notice how the vibrant, salmon colored monkey flowers stand out from the more subtle hues of bulbs and sages.

Directions: *From Highway 101 in Westlake Village, exit on Westlake Boulevard heading south, toward the Santa Monica Mountains. Turn right at the intersection with East Potrero Road and take it about 1/2 mile. Park along the shoulder of the road opposite Foxfield Riding Stables. You will notice the trailhead kiosk on the north side of the wash formed by the Lake Sherwood spillway.*

Step by step:

1. Proceed over the wash to the map kiosk. The signed White Horse Canyon Trail starts as a single-track path heading uphill and to the north as it circles some homes. Near the back of the development the path gives way to a wide fire road. The trail gently switchbacks to the east as it overlooks a canyon before heading north.

2. As you approach the ridgeline there is a trails junction. Turn left and follow the fire road 1/2 mile to a vista of Lake Sherwood. After enjoying the view return to the junction and

continue north once more on the fire road. Another 1/2 mile of easy walking brings you to another trails junction. Take the signed, single-track White Horse Canyon Loop Trail to the left. The trail loops the canyon's back wall to join the fire road once again on the far side of the canyon.

3. Stay to the right at the fire road as it leads sharply uphill to join with the Conejo Crest Trail. There is another trails junction near a large rocky outcropping. To continue today's loop once again stay to the right. This is the main trail back to the trailhead.

33. Los Robles Trail Bike Ride
West Potrero Road to Moorpark Road

Difficulty: Strenuous, serious technical riding
Length: 5 1/2 miles one-way
Elevation: 1,000 foot gain/loss
Type of trail: fire road, single-track trail
Kid appeal: * * * * *

Los Robles Trail is made up of some of the best single-track trails open to mountain bike use in Ventura County. Even though parts of the trail system pass close to residential areas, overall the trail seems quite remote. Another unique facet of this trail is that it is a multi-use trail that actually connects several inland communities to each other as well to the ocean via the trails through Point Mugu State Park. Riding the length of the ridgeline can be a grueling experience for beginning riders and it can even be challenging to experts. But riding the entire ridge is certainly an accomplishment that can long be remembered. As several areas of the trail cross privately owned easements, trail users are requested to respect private property rights along the trail.

Directions: Today's bike ride begins at the western terminus of the Los Robles Trail in the parking lot off Potrero Road. From Highway 101 in Newbury Park, take Wendy Road south toward the mountains. Stay on Wendy until it deadends at Potrero Road. Turn left and take it to the signed Los Robles Trail parking lot on the left side of the road. This large, fenced parking lot is open daily from 9 a.m. to 4 p.m. Water and maps are available at the trailhead.

Step by step:

1. The signed Los Robles Trail begins at the north end of the parking lot. From the information kiosk the trail climbs sharply through the chaparral and then to a grassland before coming to a paved service road. The trail continues west for 1/4 mile of easy riding before coming to a trails junction. Stay to the right and head uphill to keep on the main trail. The downhill route to the left exits the trail system at Deer Ridge.

2. The next couple of miles of riding is made up of some very steep hills as you climb your way to the crest of the ridge. Before too long there is a fork in the road and the trail signs indicate that the Los Robles Trail continues on the left, once more as a single-track trail traversing the north side of the slope. Another 1/3 mile of moderate riding brings you to the private Ventu Park extension road. Cross the pavement as the Los Robles Trail continues to climb the north slope.

3. Four miles from the beginning of the ride, the trail begins its descent. The next mile or so is mostly downhill via some extremely hairpin switchbacks. At the bottom of the hill there is a junction with the signed Spring Canyon Trail going to the Lynn Oaks Park Trailhead. To continue on the Los Robles Trail stay to the right as the trail climbs a small grade to a major trail junction.

4. At this hub, the trail to the left heads to the Moorpark Road trailhead or to the Arts Council Center, both 1/2 mile away. The trail to the right leads uphill toward Westlake Village and Lake Sherwood. Once at the top of the hill the trail goes through the Upper Meadow.

5. At the far end of the meadow the trail splits: To the right, it continues two miles to Lake Sherwood via the White Horse Canyon Trail; To the left, the trail continues to Westlake Village, also about two miles away, via the Triunfo Park Trail.

6. With its numerous neighborhood trailheads, Los Robles Trail can work well for various loop rides. Riding the full length of the trail makes a good "shuttle" trip since there are large parking lots at both ends of Potrero Road and at Moorpark Road as well as at the Arts Council Center.

Please remember to use caution when mountain bike riding on multi-use trails. Watch downhill speeds and blind corners: other trail-users may be closer than you think. And of course always wear your helmet and stay on designated trails, especially because the Los Robles Trail uses easements over private property. As always, be courteous to other trail-users and respect the privacy of homeowners along the pathway.

Community Parks in Thousand Oaks with Hiking Trails

The City of Thousand Oaks has ambitiously sponsored its urban open space program and as a result there are a number of community parks with excellent hiking and nature trails. Some of these trails tie into an overall district system of trails. Most of the sites are left in a natural setting to encourage local native species of plants and animals to thrive. These community parks help reduce the need to drive out-of-town to enjoy nature, and maintain a high level of biodiversity.

Chumash life at Rancho Sierra Vista

HIKES AT A GLANCE 4
Conejo Valley Community Parks

	DIFFICULTY	FACILITIES	FEATURES	FEES	OPEN TO
COMMUNITY PARKS					
34. Conejo Valley Botanic Garden	easy	w, t, p	botanic garden, views, playground	no	H
35. Tarantula Hill	easy/moderate	none	highest spot in the Conejo	no	H, C, D, S
36. Stagecoach Inn Nature Trail	easy	w, t, p	museum, historic buildings, playground	no	H
37. Conejo Creek North	easy	w, t, p	shady stream, playground	no	H
38. Oakbrook Chumash Center	easy	w, t, p	Chumash Cultural Center, woodlands	donation	H

FACILITIES: w - water, t -toilet, p - picnic table, rs - ranger station, c - camping, m - marina, s - store
OPEN TO: H - hikers, C - cyclists, E - equestrians, D - dogs on leash, S - stroller/handicap access

34. Conejo Valley Botanic Garden

Difficulty:	Easy
Length:	1 mile loop trail
Elevation:	100 foot gain/loss
Type of trail:	fire road, garden path
Kid appeal:	*****

Surrounded by the city of Thousand Oaks, the Conejo Valley Botanic Garden offers the community a true urban wilderness. Part of the Conejo Valley Community Park, the Botanic Garden occupies one of the three hilltops that make up this small open space park and provides sheltered habitat for a number of wildlife species. Attracted to the garden's various native shrubs and wildflowers as well as exotic fruit orchard, song birds and butterflies are a welcome addition to walks here.

Strolling down the cultivated garden path is but one way to enjoy your time at this beautiful park. There also is an excellent nature trail that explores the park's less formal side. This single-track footpath has recently been realigned and it is in great condition for a family hike. The nature trail is an enjoyable outing any time of year as it follows a stream through a shady, oak canopied canyon just down the hill from the Botanic Garden.

Directions: *From Highway 101 in Thousand Oaks, exit at Lynn Road and take it north about 1/2 mile to Gainsborough Road. Turn right onto Gainsborough and take it 1/4 mile to the signed Conejo Valley Botanic Garden on the right side of the road. Follow the driveway and park in the large lot near the ball fields.*

Step by step:

1. Follow the dirt jogging path from the parking lot to the entrance of the Botanic Garden. From here follow the main Garden Path uphill, staying to the left on the way to the garden's south entrance. At the sign, continue straight ahead about 100 yards on the unsigned path that leads west along the garden's cultivated perimeter. Again staying to the left, take the signed Nature Trail as it continues west through the chaparral, traversing the easy contour of the slope.

2. After an easy 1/4 mile of level walking, a flight of steep wooden steps brings the path down into the canyon. The narrow trail seems to make a tunnel along the edge of the canyon's north wall as it overlooks the stream below. The nature trail explores the stream along the canyon bottom for about 1/2 mile before leaving the creekbed to climb out of the ravine and returning to the main garden area.

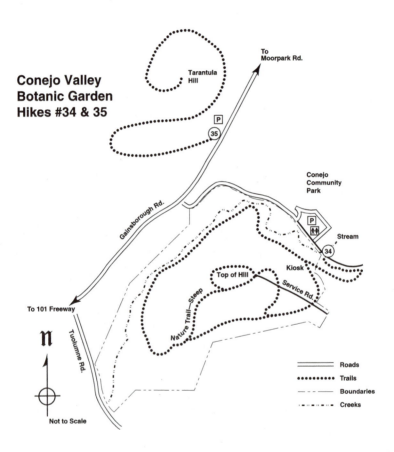

**Conejo Valley
Botanic Garden
Hikes #34 & 35**

Tarantula Hill

To Moorpark Rd.

Conejo Community Park

Gainsborough Rd.

Stream

To 101 Freeway

Tuolumne Rd.

Kiosk

Top of Hill

Nature Trail - steep

Service Rd.

	Roads
••••••••	Trails
– – –	Boundaries
·—··—··—·	Creeks

Not to Scale

35. Tarantula Hill

Difficulty: Easy/moderate
Length: 1 mile round-trip
Elevation: 250 foot gain/loss
Type of trail: paved service road
Kid appeal: * * * *

Part of the same open space park as the Botanic Garden across the street, Tarantula Hill is an excellent spot for a short but steep hike. At the top it has a tremendous 360 degree view of Thousand Oaks. And at 1,057 feet above sea level this is the highest spot in the Conejo Valley. Before Thousand Oaks became too civilized, the north side of Tarantula Hill was a popular hang-gliding site. While that exciting activity is forbidden now, the hilltop is still a very popular walk with the local crowd.

Directions: Follow the directions to the Botanic Garden: Tarantula Hill is directly opposite the entrance to the garden. Park along Gainsborough Road, but do not block the "authorized vehicles only" service road that leads up the hill.

Step by step:

1. Simply follow the paved service road to the water reservoir at the top of the hill. After enjoying the view, return to your car the way you came.

36. Stagecoach Inn Nature Trail

Difficulty: Easy
Length: 1/2 mile loop
Elevation: 50 foot loss/gain
Type of trail: footpath
Kid appeal: *****

The Stagecoach Inn Nature Trail offers hikers a very gentle path that explores a shady little glen located right behind the museum. This little trail is a delight any time of year because of its easy path around the creekbed. At the end of the trail is the site's tri-village complex which depicts glimpses of how early Conejo Valley residents lived. This is a great diversion for those youngsters who may be too active to appreciate the museum tour.

Directions: *The Stagecoach Inn Museum is at 51 South Ventu Park Road in Newbury Park. From Highway 101, take Ventu Park Road and follow it south 1/2 mile to the museum entrance on the right side of the road.*

Step by step:

1. The Nature trail begins behind the Museum off the patio. The path leads down some wooden steps under a canopy of oaks to the stream area. After criss-crossing the creek a few times the trail leads up through the chaparral a short way before returning to the stream.

2. The path takes you over the creek one last time before ending at the tri-village complex.

37. Conejo Creek North Park/Thousand Oaks Library Trail

Difficulty:	Easy
Length:	1/2 mile
Elevation:	50 foot gain/loss
Type of trail:	single-track trail
Kid appeal:	* * * * *

Perfect for family outings, today's hike on this shady little trail is sure to be a delight any time of year. The beautiful woodland at the start of the trail has a peaceful quality all its own. Canopied by huge ancient oak trees, there are outstanding specimens of both live-oak and valley oak along the trail. Live-oaks have small, cup-shaped leaves that remain on the tree throughout the year, whereas valley oaks lose their lobed leaves every autumn.

Creating its own natural habitat, this tiny wetland area between the manicured city park and library complex helps to maintain the Conejo Valley's biodiversity. By providing a place in the local environment for native species of plants and animals to thrive, the community is ensured of their existence for generations to come. Before your walk, be sure to stop at the library for a field guide to identify the birds and plant life that call this wooded ravine home.

Directions: From Highway 23 in Thousand Oaks, exit at Janss Road and head east, toward the Thousand Oaks Library. Turn left into the library driveway, then stay to the right and follow it to the entrance for Conejo Creek North Park.

Step by step:

1. Beginning just inside the park's entrance, the unsigned Nature Trail starts near the Willow Bend Pavilion. From Willow Bend, walk down the driveway toward the library and watch for the trail to start on the right side of the pavement.

2. The trail enters an open oakwoodland following along the streambed of seasonal Conejo Creek. As you near the library, the trail dips down to cross the creek before skirting the edge of the park's manicured lawn.

3. From here the trail returns to the creek as it makes its way through a small open arroyo. In a few hundred yards the trail makes the easy climb out of the wash before returning to the edge of the park's lawn area. It is your choice to return the half mile to the trailhead the way you came or along the jogging path.

38. Oakbrook Chumash Center

Difficulty:	Easy
Length:	3 mile loop
Elevation:	nearly level
Type of trail:	mostly fire road
Kid appeal:	* * * * *

The recently opened Oakbrook Regional Park is the home of Ventura County's new Chumash Interpretive Center. This 427-acre park, just northeast of Thousand Oaks, is made up of beautiful, ancient oakwoodlands and open chaparral covered hillsides. In contemporary times this area was a working cattle ranch. But for thousands of years before it became Lang Ranch, this was the site of a large Chumash Indian settlement.

Leased to the Chumash for eight years, with a 30-year option, the park will become a learning center about the original inhabitants of the Conejo Valley. A 5,400-square-foot museum with displays of ancient artifacts and murals as well as "hands-on" exhibits has been completed and is open to the public. Plans to build a replica of a Chumash village and nature trails through the woodlands are still on the drawing board. Guided tours of the park's interior treasure of ceremonial sites and pictograph caves are available on weekends by museum personnel, Alfred and Philip Valenzuela. Scheduled tours are at 1pm on Saturday and 2pm on Sunday.

Situated in this picturesque setting among the ancient oaks, the museum's displays and artifacts provide a modern window onto the world of the Chumash people. Oakbrook Regional Park also will house Jerry Thompson's Raptor Rehabilitation and Release Program. For a program schedule or more information call the museum at: 805-492-8076.

What you'll need: Donations are accepted at the museum's front desk.

Directions: *The Oakbrook Regional Park is at the end of Lang Ranch Parkway in Thousand Oaks. From Highway 101 in Westlake Village, exit on Westlake Boulevard and head north about 3 miles to Lang Ranch Parkway and turn right. From the 23 Freeway in Thousand Oaks, exit on Avenida de Los Arboles and head east about 3 miles to Westlake Boulevard and turn right. In about 1/4 mile turn left onto the Parkway and follow it to the park.*

Step by step:

1. Meet your guide and fellow hikers at the museum's front desk. After a brief introduction, the hike begins in the oak-woodland behind the museum. The tour follows an old ranch road around the woodland, which will be open to the public for picnicking, for about a quarter of a mile before coming to a locked gate. From this point on, you will need a guide before going deeper into the park.

2. As you near the stream, your guide will tell you about the Chumash village that occupied this tree-canopied canyon. Leaving the road behind, the guide will take you along a narrow path to what is known as "elephant rock." Nearby is a huge, bedrock mortar stone that was used for centuries by Chumash women to grind their bountiful acorn crop into meal before leaching it. See if you can detect a pattern in the grinding holes before your guide reveals their secret.

3. Returning to the road once more, your path leaves the woodland for the chaparral. On the opposite side of the creek, the canyon wall rises steeply to expose a number of small caves on the hillside. A little farther along, and the trail once again leaves the road to follow a narrow path down to the creekbed and to a cave with pictographs that are thought to be several thousands of years old. This is the end of the tour and your guide will bring you back to the museum the way you came.

Legendary Elephant Rock.

Chapter Three

Oak Park to Malibu

Oak Canyon Trail

HIKES AT A GLANCE 5
Oak Park to Malibu Canyon

	DIFFICULTY	FACILITIES	FEATURES	FEES	OPEN TO
OAK PARK					
39. King James Ct. to China Flat	moderate/strenuous	none	views, chaparral, oak savanna	no	H, C, E, D
40. Oak Canyon Park Nature Trail	easy	w, t, p	self-guided nature trail, shady stream, playground	no	H, D
41. Medea Creek Trail	easy	none	jogging path, stream	no	H, C, D
NATIONAL PARK SERVICE SITES					
42. Paramount Ranch	easy	w, t, p	"western town" movie set, shady trail	no	H, C, E, D, S
43. Peter Strauss Ranch/ Lake Enchanto	easy	w, t, p, rs	nature trail, playground, shady trail stream	no	H, E, D
MALIBU CREEK STATE PARK		at Campgrounds & Nature Center			
44. Old Reagan Ranch Trail	easy	t, p	oak woodlands, wildflowers, pond	no	H, E
45. Rock Pool Trail	easy	w, t, p, c	rock pool, nature center	yes	H, C, E, S
46. Malibu Lake Bike Ride	easy	w, t, p, c	lakes, M*A*S*H set, woodlands	yes	H, C, E, S
AGOURA HILLS					
47. Cheeseboro Canyon Bike Ride	easy/moderate	none	oak savanna, wildflowers, views	no	H, C, E, D, S
48. China Flat Bike Ride	strenuous	none	oak savanna, wildflowers, views	no	H, C, E, D,

FACILITIES: w - water, t -toilet, p - picnic table, rs - ranger station, c - camping, m - marina, s - store
OPEN TO: H - hikers, C - cyclists, E - equestrians, D - dogs on leash, S - stroller/handicap access

113

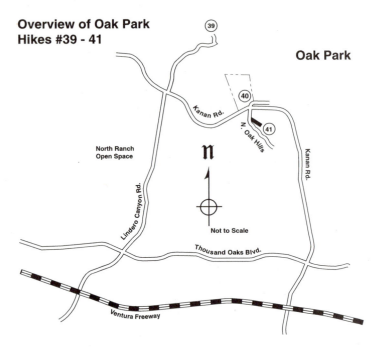

Oak Park to Malibu Canyon

Starting just north of the community of Oak Park, the Simi Peak watershed drains to the ocean via Medea Creek and its connection with Malibu Creek on its way through Malibu Canyon. This was a prime area of Chumash Indian habitation. Attracted to the inland valleys for the abundant game and useful plant materials, the Chumash people and their culture thrived in this area hundreds of years before being "discovered" by Europeans. As it is the only canyon to cut a path through the Santa Monica Mountains, Malibu Canyon served the Chumash as a major transportation route. Even today, Malibu Canyon is the most used commuter route between the valleys and the sea. In a more global sense, Medea Creek connects Oak Park and Agoura to the rest of the world by way of the ocean.

One of the best things about the hiking in this area is the diversity of the trails that are available. Many of the trails offer a mixture of scenery as well as varying degrees of difficulty. There are some trails that provide gentle family excursions into natural areas and there are some trails that can provide a strenuous physical challenge.

114

39. King James Court to China Flat

Difficulty: Moderate/strenuous
Length: 4 mile loop
Elevation: 1,300 foot gain/loss
Type of trail: fire road
Kid appeal: * * *

Hikers who enjoy a challenge in a rugged natural setting will be impressed with the China Flat Loop hike. Starting just north of Oak Park, the trail winds its way up through the open scrub chaparral to the ridgeline formed by the Simi Hills. Unfortunately, at the time of this writing, trail users are to avoid entering the National Park Service's spectacular China Flat area from the King James Court trailhead as there is no inter-agency agreement between the NPS and the Rancho Simi Park and Recreation Department. Hopefully this situation will be rectified soon and the trails between the two agencies will be officially connected.

Directions: From Highway 101, take Lindero Canyon Road and proceed north on Lindero approximately 4 miles. Turn left onto King James Court and park in the cul-de-sac near the signed China Flat trailhead.

Step by step:
1. Proceed on foot through the gate and follow the paved service road steeply uphill to the trail along the rough, dirt fire road. After 1/2 mile of steady uphill walking, there is a trails junction near the second switchback. Leave the road and take the single-track trail to the right, heading east.
2. This transverse trail follows the contour of the south face of Simi Peak and crosses several seasonal streams. A mile of easy walking brings you to another trails junction. Take the trail to the left as it heads uphill toward the ridgeline. This portion of the hike requires some serious climbing.
3. At the top of the hill, the trail follows the ridgeline west along the perimeter of China Flat. Before too long, there is a wide flat area where the ridgeline trail joins with the rough service road in step #1.
4. Turn left and take the fire road downhill 1 1/2 miles to complete the loop back to King James Court.

40. Oak Canyon Park Nature Trail

Difficulty: Easy
Length: 1 1/2 mile loop
Elevation: nearly level
Type of trail: mostly footpath
Kid appeal: * * * * *

Oak Canyon Community Park, opened to the public in October 1992, is a shining example of a recreation area designed to have minimal impact on the natural community surrounding it. Taking advantage of existing features such as a natural streambed and oakwoodland, the planners have created a park that encourages the public to get out and explore nature in their own backyard.

Located in the middle of Oak Park, the nature trail through Oak Canyon Community park links the Medea Creek Trail with the China Flat Trail. This system of trails gives residents in this area unprecedented access into the open space for which the Conejo Valley is so well known.

Directions: From Highway 101 in Agoura, take Kanan Road north about 3 miles to the intersection with Hollytree Drive. Turn right onto Hollytree and the park is immediately on the left. Follow the service road into the park.

Step by step:

1. From the parking lot proceed to the trails information kiosk near the restroom building. Be sure to pick up a Nature Trail guide, which was prepared by the students and staff of the Medea Creek Middle School. From the kiosk, proceed to step #2 located near the pavilion overlooking the creekbed. The beginning portion of the hike is along the sidewalk as the guide directs you counter-clockwise around the loop trail.

2. At station #8 on the guided tour, the trail leaves the pavement to continue on the dirt trail to the left, heading toward the oak canopied creekbed. The well-defined trail follows the creek downstream where the two crisscross each other several times in this riparian habitat.

3. At the stream crossing at station #15, there is easy access back to the sidewalk via a service road leading from the park's

main lawn area to an undeveloped, chaparral area of the park. From here the trail continues south, toward Kanan Road, through an open grassland.

4. At the park entrance the trail crosses the driveway and continues along the paved bikeway to view the planted wildflowers. Soon you will be back at the kiosk at the start of the trail.

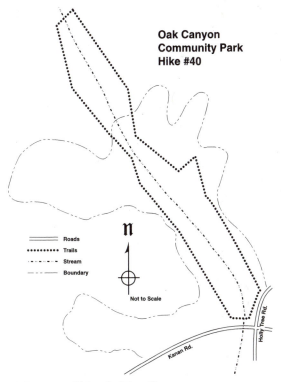

Oak Canyon
Community Park
Hike #40

Roads
Trails
Stream
Boundary

𝔫

Not to Scale

Holly Tree Rd.

Kanan Rd.

41. Medea Creek Trail

Difficulty: Easy
Length: 2 mile loop
Elevation: nearly level
Type of trail: paved jog path, single-track trail
Kid appeal: * * * * *

Medea Creek Park is another gem of a neighborhood park. Surrounded by homes and a school, it is a perfect example of an urban wilderness. The community of Oak Park has done an excellent job of enhancing, while at the same time preserving, the natural landscape of the sensitive wetlands around Medea Creek. The result is a beneficial habitat for wildlife as well as outdoor recreation for the neighborhood. As Medea Creek flows year-round and it has a good cover of native trees, it is a haven for local wildlife.

Directions: From Highway 101, take Lindero Canyon Road north to Kanan Road and turn right. Take Kanan to North Oak Hills Drive and turn right again. In half a block on the left is Calle Rio Vista; turn in and park in the cul-de-sac.

Step by step:

1. The trail follows the jogging path as it heads southeast from the cul-de-sac, skirting the Oak Park High School playing fields. In about 50 yards a dirt trail forks off the jogging path to parallel the creek. It is your choice which path to take as they both go in the same direction and crisscross several times during the course of the hike.

2. A half-mile of easy walking brings you to a residential area where the trail comes close to some homes as it crosses Medea Creek Lane. It then crosses the street and comes to a fork where you stay to the right as it heads south. (The trail to the left will eventually bring you to Oak Canyon Community Park, which is Hike #40)

3. The trail then crosses North Oak Hills Drive. Stay on the paved path on the left side of Medea Creek, which widens a bit as you approach Conifer Street. If the water level is low, you may follow the creek through the underpass, otherwise return to street level and cross there. Continue on the trail a short way and you will come to a footbridge across the creek. Take the bridge and your return leg of this hike will be on the paved jogging path.

4. When you get back to North Oak Hills Drive, sidewalk replaces the tarmac path. Before too long you will be back at the Calle Rio Vista cul-de-sac.

Medea Creek Trail
Hike #41

Community Park

Holly Tree Dr.

40

Kanan Rd.

Oak Park High School

Calle Rio Vista

41

Medea Creek Ln.

Medea Creek Mid. School

N. Oak Hills

Oak Park

Conifer St.

Rancho Simi Park Land

Ventura County

Los Angeles County

Medea Cir. Ln.

Tamarind St.

Fountain Wood St.

Medea Creek Park

Thousand Oaks Blvd.

n

Not to Scale

——— Road

•••••••• Trails

— — — Boundary

★ Structures

42. Paramount Ranch/Stream Terrace Trail

Difficulty: Easy
Length: 3/4 mile loop
Elevation: 150 foot gain/loss
Type of trail: single-track trail
Kid appeal: * * * * *

For centuries the diverse and scenic landscape of Paramount Ranch has made it a prized location in the Santa Monica Mountains. To the Chumash Indians, the area provided water and food as well as direct access through the mountains. Early Spanish settlers favored the area's rolling grasslands for their vast herds of cattle. Hollywood production crews, for whom the ranch is named, featured the scenic landscape in hundreds of movies, television shows and commercials.

The site's 436 acres are made up of grasslands, chaparral, oakwoodlands and streamside plant communities. Maintained since 1980 by the National Park Service, the park's easy trails provide several routes from which to explore these natural areas. Medea Creek, which flows through the property, and Triunfo Creek unite just south of the ranch to form Malibu Lake.

When I bring visitors to Paramount Ranch, they usually enjoy the ready made "Western Town" set located on the far side of the creek from the parking lot. Currently the popular television show, "Dr. Quinn, Medicine Woman" is being filmed at the ranch. The National Park Service keeps the park open during filming and even conducts tours of the set making this park a favorite for out-of-towners.

Directions: *From Highway 101 exit Kanan Road heading south toward the mountains. Stay on Kanan for 3/4 mile and then turn left onto Cornell Way, which turns into Cornell Road. Travel 2 1/2 miles on Cornell and the entrance to the park is on the right side of the road. Follow the road into the ranch and park near the "Western Town".*

Step by step:

1. From the parking lot, follow the paved service road 500 feet south of the information kiosk. Once you pass the picnic tables, look for the signed Medea Creek Trail to the right of the road as it leads into a wooded area. The shady trail climbs gently through the woods as it continues along the north side of the hill overlooking the old racetrack.

2. The trail leaves the woodland for the chaparral where it is joined by the Run Trail. Stay to the right as our trail heads toward the boundary of the park along Mulholland Highway. Stay on the Run Trail as it veers to the right following the perimeter of the park. A spur trail on the right loops up to a vista point at the top of the knoll and then drops down to join up once again with the Run Trail.

3. Stay to the right as the trail heads north, away from the Highway, on its way back to Medea Creek. As the trail approaches the creek it enters the woodland again. After a short stroll you are back to the service road close to the signed trailhead and the picnic tables.

The "old west" comes alive at Paramount Ranch.

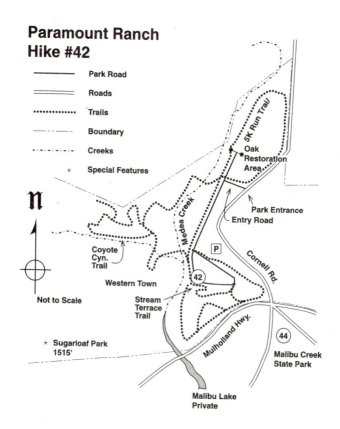

Paramount Ranch
Hike #42

————	Park Road
════	Roads
••••••••••	Trails
—·—·—	Boundary
—··—··—	Creeks
★	Special Features

5K Run Trail

Oak Restoration Area

Park Entrance

Entry Road

Medea Creek

Cornell Rd.

P

Coyote Cyn. Trail

42

Western Town

Stream Terrace Trail

Mulholland Hwy.

44

Malibu Creek State Park

Not to Scale

★ Sugarloaf Park 1515'

Malibu Lake Private

43. Peter Strauss Ranch/ Lake Enchanto Nature Trail

Difficulty:	Easy
Length:	2/3 mile loop
Elevation:	200 foot gain/loss
Type of trail:	single-track trail
Kid appeal:	* * * * *

The charming stone structures, wide green lawn and resident flock of peacocks at Peter Strauss Ranch evoke memories of a time long past: when Big Bands played and couples danced under the stars.

In its heyday, the 1940's and 1950's, this area was known as Lake Enchanto. It was a mountain resort for the wealthy Hollywood crowd, featuring a man-made lake and the largest swimming pool on the West Coast. The dam across Triunfo Creek was breached during the floods of 1969, so there is no longer a lake on the site. It does, however, still seem to be enchanted.

Directions: *From Highway 101 in Agoura, take Kanan Road south, toward the beach, about 3 miles to Troutdale Drive. Turn left onto Troutdale and take it 1/4 mile to Mulholland Highway. Turn left onto Mulholland, then immediately past the bridge turn right under the archway into the parking lot. Walk back over the bridge on Mulholland and enter the main gate into the ranch. There is handicapped parking near the house.*

Step by step:

1. The Strauss Ranch Nature Trail starts by the stone amphitheater built into the hillside in the oak grove south of the ranger station. Climbing steadily, the trail uses steps and small switchbacks as it makes way up and out of the grove.

2. Crossing some open chaparral, the trail soon levels out and returns to the oak forest. Before too long the trail heads down the hill again via another set of steps and switchbacks. It ends by the large wooden play structure in the picnic area near the dam site.

3. When Triunfo Creek is low, there is a trail that leads across it to the parking lot. Otherwise, this is a good area for a little streamside exploration as well as a good winter birding spot.

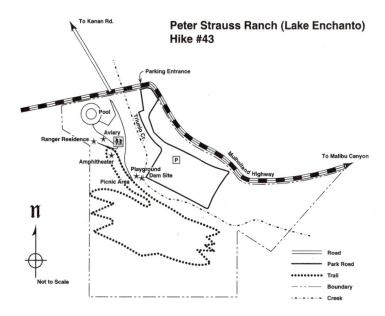

To Kanan Rd.

Parking Entrance

Pool

Aviary

Ranger Residence

Amphitheater

Playground
Dam Site

Picnic Area

Triunio Cc.

Mulholland Highway

To Malibu Canyon

n

Not to Scale

Road
Park Road
Trail
Boundary
Creek

Malibu Creek State Park

    Centrally located, Malibu Creek State Park is the linch-pin in the network of national, state and local parks that make up the Santa Monica Mountains National Recreation Area. The multi-use trails through the park provide access south to the ocean as well as east and west via the Backbone Trail

    In the early 1900's, the area surrounding Malibu Creek was developed as an outdoorsman's resort featuring hunting and fishing. At this time the creek was dammed up to form Century Lake. The lake was (and still is) stocked with fish and helped to attract more water birds to the canyon area. A few small cabins were built near the lake and along the creek as weekend retreats. Look for the ruins of these dwellings as you walk or ride through the park.

    The park's diverse land forms and unique plant communities made it a favorite location spot for many Hollywood productions. In 1946, 20th Century Fox bought the property and filmed hundreds of movies, television shows and commercials there. Besides playing host to the "M*A*S*H" 4077 unit, Malibu Creek has been "The Planet of the Apes", Africa for several Tarzan movies and home for the "Swiss Family Robinson". A complete list of films and other park activities can be found at the park's docent-run visitor center. It is open on weekend afternoons and is located inside the park about a mile from the main parking lot. See Hike #44.

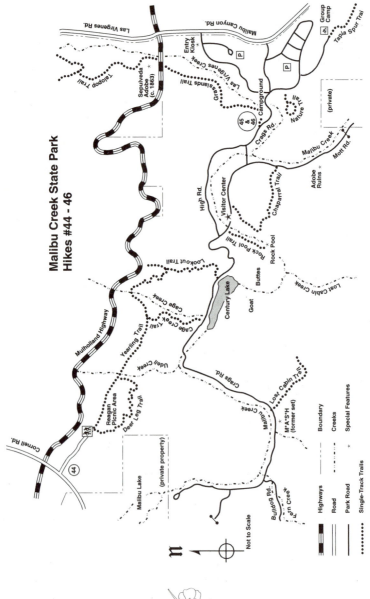

44. Old Reagan Ranch Trail

Difficulty: Easy
Length: 2 1/2 mile loop
Elevation: 150 foot gain/loss
Type of trail: single-track trail
Kid appeal: * * * *

Not to be confused with the present-day Reagan Ranch in Santa Barbara, today's hike is through the former Reagan Ranch of the late 1950's. Used by the Reagan family as a weekend retreat and to wine and dine Hollywood notables, the 360-acre ranch was located in what is now the northwest corner of Malibu Creek State Park. The ranch changed hands in the late 1960's when Reagan was elected governor of California and moved to Sacramento.

Because it is so far from the main park entrance and is closed to bicycles and park vehicles, this section of the park sees less human activity than animal activity. And while this doesn't guarantee sightings of deer and coyotes, you definitely will see much more evidence of animal usage on these trails than in most areas of the park.

Both trails that make up this loop hike are quite easy. There is a good mix of open grassland and oak groves to walk through. The Yearling Trail through the grassland is the more level of the two trails and provides a memorable wildflower walk in the spring. The Deer Leg Trail, while still easy, does have a few small grades to negotiate. The trail is shady and there are a couple of picnic tables near the halfway point that make a beautiful lunch site.

Directions: *The Reagan Ranch entrance to Malibu Creek State Park is at the southwest corner of Cornell Road and Mulholland Highway, about 3 miles south of Agoura Hills. From Highway 101 exit on Kanan Road heading south. Turn left onto Cornell Road and take it to the intersection with Mulholland Highway. Park your car along the road and follow the dirt service road into the park to the old ranch center, which is now a park maintenance center.*

Step by step:

1. Follow the park road beyond the ranch buildings to the signed trailhead. Today's loop hike begins on the Yearling Trail as it heads east through a meadow toward an old duck pond. Just past the pond, on the right side of the path, is the junction with the signed Deer Leg Trail. This will be the return trail for today's loop hike. Continue straight ahead on the Yearling Trail as it strikes out through the grassland.

2. Several unsigned trails cross the Yearling Trail, cutting the distance to the oak grove, but on the far side of the meadow there is a signed junction with the Deer Leg Trail. At this junction, take the Deer Leg Trail uphill and to the right through the chaparral to a vista point of Century Lake and Goat Buttes. After enjoying the view, take the Deer Leg Trail to the right as this high trail traverses the oak-canopied hillside on its way back to the ranch center.

45. The Rock Pool Trail

Difficulty:	Easy
Length:	2 1/2 mile loop
Elevation:	nearly level
Type of trail:	mostly fire road
Kid appeal:	* * * * *

The spectacular, craggy peaks that surround the Rock Pool in Malibu Creek State Park are best known to most of us from the opening credits of the television show "M*A*S*H". The park has a long history of being featured in numerous film productions and is still in use.

The huge, rugged rock formations that surround the Rock Pool are welded volcanic conglomerates that display a swiss-cheese look. These holes were formed by gas bubbles trapped in molten rock which then burst open when exposed to the Earth's surface millions of years ago. The results are amazing: some cavities formed caves large enough to be used as homes for animal, while others make wonderfully grotesque faces in the rock. Rock climbers often come here to practice on the rock walls that line the trail to the pool. Be careful not to disturb them. The visitor center, maintained by the Malibu

Creek docents, is in an old ranch house and is open to the public on weekend afternoons. Beautiful displays and "hands-on" exhibits of Native American culture as well as the flora and fauna make this an interesting stop for all ages.

Directions: *For this hike it is best to use the main park entrance off Malibu Canyon Road. From Highway 101 exit onto Las Virgenes Road heading south. In 3 miles you will come to the intersection with Mulholland Highway. The park entrance is straight ahead on Las Virgenes/Malibu Canyon Road, 1/4 mile past the intersection, on the right side of the road.*

Step by step:

1. From the second parking lot, follow Crags Road (across the campground road from the restroom) over Las Virgenes Creek and into the park. In about 1/3 mile there is a fork in the road. Stay to the left on Crags Road and take it over Malibu Creek. The road to the right is the High Road/Backbone Trail and will be the way to return from the Rock Pool.

2. A little past the Texas-style creek crossing, the road comes to a "T" near a row of eucalyptus trees. Stay to the right and after an easy quarter mile of walking the sycamore shaded visitor center is a welcome sight.

3. From the visitor center, follow Crags Road over the bridge. On the left side of the road is the signed Rock Pool Trail. The first part of the trail is along a service road that gives way to a single-track trail near the climbing wall. After exploring the area, return to Crags Road the same way. To make this a loop, turn left on Crags Road and proceed about 25 yards to where it is joined on the right by the oak-canopied High Road. Take the High Road back along the far side of the creek to Crags Road and then back to the parking lot.

46. Malibu Lake Bike Ride

Difficulty: Easy/moderate
Length: 7 mile round trip
Elevation: 200 foot gain/loss
Type of trail: fire road
Kid appeal: * * * * *

     Noted for its spectacular rocky peaks, creeks and year-round lakes, the diverse terrain of Malibu Creek State Park makes a popular site for mountain bike riding. The main trail through the broad canyon that parallels the creek makes for an easy ride as well as provides access up to Castro Crest via the grueling Bulldog Motorway. This variety makes the park a terrific, traffic-free site for cyclists of all levels of expertise.

Directions: *Same as for the Rock Pool - Hike #45.*

Step by step:

1. As with the hike to the Rock Pool, follow Crags Road into the park from the second parking lot. In about one-third mile, stay to the left at the junction with the High Road and take Crags Road over Malibu Creek to the visitor center.

2. From the visitor center, follow Crags Road over the bridge. On the far side of the bridge, High Road joins us from the right, (This road will be used for the return trip to the parking lot at the end of today's ride.) At this point, Crags Road climbs its only major grade with the help of two switchbacks. At the top of the hill, trapped behind Goat Buttes, is Century Lake.

3. After a half-mile of easy riding the road bridges over Malibu Creek once again. The trail becomes rather rocky as it parallels the creek through a heavily wooded area. Before too long you come to the former M*A*S*H set, which is marked by a couple of old Army vehicles burned in a wildfire that swept the area in 1982.

4. Another quarter mile of easy riding brings you to the junction with the Bulldog Motorway on the left. Castro Crest is about 4 miles away via the Motorway; however for today's easy ride we will continue straight ahead on Crags Road the remaining one-third mile to Malibu Lake where the road dead-ends in private property at the edge of the lake. After enjoying

the view, return the way you came, but take the shadier High Road along the north side of the creek mentioned in step #2.

Agoura Hills

Cheeseboro Canyon National Recreation Area is an exciting place for nature exploration. Being adjacent to a major animal corridor between the Santa Monica Mountains and the Simi Hills, the park's multi-use trails always hold evidence of animal usage. Because of the length of the trail through the open grassland in this long canyon I prefer biking to hiking in this park.

Palo Comado Canyon has followed a controversial path to becoming part of the National Park Service. Also known as Jordan Ranch/China Flat, this canyon was previously owned by entertainer Bob Hope and was slated to become another luxury housing development. A complicated land deal involving interagency cooperation and extremely far-sighted efforts made by a number of individuals have made this pristine canyon public open space in perpetuity. When combined with Cheeseboro Canyon the two parks make up almost 4,500 acres of open space, making this the largest federally operated site in the Santa Monica Mountains National Recreation Area.

47. Cheeseboro Canyon Bike Ride

Difficulty: **Easy/Moderate**
Length: **12 mile round trip**
Elevation: **nearly level**
Type of trail: **mostly fire road**
Kid appeal: *** * * * ***

Bike rides through the oak-studded savanna found in Cheeseboro Canyon are great for families just getting started in off-road riding as well as for more experienced riders. Today's ride begins on the wide park service road and is gentle enough for the most casual rider. The latter portion of the ride along the Sheep Corral Trail provides more technical, open single-track trail riding.

Directions: *Cheeseboro Canyon National Recreation Area is in Agoura. Heading toward Los Angeles, from Highway 101,* *use the Chesebro Road exit. At the end of the ramp, go straight at the first stop sign and then turn left onto Palo Comado Road and take it over the freeway to Chesebro Road. Turn right onto Chesebro Road and follow it about a mile to the signed Cheeseboro Canyon National Recreation Area on the right side of the road. Heading from Los Angeles, simply exit onto Chesebro Road from Highway 101 and follow the directions above. Turn into the park and follow the dirt road 1/4 mile to the main parking lot.*

Step by step:

1. The Sulphur Springs Trail begins near the information kiosk under two large oak trees at the northeast end of the parking lot. Follow the trail across the grassland toward an oak grove. The trail goes through several oak groves as the trees follow a seasonal stream along the canyon bottom.

2. In a few places the main trail is intersected by lateral trails that lead to the ridgetops surrounding the canyon. These trails offer more challenging rides along with bird's-eye views of the bottom land.

3. After 4 miles of easy riding through the rolling savanna on the fire road, the trail crosses the stream in a dense woodland area. At this point the trail changes to a single-track and continues with a new name, Sheep Corral Trail, through the chaparral plant community.

4. The Sheep Corral Trail follows along a seasonal streambed for another 1 1/2 miles. It then comes to a "T" junction; the trail to the right leads to deadend turnabout after 1/2 mile. The trail to the left eventually leads to China Flat. For today's ride, after enjoying the view, return to the Cheeseboro Trailhead from the "T" junction the way you came. First along the Sheep Corral Trail and then the Sulphur Springs Trail.

Optional: On your return to the trailhead, watch for the signed Modelo Trail on the right side of the Sulphur Springs Trail about 2 miles from its junction with the Sheep Corral Trail. This trail follows the ridgeline back to the trailhead so there is some moderate uphill riding on this portion of the ride.

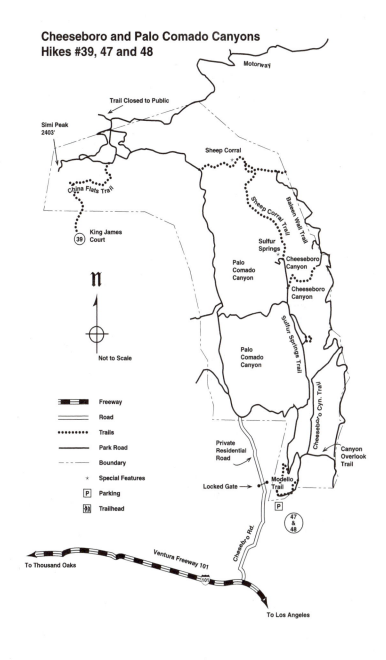

Cheeseboro and Palo Comado Canyons
Hikes #39, 47 and 48

Motorway

Trail Closed to Public

Simi Peak
2403'

Sheep Corral

China Flats Trail

Sheep Corral Trail

Baleen Wall Trail

39 King James
Court

Sulfur
Springs

Cheeseboro
Canyon

Palo
Comado
Canyon

Cheeseboro
Canyon

N

Not to Scale

Sulfur Springs Trail

Palo
Comado
Canyon

Cheesebo'o Cyn. Trail

Freeway

Road

Trails

Park Road

Boundary

Special Features

P Parking

Trailhead

Private
Residential
Road

Canyon
Overlook
Trail

Locked Gate →

Modello
Trail

P

47
&
48

Ventura Freeway 101

Chesebro Rd.

To Thousand Oaks

101

To Los Angeles

48. China Flat Bike Ride

Difficulty: Strenuous
Length: 12 mile round trip
Elevation: 1,040 foot gain/loss
Type of trail: mostly fire road
Kid appeal: * * * * *

Newly opened to the Cheeseboro National Recreation Area, Palo Comado Canyon Park also is an excellent site for mountain bike riding. Together, Cheeseboro and Palo Comado offer a 27-mile network of multi-use trails that are open to hikers, cyclists and equestrians. Just recently opened to the public, Palo Comado Canyon Park is a primitive recreation area that doesn't even have its own entrance. At this time, entry to the park is through the Modelo Trail from Cheeseboro Canyon. Other entry points to this park are still being worked out between the National Park Service and local open space agencies.

Step by step:

1. From the parking lot in Cheeseboro Canyon, proceed north on the signed, single-track Modelo Connector Trail that leads uphill and through the grassland. About 1/2 mile of riding will bring you to the top of the hill where the connector trail joins the Modelo Trail. (Using this connector trail shaves about a mile off the distance to the top of the hill via the main Sulphur Springs Trail.)

2. Continue north on the signed Modelo Trail as it travels along the top of the grassy ridgeline. A half-mile of ridge riding brings you to a "Y" trails junction near the base of a steep uphill grade. Take the trail to the left as it heads west toward Palo Comado Canyon. The trail drops several hundred feet in elevation over the next 1 1/2 miles as it traverses several hilltops on its way down the canyon.

3. Once in Palo Comado Canyon our trail joins the main park road as it heads north along the shady canyon bottom for almost a mile before coming to an abandoned ranch center. Just before the ranch, look for the junction with another park road. This unnamed road on the right is about a mile long and will be your return route to Cheeseboro Canyon.

4. The main service road through the canyon leads directly to the much renowned China Flat area. You will find several miles of easy

trails that crisscross this outstanding beautiful oak-dotted savanna. To reach China Flat continue north on the park road from the ranch center. The service road follows the streambed through the oak-canopied canyon another 3/4 mile before beginning the long climb through the chaparral up to China Flat. There is an elevation gain of 800 feet over the next mile or so but once you get to the flat, it will be worth the effort. Going all the way to China Flat will add another 4 or 5 miles to this bike ride.

China Flat vista.

Chapter Four

Simi Valley and the Rim of the Valley Corridor

Rocky Peak fun.

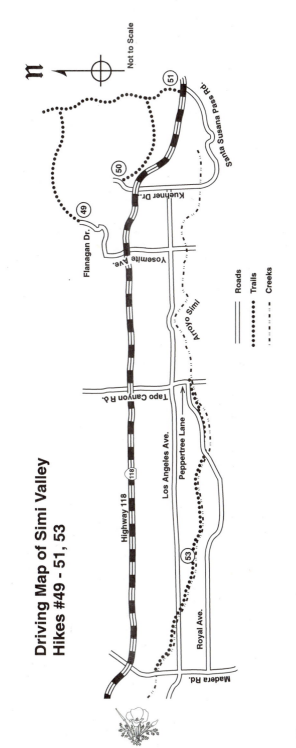

Driving Map of Simi Valley
Hikes #49 - 51, 53

Not to Scale

Roads
Trails
Creeks

Highway 118

Flanagan Dr.

Yosemite Ave.

Kuehner Dr.

Santa Susana Pass Rd.

Tapo Canyon Rd.

Arroyo Simi

Los Angeles Ave.

Peppertree Lane

Royal Ave.

Madera Rd.

HIKES AT A GLANCE 6
Simi Valley and the Rim of the Valley Corridor

	DIFFICULTY	FACILITIES	FEATURES	FEES	OPEN TO
ROCKY PEAK PARK					
49. Chumash Trail	easy	none	views, wildflowers, caves	no	H, C, E, D
50. Hummingbird Creek Trail	moderate/strenuous	none	views, wildflowers	no	H, C, E, D
51. Rocky Peak Trail	strenuous	none	views, wildflowers, rigorous workout	no	H, C, E, D
52. Sage Ranch Walkabout	easy	w, t, p, rs	views, rock forms, open woodlands	no	H, C, E, D, S
53. Arroyo Simi Bikeway	easy	w, t, p	playgrounds at parks, arroyo life	no	H, C, E, D, S
MOORPARK					
54. Happy Camp Canyon	easy/moderate	w, t, p	woodlands, wildflowers	no	H, C, E, D

FACILITIES: w - water, t -toilet, p - picnic table, rs - ranger station, c - camping, m - marina, s - store
OPEN TO: H - hikers, C - cyclists, E - equestrians, D - dogs on leash, S - stroller/handicap access

Simi Valley and the Rim of the Valley Corridor

Simi Valley and the surrounding Santa Susanna Mountains provide outdoor enthusiasts with access to numerous multi-use trails to some remote, yet close in, areas of Ventura County. One of the most popular sites is Rocky Peak Park at the valley's east end. Situated high above Simi Valley at an impressive 2,714 feet above sea level, Rocky Peak is a spectacular vista point for those willing to make the 6 mile round trip to the summit.

Besides providing some much needed breathing room between the communities of Simi Valley and the San Fernando Valley, Rocky Peak Park is a vital link in a system of wildlife corridors. These corridors, established along animal migration and foraging routes, allow wildlife to travel between the Santa Susanna Mountains, the Simi Hills and the Santa Monica Mountains without contacting humans.

The chaparral covered slopes and oak-lined canyons are home to numerous species of wildlife. In the more remote parks recent sightings of black bears and mountain lions have been reported. Some of the more common species such as mule deer, coyote and bobcat as well as road runners, horny toads and snakes also frequent these open space parks.

The general area surrounding the Santa Susanna Pass has a rich historical background. This route was used for centuries by Chumash Indian hunting and trading parties as the easiest passage from Simi Valley into the inland mountain ranges. Following the Chumash lead, white settlers developed a stage route through the pass to connect a growing Los Angeles with croplands in Simi Valley. Not to be outdone in modern times, today Amtrak trains traverse the pass in the same tunnel built by Chinese immigrants in the late 1890's. Listen for the eerie sound of the train's whistle as it leaves this tunnel.

49. Chumash Trail

Difficulty: Moderate/strenuous
Length: 5 mile round-trip
Elevation: 1,060 foot gain/loss
Type of trail: single-track trail
Kid appeal: * * *

Whether you hike, bike or ride horseback, the Chumash Trail will provide a challenging ascent into the Santa Susanna Mountains overlooking Simi Valley. Completed and opened to the public in May of 1991, this excellent single-track trail connects with the Rocky Peak Trail to explore the 4,369 acres of open space that make up Rocky Peak Park.

Starting in Chumash Park just north of Simi Valley, the Chumash trail climbs steadily into the vast backcountry area, once known as Runkle Ranch, that straddles the Ventura/Los Angeles County Line. The trail goes past some huge outcroppings of sandstone and several seasonal stream beds as well as some highland meadows while making its way to the ridgeline high above Simi Valley.

Directions: From Highway 118 exit on Yosemite Avenue in Simi Valley. Take Yosemite Avenue 1/2 mile north toward the Santa Susanna Mountains. Turn right on Flanagan Drive and follow it 3/4 mile to where it deadends and park along the street.

Step by step:

1. The trailhead for the Chumash Trail is located just north of the end of Flanagan Drive and the trail heads north through an open grassland. About 100 yards from the start the trail splits; follow the path to the left and within another 50 feet you will see an arrow signpost on the right side of the trail indicating that you are on the proper path.

2. The trail climbs gradually as it parallels a seasonal creek, making a half circle around a small hill. The trail then crosses the saddle between the hilltops and continues to follow the arrow signposts on the right side of the path as you head toward the mountains.

3. The trail enters a highland meadow dotted with wildflowers and huge sandstone outcroppings before it begins its steady

climb to the top of the ridge and its junction with the Rocky Peak Trail. The next mile of hiking crosses several seasonal streams as it switchbacks up to the ridge. At the trails junction there is an excellent vista of Blind Canyon. After enjoying the fine view, return to your car the way you came.

Optional: If you are still in a hiking mood you can continue on the Rocky Peak Trail. To the south (right) about 1 mile away is Rocky Peak and to the north (left) the trail leads through some oak savannas to the park boundary about 2 1/2 miles away.

50. Hummingbird Creek Trail

Difficulty: Moderate
Length: 5 mile round trip
Elevation: 1,100 foot gain/loss
Type of trail: single-track trail
Kid appeal: * * * * *

Located at the base of Rocky Peak and tucked into a rugged canyon just north of Simi Valley Freeway, Hummingbird Creek Trail is one of the newest trails through the vast open space area maintained by the Rancho Simi Recreation and Park District. It opened in April of 1993. The trail is unique in that it is being built solely by volunteers on "Trail Work Days" sponsored by the Rancho Simi Trail Blazers.

The large sandstone outcroppings that surround Hummingbird Creek make this an interesting area to explore. Water and wind have been carving away at these rocks for centuries creating nature's artwork. Much of the wildlife, as well as migrating species of birds, depend on the mini-wet-land around the creek for food, water and overnight shelter.

Just before the trail drops down to the level of the creek it crosses a low stone wall. In the not too distant past this wall was part of a dam spanning a re-routed Hummingbird Creek. Developers anxious to promote the area built a reservoir to create a lush tropical ambiance. The ill-fated venture went bankrupt within a year and now all that remains is the stone wall. Chumash artifacts were uncovered during the construction of the wall resulting in an archaeological dig set up by Los Angeles Pierce College. Artifacts found at the site led experts to believe this may have been a burial site. Under direction of an on-site Chumash Indian representative, many artifacts were safely removed.

Trail Work Days, celebrating Earth Day and California Trails Day 1995, enabled the Trail Blazers to complete another mile on this single-track trail through the chaparral. Rising steeply up the mountainside from Hummingbird Creek, the trail was completed to a "giants' playground" of sandstone caves and rock forms about a mile from the ridgeline. Following the 1996 Trail Work Day, the Hummingbird Creek Trail is now complete all the way to its junction with the Rocky Peak Trail.

Directions: *The Hummingbird Creek Trail is off the north end of Kuehner Drive in Simi Valley. From Highway 118, exit at Kuehner Drive and turn left, under the freeway. Continue north on unsigned Kuehner Drive about a mile and look for the map kiosk marking the trailhead on the right, or south side of the road. Park along the edge of the road. Since parking is very limited at the trailhead, many people park at the "park 'n ride" near the freeway and start their hike at the oakwoodland by the creek. From Los Angeles: follow the directions given above except turn right upon exiting the freeway.*

Step by step:

1. The signed Hummingbird Creek Trail begins near the map kiosk. Take the single-track trail heading north, paralleling the road briefly before turning south and heading down toward the creek.

2. As the trail heads deeper into the canyon it passes by some interesting sandstone formations before leveling out near an old stone wall. The trail then makes its way down to the streambed and parallels the creek for about 50 yards before coming to a small oakwoodland.

3. The trail crosses the creek and heads aggressively up the mountain. A mile into the chaparral brings you to the first of the sandstone outcroppings and a little further up the trail are a number of caves to explore. From here the path crosses a small highland flat before heading seriously uphill once more.

4. Several steep switchbacks bring you up the canyon's back wall to a low saddle with a view of the San Fernando Valley. The trail levels out somewhat but still trends generally uphill around a number of huge, wind sculpted rock outcroppings before its junction with the Rocky Peak Trail. To continue to Rocky Peak or to make a loop with the Chumash Trail, turn left and head north along the Rocky Peak Trail that follows the fire road. To the right the trail leads to the Rocky Peak Trailhead off Highway 118 and the Santa Susanna Pass Road.

Optional: If you want to make this an all day hike, the Rocky Peak Trail continues north for another 3 miles past Rocky Peak to the park's boundary where there is a large, beautiful oak savanna to explore. For an excellent car-shuttle hike, combine the Rocky Peak Trail with the Chumash Trail. Park one car at Chumash Park at the end of Flanagan Drive. Drive the hiking party to the Rocky Peak trailhead in another vehicle and start hiking from there. Continue north on the Rocky Peak Trail a mile past Rocky Peak and you will come to the trails junction with the signed Chumash Trail on the left. Take the single-track Chumash Trail 2 1/2 miles back down the mountain to the car parked at Chumash Park.

New age cave family.

51. Rocky Peak Trail

Difficulty: Moderate/strenuous
Length: 6 mile round trip
Elevation: 1,200 foot gain/loss
Type of trail: fire road
Kid appeal: * * *

The view from Rocky Peak makes it one of the best vista points in all of Ventura County. On a clear day your 360 view might include: Anacapa Island, the Topa Topa Mountains, the Santa Susanna Mountains, the San Fernando Valley, Simi Hills, Simi Valley and the Santa Monica Mountains.

However, you don't have to hike all the way to the top of Rocky Peak to get a great view. This trail is made up of one spectacular vista point after another. Besides the long-range landmarks to pick out of the scenery, there are some tremendous rock formations to explore right along the trail. These dramatic outcroppings were formed more than 65 million years ago and are the park's namesake.

Directions: To reach Rocky Peak Park from Simi Valley, exit Highway 118 on Kuehner Drive and turn right, or south onto Kuehner. Stay on Kuehner Drive/Katherine Road until it deadends at the Santa Susanna Pass Road and turn left. Take the Santa Susanna Pass road about two miles to the bridge over the freeway. Turn left and take the bridge to where it deadends in a small parking area at the entrance of Rocky Peak Park.

From Los Angeles: *Simply exit the Simi Valley Freeway at the Rocky Peak Park off-ramp and the above mentioned parking area is on the right.*

Step by step:

1. The signed Rocky Peak Trail begins just north of the parking area. The trail heads aggressively uphill passing several small, rock-strewn plateaus on either side of the path. After almost 2 miles of steady uphill walking, you will come to the trail's only shade tree, a lone live-oak. The fire road continues to climb, even more steeply, for another 1/2 mile to the top of the ridgeline.

2. At the ridgeline you will see a single-track trail leading off

to the right. This is the trail to Rocky Peak, its own bad self, about 1/2 mile away.

3. This unsigned summit trail becomes somewhat indistinct near the peak and some rock scrambling is involved to make it all the way to the top of the promontory. Be sure to wear good hiking boots if you plan to make this part of the hike. After enjoying the view, return to your car the way you came.

52. Sage Ranch Walkabout

Difficulty: Easy
Length: 2 mile loop
Elevation: 300 foot loss/gain
Type of trail: fire road
Kid appeal: * * * * *

This charming new addition to Ventura County's public open space sits atop the western Rim of the Valley Corridor. At an elevation of 2,000 feet above sea level, the park's perimeter Loop Trail offers scenic vistas of: Simi Valley against a dramatic backdrop of the Santa Susanna Mountains, the entire San Fernando valley, as well as Simi Hills and the Santa Monica Mountains. On clear days the ocean and Channel Islands also are visible from this beautiful 625-acre highland park.

The ranch's citrus and avocado groves are reminders of this area's recent agricultural past. A pine-shaded picnic site on the park's south side was once a small homestead.

Besides the spectacular urban and mountain views, Sage Ranch has some of the largest and most rugged sandstone outcroppings in the Southland. All of the park's formations are impressive, but some of the most unique are found near the Rocketdyne Field Laboratory's defunct Liquid Oxygen Plant, about halfway around the Loop Trail. Here a sheer sandstone wall stands sentinel some 300 feet above the plant's remaining foundation. The cliff's numerous small caves and other openings provide protected homes to a large number of birds. Be sure to bring your binoculars to better see their aerial displays as they feast on flying insects. A smaller rock formation on the side of the trail near the plant appears to be the face of one of the Ninja Turtles, or an extremely fierce monster.

Today's hike around the ranch's perimeter is a sample of several of the plant communities in our mountains. The park's beautiful open oakwoodlands, chaparral and streamside areas contrast the even-rowed symmetry of the homesteaders' orchard as well as the Rocketdyne test facility. The last of which is a stark reminder of the technological changes that have had their effect on the natural world around us. The Rocketdyne plant is fully operational and infrequent rocket engine tests still occur. If you are in the park during a test, a whistle will precede each test to warn of the short bursts of loud noise.

Directions: *From Highway 118 in Simi Valley, exit on Kuehner Drive heading south to Katherine Road and turn right. Cross the railroad tracks and turn right onto Black Canyon Road. Follow this narrow twisting road to the ridge top and Sage Ranch Park is on the west side of the road.*

From the Conejo Valley: *Take Highway 101 east to Hidden Hills and exit at the Mulholland/Valley Circle off-ramp. Turn right and take Valley Circle north about 6 1/2 miles to the three-way stop at Woolsey Canyon Road. Turn left, west, onto Woolsey Canyon Road and take it all the way to the top of the hill to Black Canyon Road opposite the Rocketdyne Field Laboratory and turn right. The entrance to Sage Ranch is about 1/2 mile from the intersection on the left side of the road.*

Step by step:

1. Today's hike on the signed Loop Trail begins at the northern end of Sage Ranch's parking lot. Follow the park's gravel service road to the top of the hill. Turn right and continue to walk along the road heading north as it passes several service driveways that lead to areas of the park that are restricted from public use. (These are old equipment graveyards that will be cleaned up and eventually removed from the park.) About a half-mile from the trailhead, you will come to a small grove of eucalyptus trees and some excellent vistas of Simi Valley.

2. Another few hundred yards of easy, nearly level walking brings you to a part of the trail that leads down into a shallow basin area. Huge rock outcroppings, complete with caves and hidden faces, are strewn about in a large meadow. The road

follows an old fence line as it heads west, wending its way downhill as it approaches the cement foundation of Rocketdyne's former Liquid Oxygen Plant.

3. This facility was dismantled in the 1970's after supplying liquid oxygen for Air Force programs during the 1960's. This and all other Rocketdyne property is NOT open to the public. Be sure to stay on the designated trail through this area and not wander onto private property. At this point, you are half way around the perimeter loop; the trail begins to climb out of the wash as it heads east, back toward the front of the park.

4. Near the top of a small grade, about 1/2 mile from the LOX plant, look for a pine grove to the left of the trail. Follow the indistinct path a short way to a shady, hilltop picnic area. This is the old homestead site.

5. Back on the main trail, continue east paralleling a long row of eucalyptus trees. As you do, the trail views the Rocketdyne facilities on the opposite side of the canyon. Before long you will see Black Canyon Road and soon you will be at the lower parking lot near the entrance gate.

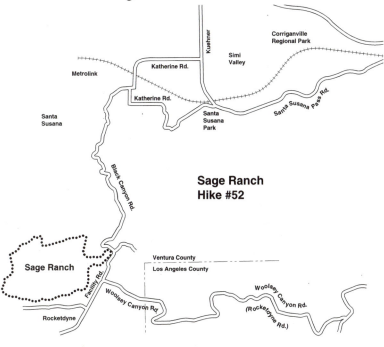

53. Arroyo Simi Bikeway/Equestrian Trail

Difficulty: Easy
Length: 10 miles round-trip
Elevation: nearly level
Type of trail: separate use paved bikeway
Kid appeal: * * * * *

Arroyo Simi could have ended up like the Los Angeles River. It could be a wide, sterile concrete ditch slicing through the heart of Simi Valley. But thanks to a total community effort, the Arroyo Simi is anything but a barren ditch.

The arroyo is actually the site of a separate-use bicycle path and equestrian trail. The bicycle path uses the north side of the arroyo, running between Madera Road and Tapo Canyon Road and is 5 miles long. The equestrian trail is about 1 1/2 miles long and uses the south side of the arroyo to connect the Equestrian Center to Rancho Simi Community Park. Cutting through the middle of town, these trails can be used as safe alternative transportation routes as well as for recreational riding.

Considering that the trail goes right through the heart of Simi Valley, there is a surprising amount of wildlife in the arroyo. Not only are there numerous ground critters such as squirrels and gophers, but great egrets, large snow-white water birds, and herons often fish the arroyo near Madera Road. Cattails growing in the waterway provide homes to many types of birds. The plants also shelter water-loving creatures such as crayfish which make tasty meals for the predators that prowl the arroyo.

You also will see more than 1,200 trees planted on both sides of the water-way making up the Heritage Tree Trail. This trail is dedicated to the men and women from Simi Valley who served in the armed forces to defend our country. Made possible by the Freedom Tree organization, from a Small Business Administration grant, the Rancho Simi Recreation and Park District purchased and planted the trees in 1991. The grant, written by the district, specified that the trees be purchased from local nurseries.

The whole community has been behind this beautification program. For the first year after planting, the trees were watered by hand by volunteers. Local businesses, such as Wang Tech (located along the arroyo), do their part by watering the trees near their facilities. Now that the trees are larger and need more water, 7 miles of drip irrigation already have been installed to deliver water to them. For information as to how you can participate in this beautification process, call the Rocky Peak Mountain Bike adopt-a-trail program at: 805-584-4400.

Directions: For out-of-towners the best place to start the bicycle path is at Rancho Simi Community Park, which is on the northwest corner of Erringer Road and Royal Avenue. From Highway 118 exit on Erringer Road heading south. The Rancho Simi Community Park is located on the west, right, side of Erringer Road between Los Angeles Avenue and Royal Avenue. To reach the Equestrian Center also exit on Erringer Road heading south. Continue on Erringer to the intersection with Royal Avenue and turn left. Proceed east on Royal for another mile to the signed turn-off for the Equestrian Center. Turn right and follow the dirt service road into the complex.

Step by step:

1. The Arroyo Simi Bicycle Path is at the north end of the Rancho Simi Community Park. From the parking lot proceed north along Erringer Road and take the path to the left, or west. The bicycle path ends near Madera Road; turn around at this point and return the way you came. Walk your bike across Erringer Road and head east toward Sycamore Drive where you must once again walk your bike across the street.

2. After another mile of easy riding, there is a fork in the bicycle path. The path to the left follows a tributary of the Arroyo Simi slightly north, angling toward Los Angeles Avenue. To the right, a wooden bridge takes the bicycle path over the tributary to continue along the north side of the arroyo toward Royal Avenue.

3. Once at Royal Avenue you must cross the street to continue your ride the rest of the way to Tapo Canyon Road. At this point you can again turn around and retrace your route back to the Community Park.

54. Happy Camp Canyon Bike Ride

Difficulty:	Easy/moderate
Length:	10 miles round-trip
Elevation:	600 foot gain/loss
Type of trail:	mostly fire road
Kid appeal:	* * * *

Like an oyster and its pearl, Happy Camp Canyon Regional Park's true beauty lies deep within its boundaries. When viewed from the trailhead, a treeless sea of grass fills the wide wash below. That, and the dry, chaparral covered slopes all around give little hint of the large shady canyon just a mile away. Located north of the city of Moorpark, this large open space is a vital link in the wildlife corridor between the Santa Susanna Mountains and the Simi Valley.

Sheltered between Big Mountain and Oak Ridge, beautiful oak-lined Happy Camp Canyon provides habitat and forage for a wide range of wildlife. This remote park is home to a number of animals no longer residing in the Santa Monica Mountains. While hiking and riding the multi-use trail, alert trail-users may even be able to pick out the tracks of bears and mountain lions along the dirt fire road. Sandwiched between these two ridgelines, this canyon seems to be a world unto itself; providing food, water, shelter and space for area wildlife in a natural setting.

Happy Camp Canyon's dedicated team of docents lead a number of specialty and family hikes into the canyon each month. To obtain a schedule of these docent-led hikes call Cheryl Reynolds at the Mountains Recreation and Conservation Authority at: 310-456-5046, ext. 101.

Directions: For today's mountain bike ride through Happy Camp Canyon Regional Park in Moorpark, take Highway 101 in Thousand Oaks to Highway 23. Go north 10 miles and exit at New Los Angeles Avenue (Highway 118 West) in Moorpark. Head west for about 1 1/2 miles and then turn right, north, onto Moorpark Avenue. Continue north for about 2 1/2 miles to a large bend in the road. Moorpark Avenue (called Walnut

Canyon Road outside of town) turns left and heads west. To get to the park, head straight at the bend on Happy Camp Canyon Road. A few hundred feet down the road and take a quick right onto Broadway Road and follow it into the park.

Step by step:

1. Today's bike ride begins on the single-track trail that heads north from the information kiosk at the northeast end of the parking lot. This easy trail meanders downhill through the grassland below. It joins with the Happy Camp Canyon Fire Road about a mile from the trailhead. At this junction, continue north, left, on the fire road. Follow the sandy road about 1/8 mile until you come to a small group of oak trees just before the locked gate near the spring.

2. At the gate there is another information kiosk with a copy of the park's self-guided nature trail. A little beyond this, the trail divides and creates a high-low loop around the interior of the park. Stay to the left, on the fire road and follow the nature trail along the low route. The roadway to the right, that leads uphill and along the ridge (the high route), is the Middle Ridge Fire Road and may be used as an alternative route on the way back.

3. The next 4 miles of riding takes you in and out of the canyon's shady oakwoodland that lines either side of the intermittent stream. The trail passes through a gateway and enters a large oak grove just before coming to the Bill Mower Memorial Trail Rest. Dedicated in 1993, this primitive picnic area occupies an old ranch center near the back of Happy Camp Canyon.

4. After enjoying the shade of the oaks, it is time to return to the main trailhead 5 miles away. The easiest way to do this is to turn your bike around and return the way you came. The trip back is a real coast, as it is downhill all the way. You may want to take Happy Camp Canyon Fire Road all the way to the junction with the Broadway Fire Road that leads to the parking lot. About a mile from the lot you will come to a "Y" junction. From here take the Broadway Fire Road to the right as it leads uphill to the parking lot. Continuing south on the Happy Camp Canyon Fire Road leads to Moorpark College.

Optional: More vigorous riders may enjoy the challenge of returning to the trailhead via the Middle Ridge Fire Road. For this route, continue east beyond the Mower Memorial Rest and in a short distance, you will come to the junction with the Switchback Road. Take it uphill and to the right, to the top of Big Mountain. This is a steep road that climbs 500 feet in a half-mile to the junction with the Middle Ridge Fire Road. At this junction, turn right again and head west along the ridge-line before dropping back down to the canyon floor near the locked gate in step #1. From the gate, it is 1 1/2 miles back to the parking lot.

Happy Camp Canyon
Hike #54

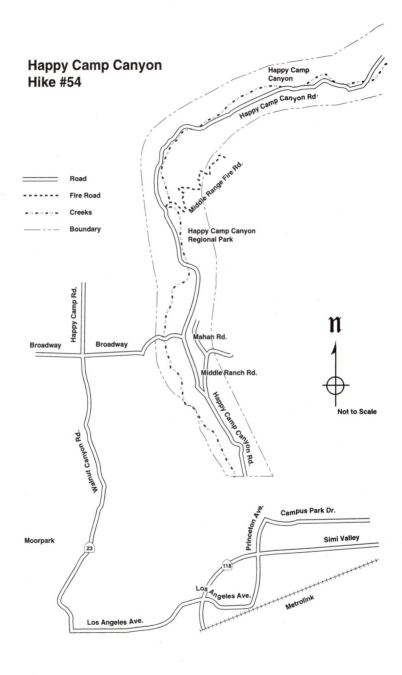

Happy Camp Canyon

Happy Camp Canyon Rd.

Road

Fire Road

Creeks

Boundary

Middle Range Fire Rd.

Happy Camp Canyon Regional Park

Happy Camp Rd.

Broadway

Broadway

Mahan Rd.

Middle Ranch Rd.

Happy Camp Canyon Rd.

n

Not to Scale

Walnut Canyon Rd.

Moorpark

23

Princeton Ave.

Campus Park Dr.

Simi Valley

118

Los Angeles Ave.

Los Angeles Ave.

Metrolink

151

Chapter Five

Ventura and Channel Islands

Our very own Pacific Ocean jewels.

HIKES AT A GLANCE 7
Ventura and the Channel Islands National Park

	DIFFICULTY	FACILITIES	FEATURES	FEES	OPEN TO
VENTURA					
55. McGrath State Beach	easy	w, t, p, c, rs	beach estuary, nature center	yes	H
56. Arroyo Verde Park	easy/strenuous	w, t, p	nature center, views, playground	holidays/weekends	H, D
57. Coastal Bikeway	easy	w, t, p, c, m, s	beach, pier, promenade	no	H, C, S
58. Emma Wood State Beach	easy	w, t, p, c	beach, tidepools	yes	H, C
59. Grant Memorial Park	easy	t, p	ocean and downtown views	no	H, C, S
CHANNEL ISLANDS					
Anacapa Island					
60. Eastend Walkabout	easy	w, t, c	museum, lighthouse	no	H
61. Frenchy's Cove	easy	none	tidepools, views	no	H
Santa Cruz Island & Scorpion Ranch		located at ranch center			
62. Cathedral Point Trail	easy	t, p, c	view of whale migration	no	H
63. Potato Harbor Trail	easy/moderate	t, p, c	views	no	H
64. Smuggler's Cove Trail	moderate	t, p, c, rs	beach, views	no	
Santa Rosa Island					
65. Torrey Pine Grove	easy	t, c	pristine beach, pine grove	no	H
Santa Barbara Island					
66. Island Walkabout	easy/moderate	t, c	views, wildflowers, rookeries	no	H

FACILITIES: w - water, t -toilet, p - picnic table, rs - ranger station, c - camping, m - marina, s - store
OPEN TO: H - hikers, C - cyclists, E - equestrians, D - dogs on leash, S - stroller/handicap access

153

Ventura & Channel Islands

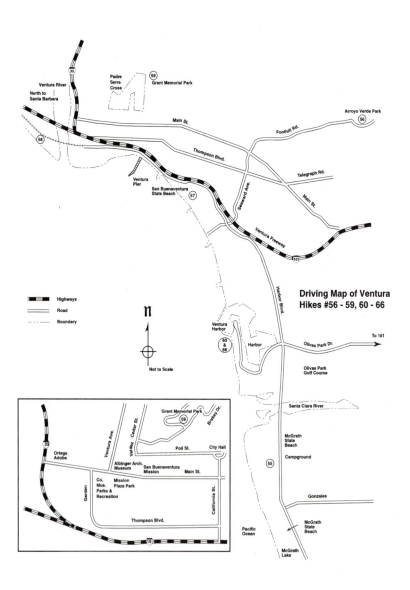

Ventura River

North to
Santa Barbara

Padre
Serra
Cross

Grant Memorial Park

Arroyo Verde Park

Main St.

Foothill Rd.

Thompson Blvd.

Telegraph Rd.

Ventura
Pier

San Buenaventura
State Beach

Seaward Ave.

Main St.

Ventura Freeway

Highways

Road

Boundary

Not to Scale

Driving Map of Ventura
Hikes #56 - 59, 60 - 66

Ventura
Harbor

Harbor

Harbor Blvd.

Olivas Park Dr.

To 101

Olivas Park
Golf Course

Santa Clara River

McGrath
State
Beach

Campground

Gonzales

McGrath
State
Beach

McGrath
Lake

Pacific
Ocean

Grant Memorial Park

Ventura Ave.

Valdez

Cedar St.

Brakey Dr.

Poli St.

City Hall

Ortega
Adobe

Albinger Arch.
Museum

San Buenaventura
Mission

Main St.

Garden

Co.
Mus.
Parks &
Recreaiton

Mission
Plaza Park

California St.

Thompson Blvd.

55. McGrath State Beach

Difficulty: Easy
Length: 1/2 mile loop trail
Elevation: nearly level
Type of trail: footpath
Kid appeal: * * * * *

Located at the mouth of the Santa Clara River in Oxnard, McGrath State Beach offers some terrific seaside family recreation opportunities. There is a large, sandy beach to play on and the northern half of the park is home to the 160-acre Santa Clara Estuary Natural Preserve. McGrath State Beach also marks the southern end of Ventura's Coastal Bike Path as well as the western terminus of the proposed Saticoy-to-the-Sea Bikeway.

With 174 developed campsites, McGrath State Beach is the largest coastal campground in Ventura County. Campsites have fire rings and tables, and drinking water and restrooms also are available. There are no day-use facilities other than parking. Besides developed sites, there also is a "hike & bike" camp for those campers who enjoy human-powered means of getting to the beach. For information about camping at any of the Channel Coast State Parks and Beaches call: 1-800-444-PARK.

McGrath State Beach is a wonderful place to explore nature and enjoy the day at the beach. The jungle-like Santa Clara Estuary Natural Preserve protects the habitat of two species of birds: the California Least Tern and Belding's Savannah Sparrow, both on the endangered species list. An estuary is where a river empties into the ocean: The fresh water from the river mixes with the salt water from the sea to produce a rich ecosystem teeming with life, making estuaries ideal nurseries and breeding spots for numerous birds and fish. Natural estuaries are particularly important today as 90 percent of California's estuaries already have been destroyed by development.

The sand dunes that you pass on your way to the beach represent another fragile habitat. These dunes are literally held in place by the plants that grow on them. Unfortunately, the plants can become dislodged very easily; even foot-traffic can kill them, and eventually the dune will blow away. So please stay on the various established trails and avoid walking over the dunes.

If you decide to walk the beach, the park stretches almost 2 miles south of the river. Along the way you will pass McGrath Lake tucked behind the dunes. This fresh-water lake helps to attract even more birds to the area. Beyond the lake is the huge old Edison power plant and then before you know it you are at the end of the park and will have to return the way you came.

Directions: *From Highway 101 in Ventura, exit at Victoria Avenue and head south to Olivas Park Drive. Turn right and head west toward the ocean. Turn left on Harbor Boulevard and head south about a mile to the park entrance on the west side of the road.*

Step by step:

1. Park your car in the extra vehicle parking lot just north of the entry kiosk. The signed nature trail begins on the right hand side at the beach end of the parking lot.

2. The trail takes on a tunnel-effect as it cuts through the dense growth of willows and reeds (and poison oak) approaching the river's edge. After a few hundred yards the trail parallels the river.

3. At sign post #8 a short spur trail leads you along the water's edge and to a bench at the next sign post. Returning to the main trail, we continue to find observation benches at various vista spots as we head through a marshy area.

4. Near sign post #14 another spur trail takes off from the main trail through the marsh toward the mouth of the river several hundred yards away. We will continue on the main trail as it loops back to the trailhead near sign post #20.

5. At this point you can return to your car and get your beach things. To get to the beach simply follow the service road around the perimeter of the campground. It begins at the end of the parking lot and it is about 500 yards to the ocean on the far side of the dunes.

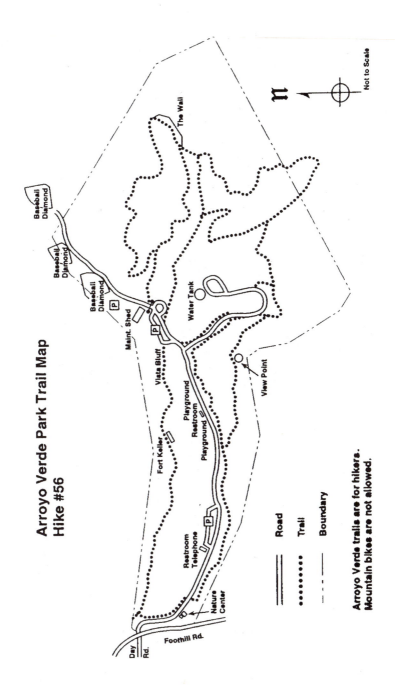

Arroyo Verde Park Trail Map
Hike #56

The Wall

Baseball Diamond

Baseball Diamond

Baseball Diamond

P

Maint. Shed

P

Vista Bluff

Water Tank

P

Fort Keller

Playground

Restroom

Playground

View Point

Restroom

Telephone

P

Nature Center

Foothill Rd.

Day Rd.

Not to Scale

Road

Trail

Boundary

Arroyo Verde trails are for hikers.
Mountain bikes are not allowed.

157

56. Arroyo Verde Park

Difficulty: Easy / moderate
Length: 3 1/2 miles of trail
Elevation: 150 foot gain/loss
Type of trail: mostly single-track trail
Kid appeal: * * * *

The city of Ventura's beautiful Arroyo Verde Park offers the public about 3 1/2 miles of hiking trails. The network of trails follows the circumference of this charming little canyon park and the trails themselves are signed as to their level of difficulty.

Serving as a buffer zone between undeveloped open space and urban sprawl, Arroyo Verde Park and the trails through it show signs of the wildlife that live in the park. The network of trails explores the coastal sage scrub plant community that hugs the rugged slopes of the hills surrounding the landscaped lawn that makes up the central portion of the park. At one time this plant community dominated the Southern California landscape. Due to the pressures of development, this eco-system is rapidly disappearing. Arroyo Verde is one of the few Ventura area parks that offers easy access to this important natural resource.

The network of hiking trails around the park is designated as follows: Trails marked with a green circle are the least difficult, those marked with a blue square are deemed difficult while those marked with a black diamond are the most difficult. There is a little more than a mile of each type of trail. The green circle trails are easy but not completely stroller accessible.

Directions: *Arroyo Verde Park is at the intersection of Foothill and Day Roads in Ventura. From Highway 101, exit at Victoria Avenue and turn right. Proceed north on Victoria for about 2 1/2 miles until it deadends at the Foothill Road intersection and turn left. Following signs for the park, continue west on Foothill Road about 3/4 of a mile. Arroyo Verde Park is on the right side of the road, opposite the intersection with Day Road.*

Step by step:

1. The easiest of Arroyo Verde's trails, which is marked with a green circle, begins near the nature center and for the most part loops around the front lawn area. Drive into the park and park in the first parking area on the left side of the road. Walk back down the road to the information kiosk located on the north side of the nature center. The green circle trail starts here and heads slightly uphill to follow the contour of the slope north toward the back of the arroyo.

2. After a little easy walking there is a trails junction on the right with a black diamond trail that climbs up to a hilltop vista point about 1/4 mile away. The black diamond indicates that this is a steep trail. From the vista point the trail continues north to meet up with the rest of the trail network a little further on. Back to the green circle trail, where 1/3 mile of easy walking brings you to another junction. At this point the green circle trail loops back to the front of the park by crossing the road and heading back to the lawn, skirting the barbecue area and starting again at the signed Vista Bluff. The path then turns south and continues through some trees along the western edge of the lawn back to the nature center.

3. A little more intense, the blue square trail continues straight ahead along the service road for a short way from the trails junction. The path then leaves the road and heads uphill to meet up with the black diamond trail coming down from the vista point in step #2. The two trails head north together from this trails junction.

4. After 1/4 mile the trails split up and the blue square trail turns left to head downhill to the west. The black diamond trail heads uphill, first to the east and then to switchback on itself and head west to the back of the arroyo. Both trails eventually lead down to the flat bottom of the arroyo. A short walk on the blue square trail brings you to another trails junction.

5. Straight ahead, a short steep section of black diamond trail takes you to the bottom of the arroyo and joins with another stretch of the green circle trail. Starting from the back parking lot this section of easy trail explores the more natural area of the park. From this trails junction, I would suggest continuing

on the blue square trail as it turns to the right to head north and make its way gradually to the bottom of the arroyo where it also joins the green circle trail.

6. Once again on the easy path, turn right and head north along the bottom land. At the end of the canyon there is another trails junction with the black diamond trail returning from step #4. This section of easy trail deadends at this junction. To continue your hike around the park, take the black diamond trail uphill and to the left a short ways. At the next trails junction you can choose to continue straight ahead on the black diamond trail for a strenuous work-out as it climbs the arroyo's back "wall"; or you can continue on the blue square trail to the left as it heads south and back toward the mouth of the canyon.

7. The black diamond trail rejoins the blue square trail and they come to an end near the maintenance shed at the back of the park. An easy 1/2 mile walk along the service road will return you to your car in the front parking lot.

Arroyo Verde's verdant meadow

57. Ventura's Coastal Bikeway

Difficulty: Easy
Length: 2 to 25 mile round-trip
Elevation: nearly level
Type of trail: mostly paved
Kid appeal: * * * * *

A bike ride along Ventura's coastal bike path is a fun family outing any time of year. This Class 1 bike path explores the beach and offers a nearly traffic-free route perfect for young or inexperienced riders. With bike rentals available at San Buenaventura State Beach, you need not even own a bicycle to enjoy this relaxing way to cruise the coast.

The path begins at McGrath State Park and then proceeds north along Harbor Boulevard to the Ventura Marina before heading to San Buenaventura State Beach. From here the Path continues north past the Pier and Promenade to Emma Wood State Beach as it follows the Old Rincon Highway toward Santa Barbara. The entire route is level, and most of it is along the waterfront. Today's easy round-trip bike ride begins and ends at San Buenaventura State Beach using Hobson County Park, near the community of Seacliff, as a turn around point.

The Coastal Bike Path brings riders eye-to-eye with Ventura's beach community. Early morning rides along the levee by the Ventura River estuary and Seaside Park are a delight for bird lovers. The pathway provides a wonderful overview of the birds' activities as many species of waterfowl flock to the mouth of the river to feed and raise their chicks.

Clear days mean outstanding views of the Channel Islands; sometimes they seem almost close enough to touch. Evening rides take on a magical quality as the sun sets over the ocean. It can be peaceful to watch the boats return to the harbor and wildlife settle down for the night. There are also some good, though not very diverse, tidepools near Seaside Wilderness Park opposite the fairgrounds. We often see dolphins playing in the surf along the Old Rincon Highway. In addition, the bike path also highlights some interesting human beach activities.

Surfers Point is aptly named: On weekends, local surfing organizations use the site for in-club competitions. Promenade Park and Ventura Pier sport a festive quality as weekend crowds bring out vendors of all kinds. San Buenaventura State Beach and Marina Park are popular picnic sites with complete facilities including tables, barbecues, bike stands and restrooms with showers.

Directions: *Because of its easy access to both the coastal bike path and Highway 101, today's family bike ride starts at San Buenaventura State Beach. From Highway 101 in Ventura, exit at Seaward Avenue and turn left toward the ocean. Follow Seaward over the bridge about a block and then make a right onto Ayala Street. Take Ayala Street all the way to the park entrance off San Pedro Street. Pull into the park and pay the parking fee at the kiosk then follow the service road around the park's perimeter to the large oceanside parking lot.*

162

58. Emma Wood State Beach

Difficulty: Easy
Length: 1/3 mile one-way (River's Edge Trail)
                  2/3 mile one-way (Ocean's Edge Trail)
Elevation: nearly level
Type of trail: fire road, footpath
Kid appeal: * * * * *

Just west of Ventura's city limits, Emma Wood State Beach is a beautiful site for a family outing. Adjacent to the city's Seaside Wilderness Park at the mouth of the Ventura River, the park's River's Edge Trail is an excellent winter "birding" venue. The estuary and surrounding wetlands constitute an extremely important and sensitive ecosystem. The area provides habitat, food and shelter for dozens of species of wildlife and several species are classified as rare or endangered. Along the Ocean's Edge Trail there are several tidepool areas along the beach just north of the estuary.

Prompted by several years of extreme winter weather and subsequent flooding of the Ventura River, the City of Ventura, the Coastal Conservancy and California State Parks have put on an all out effort to restore the Ventura River Estuary. An estuary occurs when a river empties into the sea. These unique wetland areas provide habitat and breeding grounds for a wide array of land and sea animals and birds.

Volunteers have cleaned out huge piles of debris from the river bottom as well as non-native plants that choke the river environs. Hoping to re-establish the woodland floodplain that originally covered the riverbanks, plantings of native trees have been established along the trail. Organizers hope that increased foot traffic on the park's nature trails will help to keep this area from being resettled as a squatter encampment.

The Ventura River has seen many changes in the last two hundred years. First it was the arrival of the missionaries who cleared the land near the river, then cattle herds grazed along the water's edge obliterating many of the native plants and destroying the riverside habitat. The building of the Matilija and the Casitas Dams during the mid-1900's greatly lowered the volume of water flowing down the river. Today, increased human population creates a serious pollution problem for the estuary as fertilizer runoff from agricultural lands, treated and untreated sewage and storm drain runoff all eventually end up in the river.

Directions: *To reach Emma Wood State Beach by car, exit Highway 101 at California Street in Ventura. Take California Street several blocks north to Main Street and turn left. Take Main Street through the downtown mission area until it deadends at the entrance to Emma Wood State Beach. Drive into the park, pay the day-use fee and park your car in the day-use/picnic area parking lot near the central restroom facility.*

What you'll need: There is a $6 parking fee at Emma Wood State Beach. Dogs are not allowed on the beach or on the trail.

Step by step:

1. From the day-use parking lot, follow the paved park road south around the campground. The River's Edge Trail begins at the second information kiosk. Pick up a trail guide and follow the self-guided nature trail along the wide dirt path through the woods. Near signpost #5, the trail leaves the woodland for the open soft chaparral scrub.

2. The trail becomes sandier on its way toward the river. Just before you reach the trail's end there is a small rest stop with a bench. The River's Edge Trail ends at the railroad track. From here you can cross the track to the Ocean's Edge Trail or return the way you came.

3. Carefully cross the train track and pick up the signed Ocean's Edge Trail on the other side. At first the trail meanders through an old cypress grove and then it's on to the beach. Head north to return to Emma Wood State Beach; to the south are the fairgrounds and pier. The two circular cement and iron objects at the shoreline are old anti-aircraft gun emplacements remaining from World War II. I like to use this trail during low tide as there are some good, though cobbly, tidepools along this beach. Once north of the dune area the path uses the tunnel under the train track to bring you back to the parking lot.

59. Grant Memorial Park

Difficulty: Easy
Length: 1/2 mile to vista
Elevation: 300 foot gain/loss
Type of trail: fire road
Kid appeal: * * * *

Perched high on the hilltops above historic downtown San Buenaventura, Grant Memorial Park is an excellent spot from which to view the city. These steep hillsides behind the mission served the friars as a strategic barrier between themselves and their neophytes against the heathen wilds of the New World. The restoration of the mission cross on Sept. 9, 1912, was to honor the 200th anniversary of the birth of Padre Serra, founder of the Franciscan Missions of California.

On these hills overlooking Ventura's broad coastal plain there was a large Chumash Indian settlement centuries before the Mission was established in 1782. Attracted to the foothills above the coast, the Indians discovered a protected vantage point from which they could observe the entire shoreline from the Rincon to the Santa Monica Mountains.

As the City of San Buenaventura grew and the railroad came to town, thousands of Chinese emigrants were brought to the area as a cheap source of labor. Due to a housing shortage most of the laborers crowded into this part of town, creating an instant Chinatown crammed up against the foothills.

Besides the great views of the coastal plain and city below, Grant Park is noted as a migration site for the beautiful orange and black monarch butterfly. Each year between November and March, hoards of these gorgeous creatures descend on the park's planted groves of Monterey pine and eucalyptus trees. The City of San Buenaventura celebrates this annual event by bringing busloads of school children on educational field trips to the park to view the butterflies. Some years the monarchs are so numerous that they seem to clothe the trees with orange leaves.

Directions: *There are several ways to enter Grant Park. From Highway 101 exit at California Street and take it north toward the City Hall Building on Poli Street. Near the statue of Father*

Serra, at the front of the building, turn left onto Poli Street heading west. Make a quick right onto Brakey Road, look for the small sign for the park with an arrow indicating the direction. Follow this narrow, winding road all the way to the park entrance. The park also can be accessed from Summit Drive off Kalorama Drive or from Ferro Drive off Cedar Street.

Step by Step:

1. From the park entrance, stay to the left as you drive toward the Father Serra Cross vista point. After enjoying the view from the memorial, make a turn and take the rough power line road to the hilltop opposite the parking lot.

2. At the top of the hill continue north on the narrow footpath around the public works facility. From here, take the paved road downhill to a small shady grove near the public restrooms.

3. Pick up the dirt power line road once more, on the far side of the grove, and follow it up the steep grade to another grove at the top of the next hill. From this vista point the trail continues a short ways before ending in a view point overlooking the indoor target range. After enjoying the view, return to the parking area the way you came.

Optional: History buffs doing a walking tour of downtown San Buenaventura may want to walk to Father Serra's Cross from the Mission on Main Street. Simply take Valdez Alley, located on the west side of the Albinger Museum, north to Poli Street and then walk up Ferro Drive to the park. This route passes by the remains of the Mission aqueduct, which served also as Ventura's first jail house.

Channel Islands National Park

Trips to the Channel Islands are always an adventure. The Channel Islands National Park, made up of five of the eight Channel Islands and their surrounding one nautical mile of ocean, was designated a national park in 1980. Remoteness from the mainland, provided by the unpredictable Santa Barbara Channel waters, keeps the island sanctuaries from becoming over-civilized.

The channel crossing to the islands is truly half the

fun of the trip. Every trip is an unique experience and no one can guarantee what you'll see on each voyage, but during the winter months many species of marine animals migrate through the nutrient rich waters of the Santa Barbara Channel. The most spectacular are, of course, the California Grey whales but several kinds of dolphins and pinnipeds (seals and sea lions) also inhabit these waters and numerous sea birds such as Western gulls and brown pelicans are attracted to the islands' predator-free nurseries and rookeries.

Getting there: Even though the Channel Islands National Park is only a few miles from the Ventura coast it receives fewer visitors than any other national park. Island Packers, of Ventura is the only boating concessionaire to the islands. For more information about fares and schedules call their office at: (805) 642-7688.

Directions: *From Highway 101 in Ventura exit onto Victoria Avenue heading south, toward the ocean. Follow the signs that say Channel Islands National Park and take Victoria Avenue south 1/2 mile to Olivas Park Drive. Turn right at Olivas and take it west, toward the Ventura Harbor. As Olivas crosses Harbor Boulevard it joins Spinnaker Drive. Continue into the harbor area on Spinnaker as it circles the southern end of the harbor and deadends near the visitor center, right next to the Island Packer office.*

What you'll need: The Channel Islands National Park is a primitive area. Besides hiking and/or camping gear, camera, binoculars and sunscreen you must bring all of your own everything, including water, on these excursions. Be sure to layer your clothing as weather conditions can be unpredictable and quite rough at times in the Santa Barbara Channel. Some Island Packer cruises offer galley service during the crossing. Emphasis for these tours is on "packing in" small groups. There is no cocktail service on board the cruise ship, but it's a seaworthy vessel with a first rate crew.

Camping: Due to its remote, open sea location, camping on the Channel Islands is a unique experience in primitive camping. Campers must bring their own: stoves and fuel as well as food (packed in hard plastic, mouse-proof containers), water and sturdy, windproof tents. Be sure to bring plenty of sun-screen as well as a hat or visor as the islands are fairly treeless. Campers also are responsible for the removal of all trash that they generate during their stay. For more information and camping permits call the Channel Island Visitor Center at: (805) 644-8262.

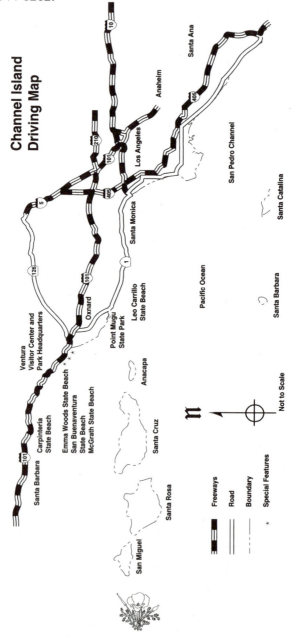

60. Eastend/Lighthouse Walkabout

Difficulty: Easy
Length: 2 mile loop
Elevation: nearly level, 152 step stairway
Type of trail: single-track trail
Kid appeal: * * * * *

The eastern end of Anacapa sports the island's only true hiking trails and finding the rocky coves with the most scenic views of the other islands and mainland is the only goal for the day as you hike around the island. At the visitor center complex there is a small museum and a sign-in log as well as a number of picnic tables and outhouses. Anacapa, being the closest to shore, is the most frequently visited of the island chain. Long noted for its spectacular display of wildflowers, this island is a showcase for the yellow blooms of the giant coreopsis.

The most difficult section of today's outing is the boat landing when you must climb up a short ladder to get off the boat. Then there is the steep stairway leading up from the landing area to the island's flat top. The staircase isn't as bad as it sounds: You can go at your own pace and even stop from time to time to enjoy the view. This is an excellent family outing as age is not a barrier to island trips, but all participants should be in good health and agile enough to go ashore. In order to keep Anacapa Island as wild as possible, there are no concessions of any kind on the island. Anything that you forget, you must do without.

Step by step:

1. Once you get to the island your boat will be met by a park ranger in the landing area. The ranger will give a short talk about the delicacy of the island habitat and ask everyone to remember to stay on the designated trails while hiking around the island. It is your option to go on a guided hike with the ranger or to strike out on your own.

2. From the top of the landing area follow the service road to

the ranger and visitor center complex. An information kiosk serves as the trailhead for the self guided nature hike. Be sure to pick up a trail guide and map.

3. The hike around the island is essentially a figure-8 loop tour that begins at the complex and can be walked in either direction. Numbered trail stops along the footpath correspond to the trail guide in pointing out cultural, geological and biological information about the island.

4. After enjoying your hike around Anacapa be sure to return to the landing area in time to catch the boat going back to the mainland.

Anacapa's famous Arch Rock.

61. Westend - Frenchy's Cove

Difficulty: Easy
Length: 1/2 mile round-trip
Elevation: nearly level
Type of trail: footpath
Kid appeal: * * * * *

Named for the hermit that called this tiny spot of land on Anacapa Island's West End home for nearly 30 years, Frenchy's Cove is the perfect destination for an exciting family adventure. Anacapa Island is made up of three individual islets that are inaccessible to each other except by boat, with the eastern end of the island being the most often visited portion of the Channel Islands National Park. Rugged, West Anacapa is a prime nesting site for the brown pelican and is a research natural area that is closed to the public except for day-use at the landing at Frenchy's Cove. The cove offers outstanding tidepools to explore during the winter as well as a great beach for swimming and snorkeling in the summertime.

Unless you have your own boat, transportation to Frenchy's Cove is provided by Island Packers of Ventura. The number of open, public tidepool excursions is somewhat limited, but it is also possible to join along on school tours. These weekday outings are quite educational; Island Packer staff naturalists present hands-on lessons in ocean navigation and sealife in the island's kelp forest habitat as well as guided tours of the tidepool area on the island's south side.

What you'll need: Be sure to wear sneakers or surf-sox for exploring the tidepools. Bring sunscreen, a hat or visor and at least two quarts of water per person. The West End of Anacapa is a primitive area; there are no facilities of any kind.

Step by step:

1. The trail to the tidepool area begins at the west end of the beach landing. The path follows some roughly hewn steps cut into a large rock outcropping and then along a short ledge. You then drop down to thread the needle through a narrow passageway between two volcanic chimneys that is only open during low tide.

2. Once on the island's south side, the indistinct path turns to the right and heads west along a cobble beach before coming to the tidepool area. Numerous rock benches extend out into the sea and provide an extensive area to explore. Use caution as the rocks can be very slippery and sometimes the swells are quite large. A good rule to remember while exploring the tide-pools is to never turn your back on the ocean. After enjoying the tidepools, return to the landing site the way you came.

Santa Cruz Island Eastend from Scorpion Ranch

62. Cathedral Point Trail

Difficulty: Easy
Length: 1 1/2 mile round-trip
Elevation: 150 foot gain/loss
Type of trail: fire road
Kid appeal: * * * *

Santa Cruz Island, the largest and most diverse of the northern Channel Islands, was thought to be the center of the universe by its ancient Chumash Indian inhabitants. Just 16 miles off the coast of Ventura, this majestic island offers many outdoor experiences for the adventurous of spirit, including today's hike to Cathedral Point.

A day trip to Santa Cruz Island is a captivating experience. The short boat ride seems to transport you back in time as well as across the channel. Scorpion Ranch's rustic setting among groves of eucalyptus and cypress trees makes it an oasis against a backdrop of amber-colored hillsides surrounding it. Because of the island's fantastic biodiversity and varied coastline, many believe it to be the crown jewel of the Channel Islands National Park. Island foxes and scrub jays are two rare, endemic species that are easily seen around the ranch center.

Day trips to the eastern end of Santa Cruz Island can be arranged through Island Packers, Inc., which provides guided boat trips for the general public to Scorpion Ranch. These trips to the island include a pleasant 1 1/2 hour boat ride each way with about four hours ashore.

Step by step:

1. Today's outing begins by crossing the always spectacular Santa Barbara Channel. Once moored at Scorpion Anchorage, landing is by a shore boat onto a rock-sand beach. (This is really much easier than it sounds.) The guided hike begins at the ranch house, usually about an hour after getting ashore. An island naturalist begins the walk with a brief history of Santa Cruz Island and the people who have lived there, from Chumash to those of recent years.

2. From the ranch, the trail follows a dirt road inland. A signpost beyond the ranch compound indicates the road to Cavern Point cuts off to the right from the main dirt service road. It leaves the valley as it heads uphill toward a centuries-old Indian midden and whale-watching spot. After enjoying the view, hikers return to the ranch the way that they came. Day-trippers may explore the rest of the Scorpion Ranch area and canyon hiking trails on their own, but be sure to be back at the pick-up spot on the beach in time for your boat's departure to the mainland.

63. Potato Harbor Trail

Difficulty: Easy/moderate
Length: 3 mile round-trip
Elevation: 200 foot gain/loss
Type of trail: fire road
Kid appeal: * * * *

To experience the total island adventure, try an overnight stay at Scorpion Ranch. Camping is available on Santa Cruz Island either on the beach or in the deep eucalyptus groves just inland from the Ranch Center. Camping on the island offers all of the convenience of car-camping, including an all terrain vehicle to transport your gear to the campsite, with all of the tranquillity of backpacking. Check with Island Packers before your outing for more information about camping rates and seasonal specials.

The adobe bunkhouse serves as a rustic B & B with kitchen privileges for those who don't like to sleep on the ground and enjoy indoor plumbing. Easy, guided kayak tours also are available from the island caretakers. The price to kayak is $25 for a two-hour excursion. No prior experience is necessary, but participants should be 12 or older.

Today's hike gives you a fine overview of Scorpion Ranch against a backdrop of Red Mountain. Undoubtedly the hike to Potato Harbor is one of the most inspirational sunset hikes you will find anywhere. Santa Cruz Island, famous for its picturesque coastline, is dotted with dramatic sea caves formed millions of years ago by molten lava that cooled quickly beneath the sea's surface. Because of this, gases that were trapped in the cooling rock formed bubbles that have been forever frozen in rock.

Step by step:

1. The hike to Potato Harbor begins on the service road that leads into the canyon. About 1/2 mile beyond the ranch compound, just before the second eucalyptus grove camping area, there is a fork in the road. Take the dirt road to the right as it leaves the canyon on its climb up to the Campo Grande Airstrip. At first, the road is rather steep, but it soon levels out to follow the contour of the slope.

2. After a mile of steady walking, you will come to the airstrip. Continue west along the runway and then take the single track path to the cliff overlooking Potato Harbor. From this vista you will see the island's rugged, northwestern shoreline. After enjoying this divine view, return the way you came. A word of caution: planes cannot land while people are on the airstrip. If you see a plane, be sure to step well away from the runway.

Optional: Another excellent day hike is to follow the ranch road up the canyon beyond the last campgrounds. About 1/2 mile beyond the grove, the road ends and you must boulder-hop your way up the streambed. There are several small seasonal waterfalls as well as some examples of the island's native trees and wildflowers on the steep walls on either side of the ravine.

64. Smuggler's Cove Trail

Difficulty: Moderate
Length: 6 mile round-trip
Elevation: 500 foot gain/loss
Type of trail: fire road
Kid appeal: * * * *

The beach at Smuggler's Cove, located on the south side of the island, is a perfect destination for a day hike out of Scorpion Ranch. The Cove's flat, sandy shoreline is great for boogie-boarding and body-surfing. There is also easy access to several snorkeling and tidepooling sites at the south end of the bay, near Yellowbanks Anchorage.

In 1869 the Santa Cruz Island Co. was formed and a number of satellite ranch centers were built around the island. Scorpion Ranch and Smuggler's Cove were two of these compounds and an overland route was built between them. This rough ranch road rises dramatically from sea level at both ends of the trail to traverse the grassy plateaus that dominate the island's eastern end, giving it a serene pastoral look.

Near the half-way point of today's hike you'll notice a number of peculiar stone mounds dotting the treeless hillsides approaching Smuggler's Cove. Thinking that these enigmas might have served a role in some ancient rite or another, I was surprised to learn that the mounds were created by Chinese laborers as they cleared the fields for cattle grazing.

Step by step:

1. Follow the ranch road out of Scorpion Ranch that heads up and over the compound's south canyon wall. At the top of the grade there is an excellent vista of the Santa Barbara Channel and the ranch below.

2. The trail crosses a sea of grass as it heads south across the island. There is only one trail through this vast open space, but there is a signpost at the trail's midpoint just to keep you in the right direction.

3. About a mile beyond the signpost you will come to the airstrip. From here, Smuggler's Cove is an easy half-mile walk down the ranch road. A shady eucalyptus grove and NPS

ranger station at the beach marks the end of the trail. After enjoying the day at the beach, return to Scorpion Ranch the way you came.

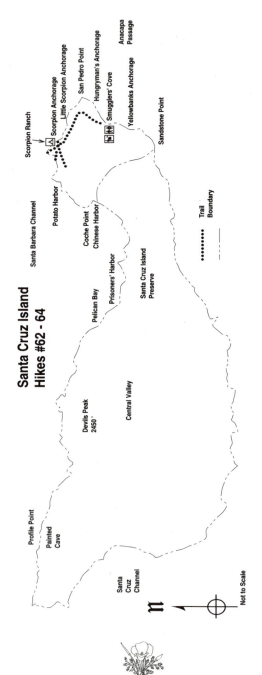

Santa Cruz Island
Hikes #62 - 64

65. Becher Beach to Torrey Pine Grove

Difficulty: Easy
Length: 3 mile round-trip
Elevation: 200 foot gain/loss
Type of trail: fire road
Kid appeal: * * * * *

Located just 25 miles from the Ventura coast, Santa Rosa Island makes an excellent destination for an unforgettable family adventure. Isolated from the mainland by one of the most unpredictable ocean channels anywhere, this is not a trip for the unprepared or the faint of heart. What awaits those willing to brave the channel crossing are miles of pristine beaches and open hiking along the old ranch jeep roads. Isolation and windy weather not only keep the crowds down, but also have created a number of biological anomalies on each of the Channel Islands. Santa Rosa Island, for instance, is known for its extremely rare Torrey Pine forest as well as several other endemic plant and animal species.

First of all, the scenic half-hour flight to Santa Rosa Island onboard the STOAL (short take-off and landing) aircraft is simply breathtaking. When viewed from the air, it's no wonder that the northern Channel Islands are considered the crown jewels of the National Park System. The flight path to Santa Rosa takes you directly up the southern coast of Santa Cruz Island, the largest of all the Channel Islands. On Santa Rosa Island, with the rancheria still in operation, it's almost as if the short plane ride transports you back in time to the days of old Spanish California.

Before European occupation, the island supported a stable population of about 2,000 people. During our tour of an ancient Chumash archeological site, our guide, Ranger Bill, told us that if we were to resurrect a Chumash Indian from 200 years ago he would not recognize the Santa Rosa Island of today. With the introduction of domestic animals such as cows and sheep came the decline of native grasses and plant communities. Not only did the imported animals overgraze the area, they also brought seeds of non-native plants with them. These introduced plants were able to beat the already stressed native plants for water and now cover most of the island.

Even though the impact of the introduced plants and animals took their toll on Santa Rosa, excellent examples of the island's unique plant communities still exist. Most of them are located on the steep slopes and canyons that cannot be reached by the livestock. The Torrey Pine groves are a good example of this. The only other Torrey Pine forest in the world is located in La Jolla, near San Diego. How these two forests came to be is a puzzle that has fascinated scientists for decades and is still being researched. Santa Rosa Island also is the only Channel Island to have a fresh-water marsh which provides habitat for its large bird population.

Directions: *Channel Islands Aviation flys to Santa Rosa out of the Camarillo Airport and you can call them directly for flight information at (805) 987-1301. You also can arrange a day-trip through the City of Ventura's outdoor adventure program.*

Other interests: Currently there is no written policy pertaining to mountain bikes on Santa Rosa Island. Permission to bike the island is handled on a case-by-case manner. Due to the remoteness of the area, you must obtain a backcountry permit before hiking, biking or camping on the island through the Channel Island National Park Visitor Center at: (805) 658-5730. If you take your own boat to the island you will need a permit to anchor, which is available at park headquarters.

Step by step:

1. The STOAL plane will land on the Becher Bay landing strip while the boat will dock at the ranch pier. Upon landing, your group will be met by the Park Service ranger and from there he will take you on a tour of the Vail & Vickers cattle rancheria and pier area. After a quick stop at the island's only outhouse, the four-wheel-drive tour of the northeastern side of Santa Rosa will begin.

2. As you explore the rough ranch roads the ranger/guide will take you to a Chumash Indian midden area (trash heap) on your way to the Torrey Pine forest. When you near the edge of the forest, the ranger will park the vehicle and take you on a

walking tour of the grove. There you will stop for a picnic lunch on a dramatic bluff overlooking the bay.

3. The next stop will be the beach, where you can explore some massive sand dunes and surf on your own for an hour or so. Later, the ranger will lead a short hike up Water Canyon. And then, all too soon, it will be time to return to the landing area for your return flight to the mainland.

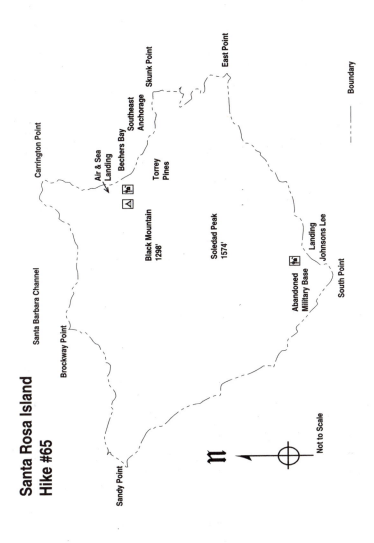

Santa Rosa Island
Hike #65

66. Island Walkabout

Difficulty: Moderate
Length: 5 mile loop
Elevation: 635 foot gain/loss
Type of trail: single-track trail
Kid appeal: * * * *

A trip to Santa Barbara Island is for the truly adventurous of spirit. Located almost 50 miles to sea, this tiny windswept island epitomizes the island experience. Extremely remote, this is the least visited, as well as the smallest of the five islands that make up Channel Islands National Park. It does, however, prove to be a sampler of the vast natural resources found on the other islands that make up this archipelago. Santa Barbara also is the only one of the southern Channel Islands included in the park; providing visitors a warmer, more temperate climate than its northern neighbors.

The island's 5 1/2 miles of hiking trails loop around to describe a large figure-eight across this marine terrace. The pathways are all single-track trails and are somewhat indistinct in spots. While most of the trails are fairly level over the open grassland, there is a 635 foot gain/loss in elevation at Signal Peak.

This arrow-shaped island with its sheltered coves and sheer sea cliffs is tremendously important as a "haul-out" and rookery site. Sea mammals including California sea lions, harbor seals and elephant seals breed and pup on Santa Barbara throughout the year. The endangered California brown pelican also has re-established nesting sites on the island. Spring is an excellent time to visit the island to view the rookeries as well as to take in the spectacular wildflowers. Goldfields carpet the island's arch point while giant coreopsis and buckwheat line the steep, rocky cliffs and canyons.

The island was named by explorer Sebastian Vizcaino on December 4, 1602 - St. Barbara's Day. This tiny island has been an important navigational site for thousands of years. While never permanently settled by the Gabrielino Indians that lived on the other southern Channel Islands, Santa Barbara Island served them as a seasonal hunting outpost for centuries. Being centrally located among the other Channel Islands, the island played an important role in pre-Columbian shipping and trading ventures between island and mainland populations.

Fishermen and farmers settled on Santa Barbara in the mid-1800's, but the lack of water brought the idea of permanent farming to an end by the 1920's. The island was turned over to the National Park Service in the late 1930's and used as a military outpost during World War II. Unfortunately, the servicemen introduced rabbits to the island during that time. These rabbits were very destructive to the fragile ecosystem and nearly destroyed what was left of Santa Barbara's unique habitat from the farming years. Since that time the National Park Service has maintained an aggressive stance on returning the island to a native state.

What you'll need: Santa Barbara Island is a very primitive area. Besides hiking gear, camera, binoculars and sunscreen, you must supply all of your own food and water for this outing. This is a very long day trip: count on it lasting at least 12 hours.

Step by step:
1. Visitors to the island must take a skiff from the boat to a landing area and then climb a ladder to the dock. This is not difficult but participants should be agile enough to go ashore under their own power. The resident ranger will be on hand and will give visitors a short talk about Santa Barbara's fragile island habitat and remind everyone to stay on the designated trails. Some trails may be closed to respect the privacy of nesting and/or breeding species at various times of the year. It is your option to go on a guided hike with the ranger or to strike out on your own.

2. Follow the steps up from the boat landing to the single-track trail that leads to the ranger station and visitor center. This path is about 1/4 mile long and gains 200 feet in elevation. At the center there is a seasonal display outlining island life as well as a map of the hiking trails. You can pick up a copy at the sign-in box by the museum.

3. The island's best wildflower display is at Arch Point. Backtrack toward the landing area and take the cliff-top trail north from the first junction. The signed Arch Point Trail heads due north to the point about 1/2 mile away. There is an excellent vista point near the NPS weather station recording equipment.

4. The trail turns south as it leaves the vista point to climb to the top of North Peak. About a mile from Arch Point, in a grassy saddle between North Peak and Signal Peak, there is a four-way trails junction.

5. From this junction you have three hiking options open to you. The trail to the left heads east and gradually downhill to the visitor center 1/2 mile away. The trail to the right heads down the other side of the ridge and to the west toward the Elephant Seal Cove Rookery a little more than a mile away. (This trail deadends at the vista overlooking the rookery, so keep in mind that is over one mile there as well as more than a mile back to the junction.) The trail straight ahead is the Signal Peak Trail. It leads over the top of Signal Peak and then down to the island's south side before looping back to the visitor center. This loop is about 1 1/2 miles long.

6. Before heading out in any direction, be sure to check how much time until your rendezvous with the boat back to the mainland: you don't want to miss your pick-up time.

7. A short self-guided nature trail, the signed Canyon View Trail, loops around the cliffs by the campgrounds and visitor center. This trail explores the island's cliff-side plant communities. It starts near the south end of the campground and continues for a few hundred yards to the rim of a steep canyon. Here it turns toward the ocean and returns to the center via a narrow path along the palisade overlooking the landing area.

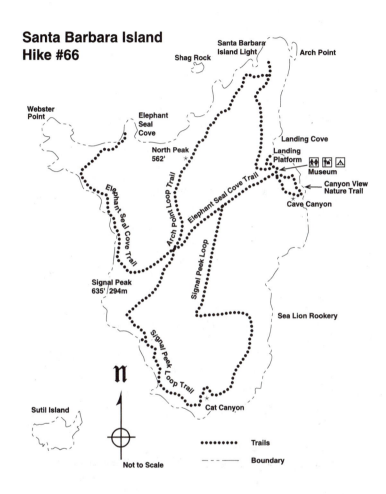

**Santa Barbara Island
Hike #66**

Shag Rock

Santa Barbara
Island Light

Arch Point

Webster
Point

Elephant
Seal
Cove

North Peak
562'

Landing Cove

Landing
Platform

Museum

Canyon View
Nature Trail

Cave Canyon

Elephant Seal Cove Trail

Arch Point Loop Trail

Elephant Seal Cove Trail

Signal Peak Loop

Signal Peak
635' / 294m

Sea Lion Rookery

Signal Peak Loop Trail

Cat Canyon

Sutil Island

N

Not to Scale

•••••••• Trails

– – – – – Boundary

183

Chapter Six

The Ojai Valley

Ojai Valley Trail

HIKES AT A GLANCE 8
The Ojai Valley

	DIFFICULTY	FACILITIES	FEATURES	FEES	OPEN TO
OJAI VALLEY					
67. Ojai Valley Trail	easy	w, t, p	views of Ojai & Ventura River	holidays/weekends	H, C, E, D, S
68. Sulphur Mountain Road	moderate/strenuous	none	ocean & mountain views, wildflowers	no	H, C, E, D, S
69. Lake Casitas Trail	easy	w, t, p, c, m, s	fishing, lake views, wildflowers	yes	H, C, S
70. Shelf Road	easy	none	views of Ojai, orchards	no	H, C, E, D, S
71. Stewart Canyon Debris Basin Trailhead	moderate/strenuous	none	connects to back country trails, wildflowers	no	H, C, E, D
72. Soule Park Walkabout	easy	w, t, p	playground, equestrian center	holidays/weekends	H, C, E, S
73. Horn Canyon Trail	easy /strenuous	none	shady canyon, hike-in camp	no	H, C, E, D
74. Santa Paula Canyon Trail	moderate	none	swimming holes, hike-in camp	no	H, C, E, D
75. Steckel Park Walkabout	easy	w, t, p, c, s	fishing & swimming in river	holidays/weekends	H, C, S
OJAI BACK COUNTRY					
76. Cozy Dell Trail	moderate	none	views, wildflowers	no	H, C, E, D
77. Matilija Camp Trail	easy	none	shady stream, hike-in camp	no	H, E, D
78. Murietta Trail	easy	none	shady stream, hike-in camp	no	H, E, D
79. Matilija Nature Trail	easy	w, t, p, c, rs	shady stream, chaparral wildflowers	no	H, D
PINE MOUNTAIN					
80. Reyes Peak Walkabout	easy	t, c	sub-alpine forest, views	no	H, C, E, D

FACILITIES: w - water, t -toilet, p - picnic table, rs - ranger station, c - camping, m - marina, s - store
OPEN TO: H - hikers, C - cyclists, E - equestrians, D - dogs on leash, S - stroller/handicap access

Ojai Valley

The Ojai area offers a wide variety of multi-use trails for hiking, bicycling and horseback riding that suit every level of expertise. The bike routes range from the nearly level and paved Ojai Valley Trail to the rough and ready single-track trails through the Los Padres National Forest. I have included a few of my family's favorite hikes and bike rides.

One of the fun aspects of going for a hike or bike ride in Ojai is browsing in the downtown art galleries, shops and ice cream parlors along Main Street after your outing.

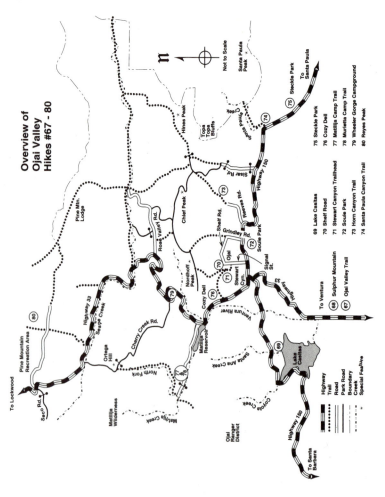

67. Ojai Valley Trail

Difficulty: Easy
Length: 18 mile round-trip
Elevation: 500 foot gain/loss
Type of trail: separate use paved bikeway
Kid appeal: * * * * *

A family bike ride is a pleasant way to wind down a long summer's day. The Ojai Valley Trail offers an easy, traffic-free venue that is just perfect for family outings. The trail connects two large Ventura County park facilities, Foster Park and Soule Park, located at either end of the Ojai Valley. Within the next couple of years the Ojai Valley Trail will be extended south to Emma Wood State Beach, making it possible to hike, bike or horseback ride all the way from Ojai to the sea.

Following the Ventura River inland, this unique separate-use trail is an excellent example of the "rails-to-trails" idea. In many areas across the United States, communities are turning out-of-use railroad right-of-ways into trails. Many people use these new trails for recreation, but a growing number of commuters also use the trails for non-motorized ways of getting to work.

The Ojai Valley Trail offers a glimpse of a slice-of-life in rural Ventura County. A bike ride along this route takes us through two very different worlds. The residential portions of the trail show how the people go about their day-to-day business of living and making a living. The open space areas, on the other hand, give us the chance to see the natural communities where the animals live. Also, the trail passes by some stunning vista points.

The Ojai Valley Trail is often cited as an outstanding example of a separate-use trail. What this means is that one side of the path is used by cyclists and hikers while the other side is used by equestrians. The two sides of the trail are separated by a wood rail fence. The surface of the horse trail is dirt while the hike and bike trail is paved.

Directions: *The Ojai Valley Trail begins in Foster County Park which is off Highway 33. From Ventura, exit Highway 101 on Highway 33 and take it north about 6 miles to Casitas Vista Road. Exit on Casitas Vista Road, turn right and follow the road under the freeway to the park on the other side. The park*

entrance is on the right side of the road just before the stone gateway to the bridge over the Ventura River. Turn into the park and drive on the service road through the picnic area and continue several hundred yards past the softball diamond to the parking lot at the end of the road.

Step by step:

1. The signed Ojai Valley Trail begins at the north end of the parking lot near a large information kiosk. The first mile of the path closely parallels the Highway, then the two are separated by a quaint residential area. About 1 1/2 miles from the trailhead there is a Texas-style creek crossing over San Antonio Creek. (The bike path is closed when the creek water is too high to cross safely.) For the next mile or so the trail goes through an open space area as it follows the contour of the Ventura River Valley hidden from the highway by an oak-dotted ridgeline.

2. The trail comes to the community of Oak View about 3 miles north of Foster Park. After skirting the perimeter of the community the trail returns once more to parallel the highway.

3. After 7 1/2 miles of relatively easy riding the trail comes to the intersection of Highway 33 and Highway 150. At this point the Ojai Valley Trail crosses the road to roughly parallel the route of Highway 150 east through Ojai, going first to Libbey Park before terminating near Soule Park Golf Course. After enjoying the sights and sounds of Ojai you can return to the Foster Park trailhead the way you came.

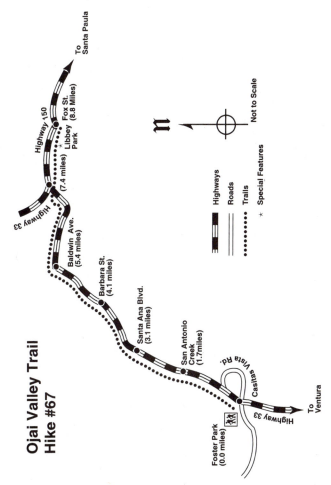

Ojai Valley Trail
Hike #67

Not to Scale

Highways
Roads
Trails
Special Features

To Santa Paula

Fox St. (8.8 Miles)
Libbey Park
Highway 150
(7.4 miles)
Highway 33
Baldwin Ave. (5.4 miles)
Barbara St. (4.1 miles)
Santa Ana Blvd. (3.1 miles)
San Antonio Creek (1.7 miles)
Casitas Vista Rd.
Highway 33
To Ventura
Foster Park (0.0 miles)

68. Sulphur Mountain Road

Difficulty: Moderate
Length: 20 mile round-trip
Elevation: 2,300 foot loss/gain
Type of trail: dirt/paved roadway
Kid appeal: * * *

Years ago, Sulphur Mountain Road was one of Ventura County's premiere Sunday drives. At an elevation of 2,600 feet, Sulphur Mountain was popular for its dramatic vista points overlooking the Ventura County coastline to the south and Ojai

Valley, Lake Casitas and the Topa Topa Mountains to the north. Later, when off-road vehicle activity became a problem, the road was closed to motorized traffic. Fortunately, Sulphur Mountain Road still remains open to those who are willing to leave their cars behind.

County-owned Sulphur Mountain Road is open to hikers, cyclists and equestrians. This beautiful old county road crosses the ridgeline of Sulphur Mountain between Highway 33 in Casitas Springs and Highway 150 through the upper Ojai Valley. Being closed to vehicular traffic, this well-graded, nine-mile road is an excellent site for a family mountain bike outing.

A real treat to outdoor enthusiasts, there is plenty to see any time of the year but late winter and early spring may be the best season for this outing. Southern views overlooking the rugged canyons of Canada Larga just north of Ventura are awash with color from the poppies, lupines and sages that grace their steep, dry slopes. These flowers, in turn, draw hummingbirds to gather in the area. Pacific breezes create air thermals that also attract birds of prey; you can often watch hawks and falcons sail on the westerlies in search of food.

Please remember to be courteous to all trail-users and also to stay on the trail at all times. Even though Sulphur Mountain Road is county-owned, the land around it is private property and off-limits to trail users. Cyclists also should control their downhill speed as many equestrians and family groups use the roadway on the weekends.

Outings on this county-owned road can be as easy or strenuous as you like depending on your choice of trailhead. One-way, downhill shuttle-trips as in today's ride are very easy. Riding the road from west to east (Highway 33 to Highway 150) will provide a great conditioning workout, as the trail gains 1,500 feet in elevation in the first 1 1/2 miles in that direction.

Directions: There are two trailheads for Sulphur Mountain Road. One is at the west end of the road, near its intersection with Highway 33 and the other is at the road's east end, off Highway 150. To reach the western trailhead take Highway 33 north from Highway 101 in Ventura, toward Ojai. Sulphur Mountain Road intersects Highway 33 in Casitas Springs, 1

1/2 miles north of Foster County Park, on the east side of the road. Watch for a maintenance turnout just before the intersection and turn right onto Sulphur Mountain Road. Follow it 1/4 mile to the locked gate at the road's end. Near the gate there is a large county sign declaring the road open to hikers, bikers and equestrians but closed to motorized vehicles. Parking is restricted in this area, so be sure to obey all parking signs. In the interest of keeping access to the trail open to future trail-users, please do not block the gate or park on private property.

To reach Sulphur Mountain Road's eastern trailhead stay on Highway 33 north until it intersects with Highway 150. Continue straight ahead on Highway 150 through the city of Ojai towards Santa Paula. Sulphur Mountain Road intersects Highway 150 outside of town, about 4 miles east of Soule County Park. There are a number of mailboxes and trash bins lined up near the intersection that make it easy to spot. Turn right and follow it 4 3/4 miles to the top of the mountain. There is a locked gate at the road's end. Park along the side of the road, once again being careful not to block the gate or park on private property.

Step by step:
1. Today's family bike ride begins at the top of Sulphur Mountain. The trail starts at the locked gate by following the paved road uphill for a short way before beginning its nine-mile descent toward Casitas Springs. After a little more than a mile of easy, mostly downhill riding, the road surface changes from pavement to dirt.

2. About 3 miles from the eastern trailhead, a major power line crosses the road near a rather sharp switchback. This is an excellent vista point of the Channel Islands. Another mile of riding brings you to an intersection with another dirt road. This marks Sulphur Road's halfway point and is near an open oakwoodland. The dirt road is off-limits to trail users, but be sure to look for the U.S. Survey Benchmark in the northwest corner of the intersection. This is about 5 miles from the road's end at Highway 33. At this point there are views in both directions as the road crosses the ridgeline to continue downhill on the mountain's shady north-facing slope for another mile or so.

3. The road returns to the ridgeline once more before heading seriously downhill. The route loses 1,500 feet in elevation over the next two miles as it descends the mountain toward San Antonio Creek and the trail's end near Highway 33.

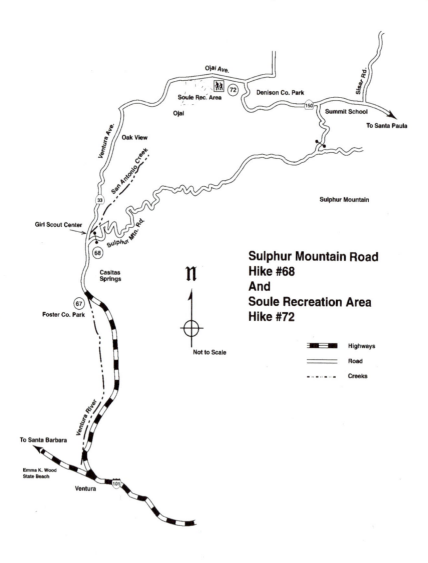

Sulphur Mountain Road Hike #68 And Soule Recreation Area Hike #72

Not to Scale

Highways
Road
Creeks

69. Lake Casitas Trail

Difficulty: Easy
Length: 1/2 mile round-trip
Elevation: 200 foot gain/loss
Type of trail: single-track trail
Kid appeal: * * * * *

Lake Casitas was created in 1956. Over the years it has proven to be a popular recreation area, best known for its first rate fishing and camping. Recurring droughts and subsequent water rationing in the Ojai area inspired civic leaders to propose the formation of the lake. From the conceptual stages it was determined that the reservoir serve a dual purpose: to be a domestic drinking water supply as well as a public recreation area.

The Lake Casitas Recreation Area includes more than 6,000 acres of rolling, oak-studded hillsides surrounding the beautiful fish-stocked lake. The public facilities are open year-round for boating, camping, fishing and picnicking. The recreation area is a very popular site on the weekends and holidays; but it sees fewer visitors during the week, especially in winter months, making it a good mid-week hiking venue. While there is no designated hiking trail through the park, there is a short, single-track trail that leads to a hilltop vista of the lake as well as the campground service roads to hike on.

Today's short hike above the lake brings you to an excellent vista point of the waterfront and the Santa Ana Valley. The chaparral covered slopes surrounding the lake are protected as part of the Charles M. Teague Memorial Watershed. Purchased with federal funds, the 3,100 acre watershed preserves the lake's water supply from the pollution of over-development.

Chaparral and oakwoodlands dominate the hills around the lake, providing suitable habitat for many wildlife species. Be sure to look for coyote scat as well as deer and raccoon tracks along the trail. The streams that feed into Lake Casitas, particularly Santa Ana Creek near the main picnicking area, serve as nurseries and feeding grounds for the lake's aquatic animals. Winter brings a number of migratory bird species to the lake, including loons, grebes, cormorants, ducks, egrets and herons.

Early spring is an excellent time for a walk around Lake Casitas as the hike to the vista point highlights several native wildflowers. This area is especially noted for its display of ceanothus, or California lilac. This large shrub grows in the transition zone between chaparral and woodland and is a prolific bloomer. Native Americans prized this plant for the delicate soap-like qualities of its beautiful blue flowers.

Directions: *To reach the Lake Casitas Recreation Area, take Highway 33 north from Highway 101 in Ventura. Exit Highway 33 about 6 miles north of Ventura onto Casitas Vista Road. Turn right, and follow the road back under the highway as it heads west over the Ventura River and passes the southwest corner of Foster Park. A quarter-mile beyond the park, turn right onto Santa Ana Road following the sign indicating Lake Casitas. Take Santa Ana Road about 3 more miles to the Recreation Area's main entrance at 11311 Santa Ana Road.*

Step by step:

1. From the entry kiosk, follow the paved park service road to the right toward the signed camp store. As you continue past the store you will see some camping and picnic areas. About 100 yards beyond picnic site #8, there is a large turnout on the lake side of the road overlooking the Marina. Park your car here.

2. Today's hike begins on the opposite side of the road. Look for a well-worn trail that heads steeply uphill through the chaparral. This unsigned trail leads to the vista overlooking the lake.

3. At the top of the hill, the trail levels out and meanders north along the oak-dotted ridgeline. The path ends above the camp store, near the Recreation Area boundary. After enjoying the view, you must return to your car the way you came.

Other interests: Lake Casitas Recreation Area is open year-round. Facilities include: camping and picnic areas, tables and stoves, restrooms with coin operated showers, boat ramps, snack bar and marina. Fishing licenses, bait and rental equipment are available at the Bait and Tackle shop. Since the lake is a source of domestic drinking water, no body contact is allowed. Horses are not allowed in the campground. There is a day fee for parking and dogs.

70. Shelf Road

Difficulty: Easy
Length: 3 1/2 miles round-trip
Elevation: 200 foot gain/loss
Type of trail: dirt roadway
Kid appeal: * * *

Ojai's Shelf Road is a quaint country lane that runs east and west along the north edge of town. This old, county owned dirt road is gated at both ends and closed to vehicular traffic. It is a perfect route for beginning mountain bike riders. Shelf Road climbs and descends less than 200 feet as it follows the contour of the foothills above downtown Ojai, making it a very easy bike ride that the whole family can enjoy.

More than likely what you'll see along this charming old country road is other people. Just a few hundred feet above the valley floor, there are a number of excellent vista points of the east end of Ojai. Be sure to watch your speed at all times as walkers, runners and horseback riders also frequent this popular trail.

Shelf Road is open year-round, but spring is probably the best time to make this ride. Not only are the hills covered with wildflowers, but the scent of orange blossoms also fills the air, making for a memorable outing. During the summer months, Shelf Road makes a lovely early morning or evening ride.

Directions: *To reach the Shelf Road Trailhead take Highway 33 from Highway 101 in Ventura, north to Ojai. Stay on Highway 33 until it turns off from Highway 150. Then bear to the right and take Highway 150 (Ojai Avenue) through the city to Signal Street. Turn left at the traffic signal and go north on Signal Street to the locked gate. Shelf Road begins on the other side of the gate and continues east 1 3/4 mile to Gridley Road. Additional parking is available at the Stewart Canyon Debris Basin; look for the turn off just before the water tank and follow the Los Padres Forest Service signs for the Foothill and Pratt Trails.*

Step by step:

1. Today's bike ride begins at the locked gate at the north end of Signal Street. Ride your bike around the gate and follow the unsigned dirt road as it curves slightly uphill to the east. After the first bend in the road, there are some fine close-in views of the city of Ojai against the backdrop of Sulphur Mountain. From this vista, the road then skirts the perimeter of a large avocado and orange orchard as it continues east.

2. Shelf Road dips and turns following the contour of this low, east-west ridgeline. Gridley Road, 1 3/4 miles from Signal Street, marks the terminus of Shelf Road. At this point, turn your bike around and return the way you came.

71. Stewart Canyon Debris Basin Trailhead

Difficulty:	Moderate
Length:	3/4 mile connector trail
Elevation:	400 foot gain/loss
Type of trail:	single-track trail
Access to trail network:	
	Foothill Trail (#22W09)
	Pratt Trail (#23W09)
Kid appeal:	* * *

The Stewart Canyon Debris Basin, just north of Ojai, serves as a convenient trailhead for the extensive network of trails through the Los Padres National Forest. Long noted for its spring display of wildflowers, the Stewart Canyon area provides access to two major National Park Service trails. They are the Foothill Trail, which is an east-west, intermediate ridgeline route above the city and the Pratt Trail, a north-south route that leads steeply up Nordhoff Ridge to the Lookout Tower.

This multi-use access-trail takes trail-users around a residential neighborhood north of the basin as it winds its way through the lower canyon to the various trail junctions. This rough and rocky single-track trail passes through two unique plant communities before joining the other trails. Stewart Canyon is a beautiful inland canyon that is graced with several wildflower meadows as well as a number of seasonal waterfalls. The eastern side of the Stewart Creek barranca is thought to be the site of a centuries old Oak Grove Indian encampment.

Directions: *To reach the Stewart Canyon Debris Basin Trailhead, take Highway 33 north from Highway 101 in Ventura to Ojai. Stay on Highway 33 until it turns off from Highway 150. At this intersection bear to the right and take Highway 150 (Ojai Avenue) through town to Signal Street. Turn left at the traffic signal and go north on Signal Street. Near the end of the street, watch for the left hand turn just before the water tank. Follow the Los Padres National Forest signs to the Debris Basin parking area.*

Step by step:

1. The unnamed access trail begins near the northwest end of the Debris Basin. The path heads through some dense brush as it follows the course of the generally dry creekbed on its way up Stewart Canyon. A few hundred yards from the trailhead, the trail crosses a couple of dirt roads before entering a large eucalyptus grove. This was once the site of the old Foothills Hotel that burned down decades ago. Be sure to stay on the well-marked pathway through this area as there is private property on either side of the trail.

2. The trail climbs steadily, approaching the homes along Foothill Drive. At this point the narrow path descends the deepening creekbed as it threads its way around the yards backing up to the creek. Trail signs placed strategically at road crossings indicate the access trail's course through the neighborhood.

3. Before too long, the trail joins upper Foothill Road Firebreak as it heads north toward Nordhoff Ridge. The first trails junction is located off this dirt road near a water tank, a little more than 1/2 mile from the trailhead. The signed Pratt Trail and Foothill Trail begin on the left side of the road. The two trails start together as a narrow path leading uphill along the west side of Stewart Creek.

4. A few hundred yards from the road, the two trails split apart. The signed Foothill Trail heads west along the ridge toward Cozy Dell which is about 2 miles away. The signed Pratt Trail

follows Stewart Creek as it begins its ascent to the Lookout Tower on Nordhoff Ridge about 3 1/2 miles away.

5. Walking north on the Foothill Road Firebreak (the old Pratt Trail), about 100 yards farther up the road there is another trails junction for the Foothill Trail. Look for it on the right side of the road. It is a single-track trail that leads slightly uphill through the chaparral around a rocky bluff. From this junction the Foothill Trail continues east for 2 miles to Gridley Road.

6. Continuing another 150 yards up the Foothill Road Firebreak near a large meadow overlooking the creek, there is a trails junction with the Ojai Front Fuel Break. The fuel break is a well-graded east-west route that is 2 1/2 miles long and ends at the north end of Gridley Road. Continuing north, straight ahead, on the Foothill Road Firebreak you will cross Stewart Creek and climb a hill before coming to the Cozy Dell Fire Road. At this point the trail continues as a single-track trail as it makes its way to the top of Nordhoff Ridge.

7. This system of trails and fire roads offers trail-users a great many day-hike options that can loop back to Stewart Canyon Debris Basin. Close in to town the trail is well signed but if you decide to strike out into the backcountry be sure to go by the ranger station and pick-up a free trails map of the area.

72. Soule Park Walkabout

Difficulty: Easy
Length: 2 mile round-trip
Elevation: nearly level
Type of trail: paved park road
Kid appeal: * * * * *

Even without a nature trail, strolling through Soule County Park's beautiful picnic grounds is an ideal family outing any time of year. The park is especially serene in the early morning hours, when the soft light reflects on the mist rising from the creekbed. This small park at the foot of Ojai's Black Mountain offers an excellent, gentle alternative to the more aggressive Forest Service trails leading into the nearby Los Padres Forest.

Surrounded by Christmas tree farms and avocado orchards, the picnic grounds occupy a large shady meadow overlooking the Soule Recreation Area Golf Course. Stately sycamore, oak and ash trees grace the landscape creating a cool, sylvan canopy in summertime and providing plenty of leaves to jump in come fall.

The juxtaposition of the rural setting around the park and the nearby arroyo creates a fabulous winter birding habitat that is right outside town. Be sure to bring a field guide with you to help identify the various migrant species of birds that make the park their winter home.

Directions: *To reach Ojai, take Highway 33 north from Highway 101 in Ventura. Stay on Highway 33 until it turns off from Highway 150. Then bear right and take Highway 150 (Ojai Ave.) east through the city. Soule County Park is located at the end of Boardman Road. From Ojai Ave. turn right, south onto Boardman Road and follow signs to park.*

Step by step:

1. Park your car near the park entrance. Today's walk follows Soule Park Drive about a mile to the gymkhana arena. Near the entrance, a small tributary to San Antonio Creek creates a wash that separates the park's meadow from the golf course. This open area is a good winter birding spot.

2. Near the halfway point of today's walk, by the park host residence, a small seasonal stream crosses the meadow. As you approach the equestrian center, the roadway opens up considerably to reveal the arena and golf course. After enjoying the vista at road's end, you can return the way you came or walk across the lawn area to get back to your car.

73. Horn Canyon Trail

Difficulty: Easy/strenuous
Length: 3 miles round trip
Elevation: 1,860 foot gain/loss
Type of trail: fire road, single-track trail
Kid appeal: * * *

Located behind Thacher School in Ojai, Horn Canyon is an excellent spot for an easy nature walk or a rigorous hike to a fantastic vista point. The trail follows a year-round stream a short way before climbing the steep chaparral covered slopes to The Pines hike-in camp. The climb is awesome, but once at the shady little camp the spectacular views of Ojai Valley are well worth the effort.

The lower portion of the Horn Canyon Trail explores the riparian plant community. This part of the hike is nearly level and the walking is easy as the path leads under a mixed canopy of sycamore, walnut and oak trees. Leaving the canyon bottom to climb nearly 1,800 feet over the next mile to the tiny ridge-line camp, the rest of today's hike provides an aerobic workout.

Horn Canyon is a beautiful spot from which to view Ojai Valley and the dramatic peaks that surround it. This rugged wilderness area is home to a number of wildlife species. Be sure to look for scat and animal tracks along the trail. See how many different ones you can identify. Besides land animals, be sure to listen for the cry of hawks in the chaparral as well as the scoldings of the scrub jays whose nests are in the trees by the creek.

Directions: *To reach Ojai, take Highway 33 north from Highway 101 in Ventura. Stay on Highway 33 until it turns off from Highway 150. Then bear right and take Highway 150 (Ojai Ave.) east through the city. Once outside town, turn left on Reeves Road, near the Boccali Pizza & Pasta House. Take Reeves Road about a mile to McAndrew Road and make another left, following the signs for Thacher School. Another mile up the road, take the road into the school grounds. Stay to the right and follow the signs to the gymkhana field. Park near, but do not block, the fire gate at the trailhead.*

Step by Step:

1. At first, the Horn Canyon Trail (22W32) follows a rough service road upcanyon. Near a shady stream crossing the trail shrinks down to a single-track path and continues to follow the creek upstream for another 1/4 mile. Soon you will come to a clearing around an old well; here the trail leaves the canyon on its way to The Pines.

2. The next mile or so of trail is very steep as it climbs the canyon wall to the ridgeline camp. Not only is the path tree-less, it is also so narrow that there is no place to stop and enjoy the view. The camp itself is nestled in a grove of pine trees that was planted by Thacher School students following the 1948 wildfires. After enjoying the vista, return to your car the way you came. (Believe me, it's much easier going down). If you feel like hiking more, the trail continues up the ridge to join the Nordhoff Ridge Road another two miles away; from here a number of trails go on to explore the Sespe Wilderness.

74. Santa Paula Canyon Trail

Difficulty: Moderate
Length: 6 mile round-trip
Elevation: 700 foot gain/loss
Type of trail: single-track trail
Kid appeal: * * * *

Santa Paula Canyon is one of the Ojai front county's most popular hiking venues. Noted for its deep, bedrock swimming holes and spectacular waterfalls, the well-used trail through this beautiful canyon is a favorite with local outdoor enthusiasts.

The trail sees its heaviest use during the spring and early summer, when fishing and swimming are the main activities along the swift flowing headwaters of the Santa Paula Creek. The stately sycamore trees that line the creekbed put on an excellent display of fall-color making this a great hike in the autumn. Also as the summertime crowds fade, there is a little more breathing room on the trail, especially if you can get away for a mid-week hike.

Located behind Ferndale Ranch and Thomas Aquinas College, Santa Paula Canyon is one of the main animal corridors between the Los Padres Forest and the Sulphur Mountain area. It is an area closed to firearms, making this a safe hiking spot even in the middle of deer hunting season. Obviously aware of this restriction, there are lots of deer in and around the canyon. Bears have been spotted from the two walk-in camps, Big Cone and Cross camp, located about 3 miles upstream from the trailhead, near the waterfalls and swimming holes.

As Ferndale Ranch and the college grants hikers and equestrians easement to cross their property to use these U. S. Forest Service Trails (21W09 and 4N03), please be sure to respect their private property rights along the first 1 1/2 miles of the hike.

Directions: *From Highway 101 in Ventura, take Highway 126 east to Santa Paula and exit onto Highway 150 toward Ojai. Look for the sign to Ferndale Ranch and Thomas Aquinas College which are located on the west side of Highway 150, about half way between Santa Paula and Ojai. There is a signed parking lot for this trailhead on the east side of Highway 150 near the bridge over Santa Paula Creek, opposite the entrance to the college.*

Step by step:

1. After parking your car in the signed lot off Highway 150, cross the road to the college entrance. Follow the hiking signs along the campus road to the trailhead. Take the Santa Paula Creek Trail about 1/4 mile across the ranch easement where you will pick up the main trail near an avocado grove.

2. The trail follows the creek up the canyon a ways before crossing it and joining a fire road that gently switchbacks up the grade making its way upcanyon. About 2 miles from the trailhead the road crosses the creek once more. The fire road climbs up and around Hill 1989 and then crosses a level saddle before entering Big Cone Camp overlooking Santa Paula Creek.

3. From the walk-in camp, a narrow footpath makes its way down to the creek to the swimming holes and waterfalls.

The trail crosses the creek and then switchbacks up the hill to an unsigned trails junction. To the right a poorly maintained trail leads east to shady Cienega Camp that occupies a large meadow about 1 1/2 miles away. To the left, the trail continues up Santa Paula Creek as it climbs another mile to Cross Camp and some even more splendid swimming holes. After enjoying this shady retreat, return to your car the way you came.

Please note: Santa Paula Creek has a very strong current following a heavy rainstorm. Please use caution when swimming here. Also rangers tell me that they do more rescues in Santa Paula Canyon than any other area of the Los Padres Forest due to trail-users diving or jumping into the creek.

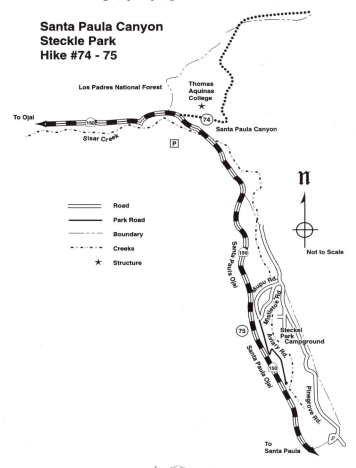

Santa Paula Canyon
Steckle Park
Hike #74 - 75

Los Padres National Forest

Thomas Aquinas College ★

To Ojai

150

74

Santa Paula Canyon

Sisar Creek

P

	Road
	Park Road
	Boundary
	Creeks
★	Structure

Santa Paula Ojai

150

Mupu Rd.

Mistletoe Rd.

Not to Scale

75

Aviary Rd.

Steckel Park Campground

Santa Paula Ojai

150

Pinegrove Rd.

To Santa Paula

75. Steckle Park Walkabout

Difficulty: Easy
Length: 2 mile loop
Elevation: nearly level
Type of trail: paved park road
Kid appeal: * * * * *

Situated along the shady banks of Santa Paula Creek, Steckle Park offers outdoor enthusiasts a quiet place to enjoy a lazy afternoon or a weekend getaway. The day-use section of the park is made up of a number of oak-canopied picnic areas; some of which are suitable for large groups and others in secluded coves are perfect for families. Campers looking for a relaxing spot along the river to pitch their tent or park their RV are welcome at the Farwest Resorts campground located up the hill, along the east bank of the creek.

Steckle Park is a popular spot for weekend gatherings but is virtually deserted mid-week, making it an ideal hike site. While there are no designated hiking trails through the park, the rustic service road winds around the wooded picnic area, providing a couple of miles of easy nature exploration. Quaint river-rock picnic tables in a sylvan setting and free-roaming peacocks add to the park's ambiance, helping to create a nostalgic feeling.

Directions: *From Highway 101 in Ventura take Highway 126 east to Santa Paula and exit at 10th Street. Turn onto Highway 150 and head north through town toward Ojai. Stay to the right on Highway 150 as it joins Ojai Road and continues north a little less than 5 miles to Steckle Park. The entrance to the day-use portion of the park is off Mistletoe Road; campers should use the Mupu Road park entrance, about 200 yards up the road.*

Step by step:

1. Once inside the park, leave your car in the lot opposite the bird aviary. Follow the service road to the right to explore the park's south end. The road heads straight through the oak-woodland about 1/2 mile before turning toward the creek. From here the road splits to parallel the creekbed in both

directions as it leads to the various picnic sites along the west bank. The park road continues another half mile before coming to an end. You may explore the creek farther south by rock-hopping the creekbed to the park's boundary.

2. From here you can follow the service road back to the central part of the park or make your own path through the shady' picnic area.

3. Walking north from the aviary along the park road will bring you to Mupu Road and its bridge over Santa Paula Creek. This road is open to residential traffic so watch for cars on the roadway. Mupu Road leads uphill to the campground area which is closed to non-campers. The well-kept camp ground has separate facilities for RV and tent camping. There is a beautiful group campsite, River's Edge, that is perfect for youth groups. For information, call (805) 933-3200.

Ojai Backcountry

Following the Maricopa Highway north from Ojai brings you to the rugged backcountry of the Los Padres National Forest. Placed under government management at the turn of the century as a way to protect the area's valuable virgin watershed, land was set aside in several locations throughout the Santa Ynez Mountains as part of the first national forest system.

There are a number of excellent trails through the area that are perfect for family outings. These mountains are wetter than their smaller cousins, the Santa Monica Mountains, to the south. Many of the mountain streams contain swimming holes that have been carved into the bedrock by eons of spring floods.

Several areas bear wilderness designation such as the Matilija and Sespe Wilderness Areas and are closed to bicyclists. In these areas, improved wilderness campsites will eventually be allowed to revert to a more natural state. Other areas such as the Pine Mountain Recreation Area are open to bicycles and allow car camping. At the time that I am writing this book, the status of the Rose Valley Recreation Area is up in the air. Some groups feel that this area should be privatized as some other Ventura County Campgrounds are, such as Steckle Park. Those that advocate this plan say that the U.S. Forest Service cannot adequately manage the property. For that reason the trails in Rose Valley have been omitted from this edition of the guide.

Directions: *To reach the Los Padres Forest Service trails in the Ojai Backcountry take Highway 33 from Highway 101 in Ventura, north toward Ojai. Just outside of town, about 12 1/2 miles from the freeway, be sure to stay on Highway 33 as it turns left and heads toward Maricopa. (This section of Highway 33 is known as the Maricopa Highway and it is a designated scenic route.)*

What you'll need: First aid supplies are a good idea for any backcountry excursion. Call the Ojai Ranger District office for information and wilderness camping permits if planning to spend the night at: (805) 646-4348.

76. Cozy Dell Trail

Difficulty:	Moderate
Length:	3 miles round trip
Elevation:	600 foot gain/loss
Type of trail:	single-track trail
Kid Appeal:	* * *

Close to town, Cozy Dell Trail makes a wonderful half-day hike, leaving you with plenty of time for sightseeing around Ojai's quaint downtown area. This is an outstanding spring wildflower hike that begins near the Friends Ranch Packing House off the Maricopa Highway. The trail starts in a shady canyon and then climbs up into the open chaparral before coming to an end in Cozy Dell Canyon. This path brings you through a whole array of wildflowers, from shade lovers near the trailhead to vast fields of morning glory and lupine on the open hilltops. Besides the flowers, this trail provides some tremendous views of Ojai Valley and the mountain peaks around it.

The first part of the trail climbs 600 feet in elevation up the shady north face of a rocky dry wash. The canopy of live-oaks that shelters this canyon provides a mild mini-ecosystem in which many shade-loving plants thrive. Fuchsia-flowered gooseberries, soap lilies and ferns are a few of the plants that grace this part of the trail. The chaparral portion of this hike is vibrant with the yellow, blue and pink hues of the flowers spreading across the hills. There are some excellent vista points of the peaks surrounding the valley: To the west is Old Man Mountain and Divide Peak, to the east Nordhoff Peak is visible. Many birders enjoy this hike as the chaparral is prime habitat for a number of different species from sparrows and towhees to hawks and other raptors.

Directions: *From Highway 101 in Ventura take Highway 33 north to Ojai. Parking for the Cozy Dell Trailhead is in a turnout on the west side of Highway 33, about 3 1/2 miles north of town. Watch for the Friends Ranch Packing House on the east side of the road and the large, brown US Forest Service sign that identifies the trail by name and number: Cozy Dell Trail #23W26.*

Step by step:

1. From the parking area, cross Highway 33 near the bridge over the usually dry Cozy Dell Creek. The trail begins along the streambed and follows it east into the canyon. The first part of the path along the streambed is relatively level. But after 1/4 mile of easy going the trail heads aggressively uphill via a number of well-worn switchbacks.

2. About a mile from the trailhead the path finally tops out near an open saddle area that sports a good view of Meiners Oaks and the west end of the valley. From the saddle, the trail dips down into the chaparral through a forest of ceanothus. Here the trail levels out once more as it follows the contour of the canyon east for another 1/2 mile. At this point there is an excellent vista of the trail's end in Cozy Dell about 1 1/2 miles away.

3. The trail drops 200 feet to enter the Cozy Dell Canyon area where it connects with the Pratt and Foothill trails. After enjoying the Dell's open oakwoodland you can return to your car the way you came, 2 miles back to Highway 33.

Optional: Many hikers make this trip a "shuttle-hike" by using two vehicles. They park one vehicle at the Stewart Canyon

Debris Basin and continue to hike east out of Cozy Dell on the signed Pratt Trail and Cozy Dell Road instead of backtracking to Highway 33. To reach this parking area, take Highway 150 through Ojai to Grand Avenue and turn left. Drive up North Signal Street to the top of the hill, turn left and follow the road to the Debris Basin Area. Park one vehicle here and drive the hiking party to the trailhead near Friends Packing House. After hiking trail #23W26 to Cozy Dell, continue straight up the hill to the Cozy Dell Road and take it to the signed Pratt Trail #23W09. Take the Pratt Trail east though Stewart Canyon to the Debris Basin area. Be sure to stay on the well-signed trail in this area as it crosses through private property.

77. Matilija Camp Trail

Difficulty: Easy
Length: 2 mile round-trip
Elevation: nearly level
Type of trail: fire road, single-track trail
Kid Appeal: * * * * *

Matilija Camp is nestled in the narrow canyon along the North Fork of Matilija Creek. This shady hike-in camp is situated on a small knoll overlooking several of the creek's medium sized swimming holes. Located about a mile from the trailhead, this is an excellent spot for a day-hike picnic as well as a perfect destination for a beginner backpacking trip.

The first part of the hike is along the dirt fire-road that goes through Matilija Canyon Ranch Wildlife Refuge, which is private property. The ranch owners grant access through their property to Los Padres National Forest to hikers and equestrians only - no bikes are allowed in the Wildlife Refuge.

Beginning backpackers will love this little camp. The hike in is beautiful, nearly level and quite short. But with five creek-crossings it is not uneventful. Even though Matilija Camp is close to town, the camp offers a solitude usually associated with more remote backpacking venues. A campfire permit is required if you would like to backpack into Matilija Camp or any of the Los Padres National Forest backcountry camp sites. A permit may be obtained in person or by mail from the Ojai District Ranger Station: 1190 East Ojai Avenue, Ojai, CA 93023. For more information call (805) 646-4348.

You will enjoy strolling through the ranch area and checking out the organic fruit stand. The ranch road through the river basin is fairly open and the huge boulders and lack of trees makes the area appear almost desert like and quite a contrast to the cool, tree-canopied canyon.

Directions: *From Highway 101 in Ventura take Highway 33 north to Ojai. Continue on Highway 33 north out of Ojai toward Maricopa. Watch for the intersection with Matilija Road on the left side of the highway about 5 miles outside of town. The intersection may be unsigned but it is easily recognizable from the large number of mailboxes along the highway. Take Matilija Road another 5 miles, passing the dammed Matilija Lake, to where it deadends at a locked gate and park in the lot provided.*

Step by step:
1. From the parking area, proceed on foot around the locked gate at the end of the road. (Be sure to stay on the road, heeding the No Trespassing signs for the wildlife refuge.)
2. Once past the ranch, there are a couple of Texas-style stream crossings in the half-mile walk to the beginning of the trail. Look for the signed Matilija Camp Trail #23W07 on the north side of the fire-road. Take the single-track trail to the right as it heads upcanyon, back over Matilija Creek to follow the Creek's North Fork upstream.
3. The path becomes quite shady as it follows the creek deeper into the narrow canyon. After another half-mile of hiking and crossing the creek three more times you will arrive at the walk-in camp. Here you can soak in the cool water in one of the many pools formed in the boulder strewn creekbed.
4. After enjoying the delights of peaceful Matilija Camp, return to the trailhead the way you came.

Optional: Only 10 miles north of Ojai, near the civilized edge of the Los Padres National Forest, Matilija Camp is on trail #23W07. This trail is about 10 miles long and continues beyond Matilija Camp to two more walk-in campsites. First to Middle Matilija Camp which is another two miles upstream and then onto Maple Camp, another 4 miles up the canyon. From here the indistinct trail climbs over Ortega Hill to the beginning of Sespe Creek and Cherry Creek Campground a few miles off Highway 33 on Cherry Creek Road.

78. Murietta Camp Trail

Difficulty: Easy
Length: 2 mile round-trip
Elevation: nearly level
Type of trail: fire road, single-track trail
Kid Appeal: * * * * *

Claimed to be the hideout of the notorious Mexican badman Joaquin Murietta, Murietta Camp is another perfect destination for a day-hike picnic or an easy overnight backpack trip. Using the same trailhead as Matilija Camp, the trail to this primitive walk-in campsite explores the south side of Matilija Canyon.

Directions: *From Highway 101 in Ventura take Highway 33 north to Ojai. Continue on Highway 33 north out of Ojai toward Maricopa. Watch for the intersection with Matilija Road on the left side of the highway about 5 miles outside of town. The intersection may be unsigned but it is easily recognizable from the large number of mailboxes along the highway. Take Matilija Road another 5 miles, passing the dammed Matilija Lake, to where it deadends at a locked gate and park in the lot provided.*

Step by step:

1. The same as for the trail to Matilija Camp, park in the dirt lot at the end of Matilija Road and proceed around the locked gate on foot. Follow the road through the Matilija Canyon Ranch Wildlife Refuge to the second Texas-style stream crossing. The signed Murietta Trail #24W07 begins on the left side of the road, just beyond the crossing. Take this single-track path south as it follows the canyon bottom to Murietta Camp.

2. Over the next 1/2 mile the trail becomes increasingly more shady and narrow, forcing the path to cross Murietta Creek.

After rock-hopping the creek, you will find that the trail continues on the other side, where it climbs a steep, small grade.

3. The trail follows the creekbed upstream through this riparian area for another mile before coming to the campsite. Here on a small sylvan flat, an open oakwoodland makes a picture-perfect walk-in camp. The trail continues a short way beyond the campsite. It grows quite indistinct as it leads upstream to several more shallow swimming holes. After enjoying your picnic or campout in peaceful Murietta Camp, return to your car the way you came.

79. Matilija Nature Trail at Wheeler Gorge Campground

Difficulty: Easy
Length: 3/4 mile loop trail
Elevation: 150 foot gain/loss
Type of trail: single-track trail
Kid Appeal: * * * * *

Continuing up the Maricopa Highway another 5 miles beyond Matilija Canyon brings you to the Wheeler Gorge area. The first landmark is the Wheeler Hot Springs and then in a few miles is the Wheeler Gorge Campground located on the North Fork of the Matilija Creek. The shady campsites next to the creek make this one of Ventura County's most popular car-camp destinations. At the north end of the campground is the Matilija Nature Trail, which is just the ticket for a short easy hike along a mountain stream to refresh your spirits.

Wheeler Gorge is a beautiful place to hike and the easy nature trail is a good path for even the youngest of hikers. The deep bedrock pools and huge boulders in the tree-canopied creekbed can provide a delightful venue for a few hours of relaxed nature exploration.

This trail passes through two outstanding examples of the plant communities common to the mountains north of Ojai. The deeply shaded riparian area around the creek is wonderful anytime of year but it is especially so in the fall when the leaves are turning. The chaparral environment along the last part of the hike puts on a tremendous wildflower display in the spring.

Many of the plants found in both of these ecosystems are also found in the Santa Monica Mountains further south.

Directions*: From Highway 101 in Ventura take Highway 33 north to Ojai. Continue on Highway 33 north from Ojai about 8 1/2 miles to the Wheeler Gorge Campgrounds. Proceed another 1/4 mile up the highway to the turnout for the signed Matilija Nature Trail on the left side of the road.*

Step by step:

1. The signed Matilija Nature Trail begins next to the parking area on the south side of the creek. The path follows the creek upstream and under the bridge on Highway 33. There are a number of footpaths that lead down to the creek in this area, but the nature trail is clearly marked with arrow trail signs. About 100 yards upstream, the trail crosses the creek and then heads east as it parallels the creekbed along the north bank.

2. After about 1/4 mile the path leaves the tree-canopied stream area and heads up into the chaparral. There is a trails junction near a lone pine tree; follow the arrow sign pointing uphill and to the left to continue on the signed nature trail. The trail going straight ahead leads back to the creek and several large pools.

3. The nature trail continues uphill to a vista point with a view of the gorge area. From the top of the grade the trail heads west as it winds its way down the north side of the slope. Before too long you are near the highway once more and the trail leads back down into the stream area near the bridge. After exploring the deep pools in this part of the creekbed, take the trail back under the bridge to the parking area.

Pine Mountain Recreation Area

Pine Mountain Recreation Area offers outdoor enthusiasts miles of prime mountain trail through a thick forest of giant pines. Off of Highway 33, this backcountry campground is just 30 miles north of Ojai. At an elevation of almost 7,000 feet, this dramatic ridgeline is like an island rising out of a sea of chaparral covered badlands. Known to locals as the best car-camping site in Ventura County, Pine Mountain is also the trailhead for a number of trails through this remote area of the Los Padres National Forest.

Just getting to the Pine Mountain campgrounds on the narrow, pot-hole filled road from Highway 33 can be an adventure. While it may appear intimidating, the roadway is okay for most two-wheel drive vehicles with good ground clearance. This ridgeline drive through the scrub chaparral gives little indication of the sub-alpine forest at the road's end, 14 miles away. At the mountaintop, a dozen or so campsites are close to the road situated among the huge conifers and numerous sandstone outcroppings that dominate the mountain landscape.

As it is an excellent destination for a picnic, day hike, mountain bike ride or overnight camping trip, Pine Mountain Recreation Area sees a good deal of human activity on the weekends throughout the summer. If at all possible, try to plan your outing to this spectacular summit for mid-week.

Pine Mountain's elevation gives it a climate all its own. Here you will see plants that are common to much colder climes, such as rhododendrons, white fir and giant Ponderosa pines. There is almost always a cool breeze on the mountaintop, even in the middle of summer. There are also some splendid views from this breathtaking elevation. Not only are the local Badlands, Lockwood Valley's irrigated agricultural fields and Sespe Wilderness visible, but clear days also provide outstanding views of the Channel Islands.

Bounded by designated wilderness on either side, Pine Mountain is home to a variety of wildlife. Deer, raccoons and chipmunks are visible near the car campsites, while bears and mountain lions have been sighted from the walk-in camps.

The mountaintop is an excellent birding venue: besides woodland species, hawks, golden eagles, and even condors have been spotted from this 7,000 foot summit.

Facilities at the campgrounds include picnic tables, fire rings and pit toilets. There is NO water at this primitive site, so be sure to bring extra water as well as first aid supplies. There are no camping fees. For more information about camping on Pine Mountain, call the Ojai District Ranger station at (805) 646-4348.

80. Reyes Peak Walkabout

Difficulty:	Easy
Length:	1 mile loop
Elevation:	100 foot loss/gain
Type of trail:	single-track trail
Kid Appeal:	* * * *

Today's short hike explores the north side of Pine Mountain. Dramatic views of Lockwood Valley's irrigated agricultural fields are a striking contrast to the sub-alpine forest on the mountaintop. Giant Ponderosa pines and white fir thrive here in the mountain's rain shadow.

The forest's diversity provides habitat for many species not commonly seen in Southern California. Chipmunks are especially abundant around the campsites. White-headed woodpeckers, nuthatches, chickadees and flycatchers cover the old trees in their search for grubs, making this an exciting birding spot for any level of expertise.

Step by step:

1. Reyes Peak Campground is the second one that you'll encounter in the Pine Mountain Recreation Area. Park your car near the Forest Service sign by the large campsite on the north side of the road. Today's loop hike begins at the northeast end of the parking lot. Look for the unsigned trail on the left as it heads into the forest.

2. Following the slope of the mountain, the narrow trail loops around the hillside under a thick covering of mixed pine and fir trees. Returning to the roadway after 3/4 of a mile of easy

walking, turn left and follow it uphill the remaining 1/4 mile back to your car.

Directions: *From Highway 101 in Ventura, exit onto Highway 33 north toward Ojai. Continue on Highway 33 for 42 1/2 miles to the Pine Mountain Summit. Turn right onto Pine Mountain Road (Forest Service Road #6N06). This rough roadway is closed in winter months because of snow. Call the Ojai District Ranger Station, 805-646-4348, to be sure that the road is open.*

Hilltop Memorial.

Chapter Seven

Santa Barbara

Rattlesnake Creek Crossing

HIKES AT A GLANCE 9
Santa Barbara

	DIFFICULTY	FACILITIES	FEATURES	FEES	OPEN TO
FRONT COUNTRY TRAILS					
81. Rincon Beach Trail	easy	w, t, p	beach views, tidepools	no	H
82. Carpinteria State Beach Walkabout	easy	w, t, p, c	tidepools, swimming beach, seal rookery	yes	H, C
83. Toro Canyon Park Trail	easy	w, t, p	views, playground, shady canyon, equestrian center	holidays/weekends	H, C, E, D
84. East Cold Springs Trail	moderate	none	views, shady canyon with stream	no	H, E, D
85. Seven Falls Trail	moderate	none	waterfalls, bedrock pools	no	H, E, D
86. San Antonio Creek Trail	easy	w, t, p	shady canyon, playground	no	H, E, D
87. Knapp's Castle Trail	easy	none	castle ruins, views	no	H, C, E, D, S
88. Painted Cave State Park	easy	none	Chumash pictographs	no	H
89. Rattlesnake Canyon Trail	moderate	none	waterfalls, shady canyon	no	H, E, D
90. Nojoqui Falls Trail	easy	w, t, p	grotto, waterfall, playground	holidays/weekends	H

FACILITIES: w - water, t -toilet, p - picnic table, rs - ranger station, c - camping, m - marina, s - store
OPEN TO: H - hikers, C - cyclists, E - equestrians, D - dogs on leash, S - stroller/handicap access

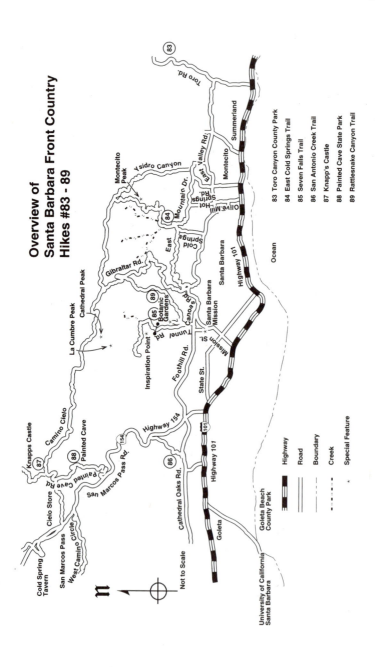

Overview of
Santa Barbara Front Country
Hikes #83 - 89

83 Toro Canyon County Park
84 East Cold Springs Trail
85 Seven Falls Trail
86 San Antonio Creek Trail
87 Knapp's Castle
88 Painted Cave State Park
89 Rattlesnake Canyon Trail

Highway
Road
Boundary
Creek
* Special Feature

Not to Scale

218

Santa Barbara Front Country

Cut-off from the rest of the world by the Pacific Ocean and the mountain wall formed by the Santa Ynez Mountains, Santa Barbara seems to epitomize the Southern California lifestyle of understated elegance. In the late 19th century the city's name grew as a health resort of the era's rich and famous. At that time the main, utilitarian trails over the mountain wall took on a new significance. The glory of their rugged, natural beauty was immortalized in the literary world, the most notable of which was Stewart Edward White's <u>The Mountain.</u>

Rising dramatically from behind the city of Santa Barbara, the Santa Ynez Mountains comprise nearly a million acres of unspoiled backcountry. The "Wall" represented an almost insurmountable barricade to Santa Barbara's first European settlers. Originally, only six main trails made their way over the steep, chaparral-covered slopes from the city to the mountain's vast natural resources and the rich agricultural lands beyond them. The trails were a necessity, as they were the only way to supplies could be moved over the Wall.

Many of these trails begin in tree-canopied canyons cooled by ocean breezes. As you leave the coast however, the gentle canyons quickly give way to steep, open chaparral-covered slopes. Most of the trails end at Santa Barbara's famous ridge-line drive, Camino Cielo, or "Sky Highway", and provide stunning vistas of the city and the immediate coastal plain as well as the Channel Islands. A hike to Knapp's Castle will give you a tremendous view of what's on the other side of the Wall -- Lake Cachuma in the center of the Santa Ynez Valley and the rugged, Sierra Madre Mountains far to the north.

81. Rincon Point Beach Trail

Difficulty: Easy
Length: 2 miles
Elevation: nearly level
Type of trail: footpath
Kid appeal: * * * * *

Straddling the Ventura/Santa Barbara County Line, scenic Rincon Point is a gateway for miles of beachcombing along the pristine shorelines of the Santa Barbara Channel. Attracted by the challenging waves at this remote, south-facing beach, Rincon Point has long been a favored surf spot among local enthusiasts. The Point's extensive and easily accessible tidepools make this a great venue for family outings.

Rincon Point is a good example of the two types of tidepools found in Southern California: cobbles and rock benches. As the name implies, cobble pools are made up of hard rock cobbles. These pools see a lot of tumbling action from the rocks; the water is very aerated and creates a great environment for anemones and other stationary filter feeders. Rock benches on the other hand, are formed by very large boulders or shelves of stabilized rock that see little movement and tend to create deeper pools of water.

Rincon Point has a rich history in the European settlement of the coastline. Long before the highway and railroad were cut out along the narrow rim of land at the foot of the coastal palisades between Ventura and Santa Barbara, today's beach walk served as the stage route between the two settlements. Very often the stage would have to wait at the Point for the tide to recede before the trip could continue.

The Santa Barbara County Beach has the picnic facilities and restrooms, whereas this part of Carpinteria State Beach offers no facilities other than parking. Both parks are open from sunrise to sunset and are free of charge. Please note that there are NO lifeguards assigned to these beaches. Dogs are not allowed on the state beach but are allowed on the county beach if they are leashed. Please remember that shoreline creatures depend on low tides for their meals, and do not allow your animal to harass them while they are feeding. It may be a long wait for their next meal. Horses are not allowed at Rincon Point.

Directions: *From Highway 101, between La Conchita and Carpinteria, exit on Bates Road and turn toward the ocean. Bates Road deadends at the entrance of a gated community. Two parks, one on either side of the residential area, provide parking and beach access to the general public. Santa Barbara County Beach at Rincon Point is on the north side of the road and Carpinteria State Beach is on the south side of the road (the Ventura County side).*

Step by step:

1. Park your car in the Carpinteria State Beach parking lot. A wide dirt path at the south end of the lot leads to the ocean. The trail is about 250 yards long and it parallels the highway as it gradually descends to the beach. At the end of the path there is a vast cobble tidepool area that extends north about 1/2 mile. (At a very low tide it is possible to walk south to La Conchita about 1 mile away.)

2. Walking north from the tidepools along the wet sand will bring you to Santa Barbara County Beach. Another mile of easy walking will bring you to another interesting tidepool area. This area consists of large rock benches that parallel the shore. In the deep pools between the benches there is a greater diversity of marine life than in the shallow cobbles we passed earlier to the south. Several other good tidepool areas occur between here and the Chevron USA Pier.

3. After enjoying the beach, turn around and head south, back to the Santa Barbara County Beach. Take the path or the 107-step stairway up to the parking lot. The park's facilities are near the top of the stairs. Crossing Bates Road will bring you back to your car parked at the state beach.

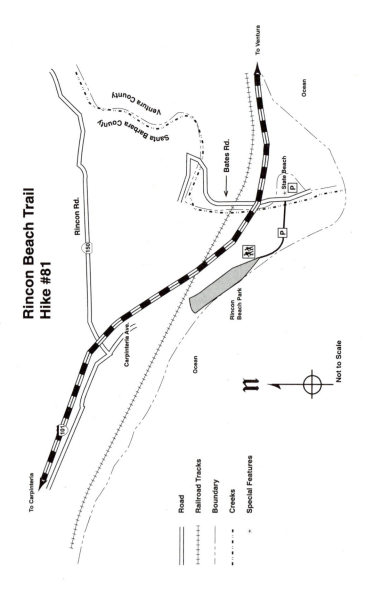

Rincon Beach Trail
Hike #81

To Carpinteria

101

Carpinteria Ave.

150

Rincon Rd.

Santa Barbara County
Ventura County

Bates Rd.

To Ventura

Ocean

State Beach

P

P

Rincon Beach Park

Ocean

N

Not to Scale

Road
Railroad Tracks
Boundary
Creeks
Special Features

Low tide explorations.

82. Carpinteria State Beach

Difficulty: Easy
Length: 2 miles round-trip
Elevation: nearly level
Type of trail: footpath
Kid appeal: * * * * *

The small, seaside community of Carpinteria offers a welcome respite from the busier harbor towns of Santa Barbara and Ventura, directly north and south of it. Carpinteria claims to be "the world's safest beach" due to the fact that the beach slopes down gradually with no sharp drop-off. The gentle slope means that there are no rip-tides or big surges to overwhelm swimmers, so this is an ideal spot for novice surfers or small children to practice their wave-riding skills.

Carpinteria has a rich history dating back thousands of years. Known as "Mishopshnown" by the native Chumash Indians, a large village grew up around several tar pits near the beach. This natural resource provided the Indians with a waterproof material used to caulk the inside of their ocean-going canoes, called "tomols". These unique vessels made the Chumash people the most sea-faring of the native Pacific Indian tribes. The Indians' huge canoe-building operation so impressed Spanish explorer Capt. Gaspar de Portola, that he renamed the area La Carpinteria, which means "the carpenter shop."

The presence of the onshore seal rookery makes this walk along the beach a unique one. California harbor seals are not an endangered species, but due to human encroachment there are only four onshore rookeries in all of California. The Carpinteria rookery is the only one that is open to the public. Chevron and city officials are working together to provide better beach access to the rookery. During winter months when the females come ashore to give birth to their "pups", docents from the Carpinteria Seal Watch Committee stay at the beach every day in shifts to educate beach walkers about the dangers of human intrusion on the seal colony.

Summertime brings the elusive grunion run. Thousands of these small silvery fish haul themselves out onto the sandy beach in the dead of night to spawn. Check at the state park's visitor center for more information and the exact times of the next run. Other winter migrating animals to watch for are California gray whales and Monarch butterflies. The whales sometimes swim so close to shore that you can see them from the campsites. The large orange and black butterflies cluster in the eucalyptus grove near the railroad track.

Carpinteria State Beach is situated on 48 acres along almost a mile of beach front. There are more than 250 campsites that can accommodate RV's, trailers or tents. All campsites include: tables, firerings and clean water, and restrooms with showers are available. Dogs on leashes are allowed in the camp-grounds but not on the beach. There is a bicycle/in-line skate/body-board rental shop near the Linden Avenue exit of the park. The state beach is the mid-way point of a six-mile stretch of beach between Rincon Point to the south and Loon Point to the north. At low tide the entire length of the beach is walkable.

Directions: *From Highway 101 in Carpinteria, exit on Casitas Pass Road and turn left toward the ocean. Go about a block and turn right onto Carpinteria Avenue. After another block, turn left on Palm Avenue and take it to the park entrance. Pay the $5 day-*

use fee at the kiosk and follow the park service road to the parking area located at the far end of the San Miguel Campground.

Step by step:

1. From the southern-most parking lot walk across the grassy area toward the ocean and take the short flight of stairs down to the beach. There are some interesting tidepools immediately opposite the steps. After examining the pools continue south toward the Chevron USA pier. As you approach the pier you will come upon the California harbor seal rookery. The area is well signed to remind us to leave the animals alone at all times, but to be especially respectful of their privacy during the winter months when the females come ashore to give birth.

2. From the pier you can continue along the beach for another mile or so until you come to Rincon Point. There are more tidepools to explore near the point but be sure to check the tide table and give yourself enough time to return without the tide cutting off your route back to the state beach.

3. If you feel like more beachcombing, you can continue north from the state beach along Carpinteria city beach to Sandy Point. Inland from Sandy Point is a large mud flat. At low tide, ambitious walkers should be able to make their way along the waterfront to the tidepools at Loon Point about 2 miles away before the tide returns.

83. Toro Canyon Park Trail

Difficulty:	Easy
Length:	1 mile loop
Elevation:	300 foot gain/loss
Type of trail:	fire road
Kid appeal:	* * * *

Santa Barbara's Toro Canyon County Park is a delightful shady picnic spot above Summerland. The park fills a bowl-shaped glen nestled between the foothills that rise up from the coastal plain south of Santa Barbara. Just a short, 10-minute drive from Pacific Coast Highway, the easy trails through this park explore a small part of the extensive Los Padres National Forest.

Rustic Toro Canyon makes for an excellent family outing. There are a number of tree-canopied picnic spots, with some large reservable areas as well as single tables placed alongside the seasonal stream that runs the length of the park. Barbecues and playground equipment add to the park's family atmosphere. The main hiking trail soon leaves the oakwoodland to climb through the chaparral to a vista point complete with an oriental-style gazebo from which to enjoy the view.

Directions: Take Highway 101 north to the Summerland area, which is a few miles south of Santa Barbara. Exit at Padaro Lane and turn right onto Via Real. Take Via Real east 1/2 mile to Toro Canyon Road and turn left, inland toward the mountains. (If you accidentally exit at the Santa Claus Lane/Padaro Lane off-ramp, simply turn left on Via Real and proceed 1 1/2 miles to Toro Canyon Road and turn right.) Follow this paved, unmaintained road to the park entrance and turn left into the park. Follow the park road to the left, deeper into the canyon, to reach the main trailhead located near a large sandstone outcropping.

Step by step:
1. The trailhead is at the end of the park road, with a single-track path leading from the lot, crossing a large meadow and heading toward the sandstone outcropping in the middle of the field. Just before you get to the rock however, the trail to the gazebo cuts off to the right.
2. Follow the bulldozed path uphill through the chaparral for about 1/2 mile until you come to a fork in the road. At this point the trail splits to loop around the small knoll before you. As the trail makes a loop, follow it in either direction to the hilltop gazebo.
3. After enjoying the splendid view, continue on in the direction of your choice back to the fork in the trail. From there return to your car the way you came.

84. East Cold Springs Canyon Trail

Difficulty:	Moderate
Length:	3 miles round trip to Montecito Overlook
Elevation:	900 foot gain/loss
Type of trail:	single-track trail
Kid appeal:	* * * * *

Cold Springs East Trail may be one of Santa Barbara's most popular front country trails. Established at the turn of the century by the Santa Ynez Forest Reserve, this trail meanders through lush Cold Springs Canyon, crisscrossing the year-round stream to reveal numerous bedrock pools and waterfalls. Located closed to town, the Cold Springs East Trail makes an exciting family adventure for an hour or an entire afternoon.

Cold Springs Canyon's beauty is evident from the moment that you leave your car along the roadside. The trail-head is permeated with the fragrance of California Bay trees mixed throughout the riparian oakwoodland. Giant alders, sycamores and big-leaf maple trees are a few of the other species that help to make the overhead tree-canopy an effective barrier against the summer heat. In contrast, the hike up to the Overlook explores the chaparral plant community as hikers climb out of the canyon for one of the finest close-in views of Santa Barbara and its stunning coastline.

Originally, the Cold Springs Trail led up the west Fork of Cold Springs Canyon. Because this trail scrambled past a 300 foot waterfall as well as traversing a bank of loose shale, the Santa Ynez Forest Reserve thought it too difficult to maintain and as a result the present Cold Springs East Trail was made. Today both trails begin at the same trailhead but the older, West Trail splits from the now main East Trail at Kevin's Bench and follows the creek's rugged west fork to Gibraltar Road about 1 1/2 miles away.

Symbolic of the conflict between trail-users and property owners, a small bronze plaque embedded in a boulder near the trailhead declares the owner's right to the land. To ensure the trail's use in perpetuity, please be sure to respect the owners' property rights near the trailhead as well as long the trail.

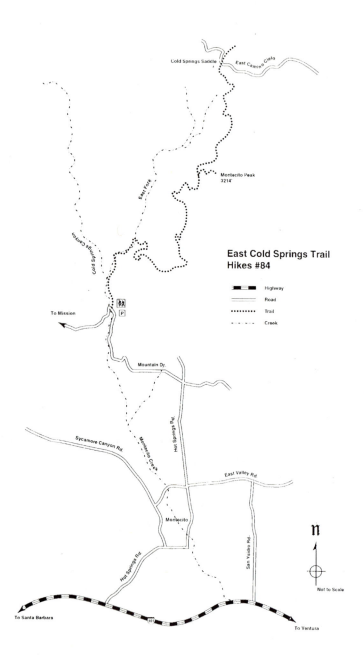

**East Cold Springs Trail
Hikes #84**

Cold Springs Saddle
East Camino Cielo

East Fork

Montecito Peak
3214'

Cold Springs Canyon

To Mission

	Highway
	Road
•••••••	Trail
-•-•-•-	Creek

Mountain Dr.

Sycamore Canyon Rd.

Hot Springs Rd.

Montecito Creek

East Valley Rd.

Montecito

Hot Springs Rd.

San Ysidro Rd.

n

Not to Scale

To Santa Barbara

101

To Ventura

228

Parking is extremely limited, so be sure to heed posted warnings. Car pooling is advised.

Due to the rocky nature of this and many other Santa Barbara Front Country trails in general and in view of trail-use conflicts between cyclists and other user groups, cyclists are urged to avoid using Front Country trails (those south of Camino Cielo).

Directions: *From Highway 101 in Santa Barbara, use the Milpas exit and take it to Montecito Street. Turn right onto Montecito Street near Scolari's Market and proceed north to Sycamore Canyon Road which starts on the far side of the roundabout traffic circle and continues north and signed as Highway 144. Continue north on Sycamore Canyon Road (it later becomes Highway 192) about 2 miles to Cold Springs Road. Turn left, north, opposite the fire station and drive 1/2 mile past Westmont College to Mountain Drive and turn right, east. The Cold Springs trailhead is about 1/2 mile up the road on the left side near the point where the creek crosses the road.*

Step by step:

1. From the parking area, the signed Cold Springs East Trail begins on the north side of the creek. Take the trail over the stream and stay to the right of the temporary trail marker. The path follows the tree-canopied creekbed upstream about 1/4 mile to where the less traveled Cold Springs West Trail crosses the creek as it splits off from the main Cold Springs East Trail. This spot is easily recognizable; look for Kevin's Bench overlooking a large pool just below a seasonal waterfall.

2. The East Trail leaves the creekbed for higher ground as it switchbacks up the canyon for about another 1/4 mile. There is some level walking for a short way before the trail comes back to cross the shady creek once more. This pattern is repeated again, and in another 1/2 mile the trail comes to its last creek crossing before heading out of the lush canyon.

3. About a mile from the trailhead, our path leaves the sheltered canyon as it heads more aggressively uphill through some very densely chaparral-covered slopes. Several steep switchbacks bring you to the top of the hill and a junction with the power line service road. Turn right and the road leads to

the Montecito Overlook. This will be the turnaround for today's hike. After enjoying the fine views of the Santa Barbara area and surrounding ocean, return the 1 1/2 miles back to the roadside trailhead the way you came.

Optional: A few hundred yards beyond the first junction is the junction with the Hot Springs Trail heading east. It is possible to make today's outing into a loop hike by returning to town via the Hot Springs Trail and walking back to the Cold Springs Trailhead along Mountain Road. This route adds about 1 1/2 miles to the total hiking distance.

Cold Springs East Trail continues all the way up the Mountain Wall to Camino Cielo (the Sky Highway) 3 miles away. From the junction with the power line road, instead of going to the overlook, turn left and proceed north to the next ridgeline. Once at Camino Cielo, the trail continues north for 6 more miles to Forbush Flat Trail camp. The trail is signed as: North Cold Springs Trail #26W10 and connects with several other Los Padres National Forest Service trails.

85. Seven Falls Trail

Difficulty: Moderate
Length: 3 miles round-trip
Elevation: 600 foot gain/loss
Type of trail: fire road, single-track trail
Kid appeal: * * * *

Today's hike up Mission Canyon is a short one to the renowned Seven Falls. Most of this hike is along the Tunnel Fire Road before joining the multi-use, single-track Jesusita Trail that leads to Mission Creek before heading up to the crest of Camino Cielo.

More than likely, especially if you do this hike on the weekend, what you'll see along this trail is other nature-lovers. Close-in to the city as it is, the Seven Falls area sees a lot of local hikers and cyclists who come to the gorge to enjoy this enchanting series of waterfalls, deep pools and cascades as well as the numerous spring wildflowers that grace the steep mountain slopes.

The first part of this hike, along the paved Tunnel Fire Road, climbs steadily through the chaparral that overlooks rugged Mission Canyon. The second part of the hike, along the single-track Jesusita Trail, explores the shadier oakwoodlands and riparian plant community. From the Seven Falls area, the Jesusita Trail continues on the far side of Mission Creek all the way to Inspiration Point and then on to Camino Cielo.

Step by step:

1. After legally parking your car along Tunnel Road, proceed up the road to the water tank. Walk around the locked gate and continue uphill on the Tunnel Fire Road. Near the top of the first small hill there is a "Y" junction in the road. Stay to the left and continue uphill.

2. About 1/2 mile from the trailhead, a sturdy wooden bridge takes the road over Mission Creek. Little Fern Canyon Falls empties into a deep pool just under the bridge, where a small path leads down to the creek. Another 1/2 mile of uphill walking and the fire road enters the oakwoodland where there is a major trails junction as well as an information kiosk. This is where we leave the paved road for the single-track Jesusita Trail.

3. The Jesusita Trail continues uphill for another hundred yards before coming to a signed junction with the Tunnel Trail. Stay on the Jesusita Trail as it will soon begin its ascent into the gorge around Mission Creek about 1/2 mile away. Once at the creek you can boulder-hop up and down the creekbed to visit each of the Seven Falls. Please note that there is quite a bit of poison oak in and around the creekbed. Also be sure to use caution if you do any rock climbing in the gorge, the rocks can be very slippery. See trail map for Hike #89.

86. San Antonio Creek Trail

Difficulty: Easy
Length: 3 mile round-trip
Elevation: 150 foot gain/loss
Type of trail: single-track trail
Kid appeal: * * * * *

Nestled in the hills of Santa Barbara's front country, Tucker's Grove County Park occupies the mouth of a lush coastal canyon. The park is the starting point of the San Antonio Creek Trail. This easy hiking and bridle trail explores the riparian and oakwoodland plant communities along year-round San Antonio Creek as it snakes its way through the canyon.

Tucker's Grove is one of Santa Barbara's oldest public parks. The park was named for Charlie Tucker who owned the property around the turn of the century and allowed the public to picnic in the sheltered oak grove free of charge. Today's hike makes for a perfect family outing. The park is still free and the easy trail along the stream can provide a couple of hours of relaxed hiking.The canyon is pleasant any time of year; springtime hikes can be filled with exploration of the countless myriads of small waterfalls in the creek and the wildflowers in the meadows.

This beautiful canyon was reduced to ashes by the Painted Cave Fire of 1990. Walking through it several years later, the canyon's comeback from the fire is quite amazing.

Directions: Take Highway 101 north to the Goleta area, which is immediately west of Santa Barbara. Exit at Turnpike Road north to where it ends at the intersection with Cathedral Oaks Road. The entrance to Tucker's Grove is opposite this intersection. Cross Cathedral Oaks Road and enter the park. Stay to the right on the service road and follow it about 1/2 mile to the last parking lot before the signed Kiwanis Meadow and park your car.

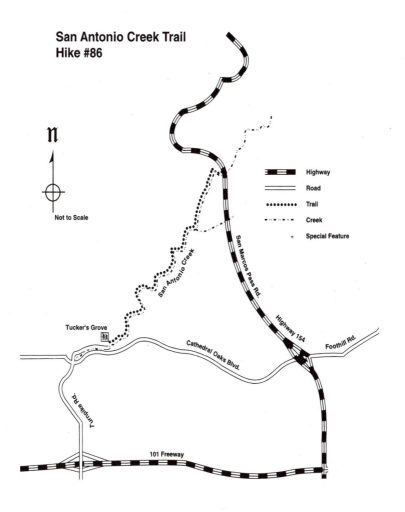

**San Antonio Creek Trail
Hike #86**

▪▪▪▪	Highway
———	Road
••••••••	Trail
–▪–▪–▪–	Creek
★	Special Feature

Not to Scale

San Antonio Creek

San Marcos Pass Rd.

Highway 154

Foothill Rd.

Cathedral Oaks Blvd.

Tucker's Grove

Turnpike Rd.

101 Freeway

Step by step:

1. The San Antonio Creek Trail begins near the creek at the northeastern end of the parking lot. It is marked with a small wooden sign that says "Bridal Path"(sic). Take the trail over the creek through the woods to another larger creek crossing that is still within sight of the Kiwanis Meadow. (In the spring, when the creek is high, many hikers forgo the crossing by walking through the Meadow and picking up the trail on the other side.)

233

2. The trail skirts the rail fence around the lawn as it follows the creek deeper into the canyon. The unmarked, but easy to follow trail meanders through an open oakwoodland. There are a number of short spur trails, but watch out for the poison oak if you decide to follow one to the edge of the creek.

3. After half a mile of walking on the east side of the creekbed, the trail makes its first of a number of crossings. For the next half mile, the trail crisscrosses the creek through a beautiful riparian area that is dominated by sycamore seedlings. Then you come to a rather open area where there is a concrete flood-control dam spanning the streambed. The trail crosses over the top of the dam and continues on to follow the creek upstream. A number of small trails cross here, but the main San Antonio Creek Trail soon turns right to cross the creek once more.

4. Back on the east side of the stream, the trail parallels a chain-link fence that is visible from the crossing. The trail continues through another open oakwoodland for its last half-mile segment. It comes to an end under the bridge where the trail intersects Highway 154 near Vista Del Mundo Ranch. At this point return to your car the way you came and enjoy a picnic at Tucker's Grove.

87. Knapp's Castle Trail

Difficulty: Easy
Length: 3/4 mile round-trip
Elevation: nearly level
Type of trail: fire road
Kid appeal: * * * * *

High above Santa Barbara, Camino Cielo Road cuts a thin ribbon along the summit of the Mountain Wall. The Wall separates the city from the Santa Ynez Valley backcountry. Taken aback by the views from this "sky highway", George Knapp built a mansion here, near the top of the Santa Ynez Mountains. Constructed during the elegant heydays of Santa Barbara's Roaring Twenties expansion era, Knapp's Castle took 20 men more than four years to complete.

As so often happens with structures in Southern California built along chaparral-covered ridgelines, this remarkable estate was destroyed by a wildfire. Following the Paradise Canyon Fire of 1940, all that remained of Knapp's skyline lodge were the stone arches, chimneys and building foundations. An easy walk from Camino Cielo, hikers can enjoy the same extensive 180° view that Knapp treated his guests to at extravagant mountaintop soirees in times long past.

Simply outstanding views of the Santa Ynez Valley unfold as you make your way around the Castle ruins. Be sure to bring your binoculars so you can watch the sailboats racing around far below on Lake Cachuma. The short, easy walk from the road makes this a great destination to watch the sunset against the ruggedly beautiful Santa Barbara backcountry.

Besides magnificent vistas, a walk to Knapp's Castle helps trail users to catch a glimpse of another, perhaps more cultivated era of human habitation in Southern California. Walking around the ruins it's almost possible to imagine what it looked like to the socialites invited to one of Knapp's elegant gatherings. Knapp's ambitious construction projects throughout the Santa Barbara area represented the pioneer mind-set that thrived on making the impossible, possible.

Directions: *In Santa Barbara, exit Highway 101 at San Marcos Pass Road (Scenic Highway 154). Take the highway north 3 1/2 miles to East Camino Cielo Road and turn right. Follow Camino Cielo for several miles, watching for a locked gate on the north side of the road about 1 1/4 miles beyond the intersection with the Painted Cave Road. There are parking turnouts on both sides of road at this unsigned trailhead for the Snyder Trail (28W02) that leads to Knapp's Castle on its way down the mountain to Paradise Road.*

Step by Step:

1. Walk around the gate and follow Snyder Trail along the dirt service road northeast to the ruins of Knapp's Castle less than 1/2 mile away. The castle is on private property and present owners currently allow access to the ruins but this could change.

Please be careful and considerate of private property during your visit to this wonderful site.

2. Snyder Trail cuts off to the left about 200 yards before the castle ruins. It starts as an old ranch road that eventually gives way to a footpath as it threads its way down the mountain to Paradise Canyon Road about 3 miles away. The trail loses more than 2,000 feet in elevation as it passes through some more chaparral, an oakwoodland and several grassy meadows before reaching the canyon bottom.

Optional: On your way home, take winding and often times single-laned Painted Cave Road to view the ancient mystic Chumash pictographs at Santa Barbara's legendary Painted Cave. The roadside, Painted Cave State Park is located about 1 1/2 miles south of Camino Cielo. Parking is extremely limited, please use patience and caution when looking for a spot. After viewing the pictographs, continue down the mountain on Painted Cave Road for another 2 miles to San Marcos Pass Road where a left turn will return you to Santa Barbara.

Knapp's Castle ruins

Knapp's Castle and Painted Cave State Park
Hikes #87 & 88

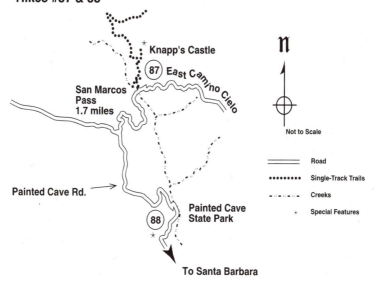

Road	══════
Single-Track Trails	••••••••
Creeks	▬ ▪ ▬ ▪ ▬ ▪
Special Features	⋆

Knapp's Castle

(87) East Camino Cielo

San Marcos Pass
1.7 miles

Painted Cave Rd.

Painted Cave State Park

(88)

To Santa Barbara

Not to Scale

88. Painted Cave State Park

Difficulty: Easy
Length: 100 yards
Elevation: 20 feet
Type of trail: footpath
Kid appeal: * * * * *

      Overlooking the ocean, Painted Cave State Park sits perched half-way up Santa Barbara's "backwall". Hardly more than a pull-over off a country lane that threads its way up the mountain, this is one of the smallest of the California State Parks. Physically small but spiritually rich, the rough walls of the Painted Cave are covered with some of the most beautiful and mysterious Chumash Indian pictographs, or "rock-art", found anywhere in the Santa Ynez Mountains.

      Tucked into a shady fold in the mountain, the Painted Cave is actually more of an overhang than a cave. Sculpted by time and the elements, the rocky promontory that makes up the park holds a commanding view of the Santa Barbara coastline.

237

Ancient Painted Cave treasure.

Perhaps it was this view that, over the centuries, drew the tribal shamans to the cave to paint these intriguing images on its walls. Some of the most extraordinary Chumash pictographs are found in the Painted Cave. Many Native Americans feel that the Cave is one of the most sacred ceremonial sites in the mountains and come here to offer prayers to their ancestors. Often times you will see burnt bundles of white sage left on the cave's floor from these visits. Unfortunately, desecration by vandals has caused the state to close off the cave with iron bars, limiting how well you can see this elaborate piece of "rock-art".

Directions: *Painted Cave State Park is located off Painted Cave Road in Santa Barbara. From Highway 101, exit at San Marcos Pass Road (Scenic Highway 154) and take it north 5 miles to Painted Cave Road. Turn right and continue up the mountain on this narrow, sometimes one-lane road less than 2 miles to the signed, Painted Cave State Park. The road continues on for another mile or so before ending at the ridgeline East Camino Cielo Road, or "Sky Highway". It makes for a nice driving loop to continue up the mountain, turn left on Camino Cielo and take it back to San Marcos Pass Road.*

89. Rattlesnake Canyon Trail

Difficulty: Easy/moderate
Length: 3 mile round trip
Elevation: 200 foot gain/loss
Type of trail: single-track trail
Kid appeal: * * * * *

In spite of its forbidding name, Rattlesnake Canyon Trail through Skofield Park, is one of Santa Barbara's most popular front country hikes. Indeed, you will be much more likely to see other hikers than rattlesnakes as this single-track trail snakes its way upcanyon. The shady pathway follows a delightful, year-round stream before coming to a wildflower meadow.

Today's outing through Rattlesnake Canyon offers a number of easy hiking possibilities. Several spur trails lead off the main trail to small waterfalls and secluded pools as you make your way upcanyon. Hikers looking for a more vigorous workout can take the Connector Trail from the meadow as it climbs through the chaparral east to Gibraltar Road, where there is an outstanding vista of the Santa Barbara coast, or west to the Tunnel Trail.

The first part of today's hike is under an exquisite mixed canopy of alder, bay, sycamore and oak trees. The stream crossings, rock pools and waterfalls make this canyon hike an exciting outdoor family adventure. As you make your way uphill through the chaparral be sure to look back toward the ocean as there are stunning views of the Channel Islands. On a recent weekend, hang-gliders over the canyon looked like huge, prehistoric birds of prey as they made their way toward the coast.

Directions: To reach the trailhead for the Rattlesnake Canyon Trail, exit Highway 101 at Mission Street in Santa Barbara. Follow the signs to the Mission and from there, proceed north on Mission Canyon Road to Foothill Road and turn right. Make a quick left, once again onto Mission Canyon Road, and take it to Las Canoas Road and turn right. Take Las Canoas about 1 1/2 miles, look for a wide turnout near the second

stone bridge and park your car along side the road. Special note: this multi-use trail is closed temporarily to cyclists and equestrians due to storm damage.

Step by step:

1. The signed Rattlesnake Trail begins at the side of the road at either end of the stone bridge. The easterly path avoids the first stream crossing and the two trails meet just a few hundred yards upstream where the trail heads steadily uphill through a planted area of rock-roses. Stay to the left at the junction with the old fire road as the main trail follows the course of the stream upcanyon. About a half mile up the trail there is another trails junction, this time proceed straight ahead as the main path continues to follow the creek.

2. A little farther upstream there is another stream crossing and the canyon once more narrows in over the creekbed. The trail switchbacks up from the creek through the open chaparral to a grove of planted pine trees that is about 2/3 mile from the trailhead. After a bit the trail leads back under the tree canopy to cross the creek once more before climbing out of the forest to the grassy meadow. At the end of the meadow, near a large live-oak tree, is the junction with the Connector Trail. This is a good turnaround point for an easy hike or rest stop for hikers continuing on the trail in either direction.

3. On the return trip, it is possible to make a loop by hiking back down the canyon's east side. Look for the trail to start in the forest below the last stream crossing. This path is much less traveled than the main trail but is in good hiking condition. The east canyon loop trail rejoins the main trail below the second stream crossing, about 1/2 mile from the roadside trailhead.

Seven Falls Trail Hike #85 and
Rattlesnake Canyon Trail Hike #89

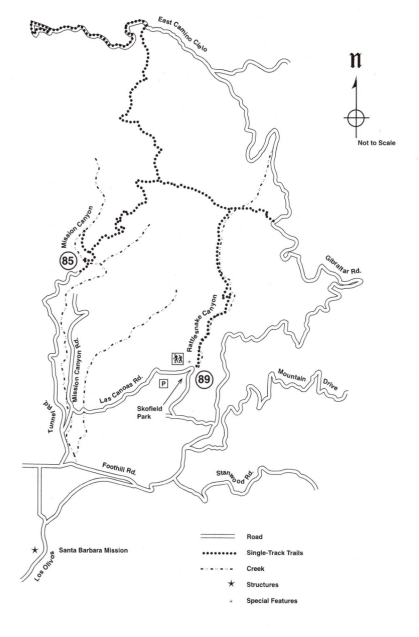

East Camino Cielo

Mission Canyon

85

Gibraltar Rd.

Rattlesnake Canyon

Mountain Drive

Las Canoas Rd.

P

89

Skofield Park

Mission Canyon Rd.

Tunnel Rd.

Foothill Rd.

Stanwood Rd.

★ Santa Barbara Mission

Los Olivos

N

Not to Scale

——— Road

•••••• Single-Track Trails

—·—·— Creek

★ Structures

⋆ Special Features

241

90. Nojoqui Falls Trail

Difficulty: Easy
Length: 1/2 mile round-trip
Elevation: nearly level
Type of trail: fire road
Kid appeal: * * * * *

Folded into the Santa Ynez Mountains, Nojoqui Falls County Park occupies a cool, shady little glen. Situated at the back of a steep canyon, the heavily forested bowl-shaped area at the bottom of the falls creates a magical setting under a canopy of bay, big-leaf maple, sycamore and live-oak trees. Beyond the forest, this beautiful Santa Barbara County Park opens onto a broad grassland that was once the site of a Chumash Indian village. The settlement was known as "Nakhuwi", which is a Chumash word thought to mean mead-ow. Located just a few miles off El Camino Real (Highway 101) between Gaviota Beach and Buellton, Nojoqui Falls is a terrific picnic destination in itself. Close to the highway, it also is an excellent rest-stop where travelers can stretch their legs far from the madding crowd. Open daily, the short stroll to the park's namesake waterfall makes a delightful family outing any time of year. This charming little canyon is cool in the summer and mild in the winter. The tree-canopy above the falls also offers a surprising display of autumn color. Additional water from seasonal rains makes Nojoqui Falls even more splendid in the spring.

The short walk to Nojoqui Falls is a real treat. Easy enough for walkers of all experience levels, the falls' beauty and unique geologic formations will delight even the most adventuresome hiker. Starting as an open oak savanna, the park closes down to the narrow, tree-canopied canyon sur-rounding the falls. A sign near the end of the trail describes the geologic process by which Nojoqui Falls where formed over 30 million years ago as a result of activity along the San Andreas Fault. The area is still slowly rising, even today. Nojoqui Falls itself continues to grow as travertine material is built up from calcium and magnesium in the rocks above the falls, similar to stalagmites found in caves.

Directions: *From Santa Barbara continue north on Highway 101. Just beyond the community of Gaviota, the highway leaves the coast and turns inland. The turn-off for Nojoqui Falls is about 5 miles from the coast. Watch for the county park sign on the right side of the road at the end of the truck parking area near the highway rest area. Turn right and proceed down the Old Coast Highway about a mile to Alisal Road and turn right. Stay on Alisal for another 1/4 mile or so and Nojoqui Falls County Park is on the right side of the road. Turn into the park and follow the driveway all the way to the last parking area. From Buellton, drive south on Highway 101 about 5 miles. Watch for a county park sign at the intersection with the Old Coast Road. Turn left on the Old Coast Road and follow it to Alisal Road and turn left again. Follow the above directions into the park.*

Step by step:

1. Follow the park signs to the trailhead at the back of the canyon. The gentle dirt trail trends slightly uphill as it makes its way through the dense woodland. Following the year-round creek upstream about a quarter mile, the trail comes to an end at the base of the waterfall.

2. After enjoying the area around the falls, return to your car the way you came. Please remember that climbing Nojoqui Falls is forbidden. Not only are the canyon's shale and rock walls very unstable, but a number of endangered plants that call this fragile environment home could be damaged by careless foot traffic.

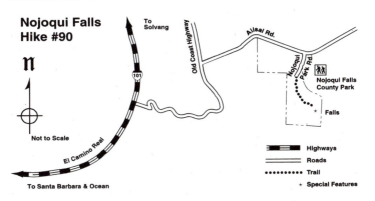

243

Organizations

Angeles Forest District fire closure 805-488-8147
Back to Blue 818-789-7866
California Department of Parks & Recreation . . 818-880-0350
   - Southern California office 916-653-6995
   - Camping reservations 800-444-7275
California Native Plant Society. 818-348-5910
Circle X Ranch 310-457-6408
Coastwalk, state office 707-829-6689
Concerned Off-Road Bicyclists' Association . . . 818-773-3555
Conejo Rec. & Park District Outdoor Unit. 805-381-2737
Conejo Valley Audubon Society 805-289-0440
County of Ventura GSA Recreation Services . . . 805-654-3951
Earth Skills 805-245-0318
Friends of Channel Islands National Park. 805-658-5700
Friends of Satwiwa 805-499-2837
Island Packers, Inc 805-642-1393
Los Angeles County Parks & Recreation 213-738-2961
Los Padres National Forest
   - Ojai office. 805-646-4348
   - Goleta office. 805-683-6711
Malibu Creek State Park. 805-499-2112
Nursery Nature Walks. 310-998-1151
Oakbrook Regional Park - Chumash Center. . . . 805-492-8076
Point Mugu State Park 805-488-5223
Santa Barbara County Parks Department
   - south 805-568-2460
   - north 805-934-6211
Santa Monica Mountains National Recreation
Area Visitor Center . . . . 818-597-9192 ext. 201
Santa Monica Mountains Conservancy. 310-456-5046
Santa Monica Mountains Trails Council. 818-222-4531
Sierra Club
   - Los Angeles Chapter. 213-387-4287
   - Los Padres Chapter. 805-966-6622
Simi Valley Park & Recreation District 805-584-4400
Wildflower Hotline 818-768-3533

Glossary

General Terms:

Chaparral -- Sometimes referred to as the elfin forest, this is the type of dense evergreen shrubs that dominate Southern California's rocky, dry slopes. It is difficult to love, especially when seen from the freeway at 65 mph, when it seems to be an indistinguishable blur of blue-green. These plants are adapted to hot, dry conditions and late summer is their dormant time of year.

Hard chaparral -- found on south facing slopes where it is hotter and drier. Soft chaparral -- found on cool, moist coastal and north facing slopes.

Endemic species -- a species unique to a given area because of its specific growing conditions such as water, temperature and sunlight. The Santa Monica Mountains has one of the largest varieties of endemic species anywhere in the world.

Estuary -- where a stream or river meets the ocean. A rich ecosystem is created as the salt water mixes with the fresh.

Fire followers -- wildflowers and herbs that are only seen the first few years following a wildfire when the larger plants are still growing back.

Fire hazard -- This occurs when weather factors such as: humidity, temperature and wind velocity combine to make prime conditions for wildfires. This usually happens September through November during the Santa Ana winds. The open space parks will close; call the Fire Hot Line 805-488-8147.

Habitat -- comprised of four elements: water, food, shelter and space in the proper arrangement for a given animal species.

Indicator species -- Some plants and animals can only survive in a particular, limited environment or community. Their presence is said to define a particular community.

Live-oak trees vs. valley oak trees -- Live-oaks retain their cup-shaped leaves throughout the year. Valley oaks are characterized by lobed leaves that drop in the Fall and rough, lined bark. Live-oaks thrive throughout Southern California whereas valley oaks are only found inland from the ocean.

Mediterranean climate -- characterized by a seasonal weather pattern of warm, wet winters and cool, dry summers. As you travel inland from the coast the climate becomes increasingly semi-arid upon leaving the rain shadow of the coastal mountains.

Native plant -- a plant that naturally grows in a given area without any help from man.

Oakwoodland -- are open forests of oak trees. These forests were nearly destroyed by over grazing in the rancho era. Oakwoodlands were treasured by the Chumash Indians as acorns were their staple food for centuries.

Plant community -- an assemblage of interacting plant species that is characterized by sharing the same growing conditions of: water, sunlight and soil type. Southern California's climate and mountainous terrain combine to make many unique plant communities.

Rain shadow -- mountains act as barriers to rain laden winds off the ocean. As the clouds rise to get over the mountain tops, they must lose some of their water through rain, leaving them lighter and drier as they make their way east.

Riparian -- streamside plant community characterized by dense vegetation surrounding a source of water, usually producing a tree canopy made up of: alder, bay, big-leaf maple, oak and sycamore trees.

Saddle -- a low area between two hilltops.

Santa Ana winds -- hot, dry blasts of wind that sweep through the mountain passes and canyons of southern California from the northern deserts.

Savanna -- a grassland. oak savanna -- a grassland with oak trees. Sycamore savanna -- a grassland with a ribbon of sycamore trees following a seasonal creekbed, indicating ground water.

Scat -- animal poop; owl pellets -- owl poop.

Scrub -- a term for dry, shrubby vegetation.

Slough -- a wetland that is a mixture of salt and freshwater.

Tidepool -- oceanside pools formed at low tide in rocky areas along the coast where water is trapped as the tide recedes. Generally, there is a high and a low tide every 12 hours.

Tidepooling -- observing sealife that is uncovered at low tide. The best time to go tidepooling is when the low tide is at 0.0 or it is negative such as -1.0. Usually the best time of year to go tidepooling is in the winter, when the lowest tides appear during daylight hours.

    - **High tide** -- the tide is at its highest point on the beach.

    - **Low tide** -- the tide is at its lowest point.

Tide chart -- computer generated for the entire year and available from local bait and/or surf shops.

Umbrella species -- going hand-in-hand with an indicator species, an umbrella species is said to define the overall health of a particular environment.

Watershed -- The area through which water drains from high ground toward the ocean.

Wetland -- bogs, swamps and marsh areas with high moisture content: can be seasonal due to flooding

Woodland -- a tree covered area.

Trail terms:

Fire road -- a road that provides access for fire trucks and other service equipment into open space. Most are closed to regular traffic.

Full-access trail -- provides access to those with physical impairments, such as smooth surfaces and ramps for wheelchairs and/or guide wires for the blind.

Indistinct trail -- a route that is not a maintained trail but is defined by use.

Loop trail --trail begins and ends at the same trailhead after circling a given area.

Motorway -- open space roads to power and gas lines but closed to regular traffic.

Multi-use trail -- one that is open to the three major user groups: hikers, cyclists and equestrians

One-way -- the distance or route from one place to another.

Round-trip -- the hike back is the same as the hike out.

Separate-use trail -- open to various user groups because the pathway is divided with a barrier of some sort such as a fence or wall.

Shuttle hike -- before the hike a vehicle is left at the end of the trail and after the hike you drive it back to the trailhead.

Single-track trail -- a narrow path.

Staging area -- stopping or resting place for migrating animals. Also refers to trailheads and camping or rest spots for humans.

Switchback -- Switchbacks help you climb a steep slope when the trail turns back on itself as it crisscrosses up the hill. Do not shorten your hike by "cutting the switchbacks": this makes tiny pathways and causes erosion on the trail.

Texas-style creek crossing -- this is a trail-level bridge over a creek or stream. In wet weather water flows over the bridge; use caution during flooded conditions.

Trailhead -- where the trail begins.

Waterbreak -- this is a barrier such as a log dug into the trail to control erosion by diverting water off the pathway.

Directional Terms:

Frontcountry vs backcountry -- in Ojai and Santa Barbara-- frontcountry refers to the hikes in the foothills around these two communities. Backcountry refers to the hikes in the more remote areas of Los Padres National Forest.

Kanan Dume Road/Kanan Road -- this road changes its name as it crosses Mulholland Highway. It's the same road, only the names have been changed to protect the innocent.

Las Virgenes Road/Malibu Canyon Road -- same as above.

Notes:

Additional Reading

Hiking References:

1. Benkiam, Arthur. Santa Barbara TRAIL GUIDE. 7th ed. n.p.: Santa Barbara Group, Los Padres Chapter, Sierra Club, 1993.

2. Ford, Jr., Raymond. Santa Barbara DAY HIKES. 4th ed. Goleta, Ca: McNally & Loftin, 1992.

3. Gagnon, Dennis. HIKE Los Angeles Vol. 1. 3rd ed. Santa Cruz, Ca: Western Tanager Press, 1991.

4. Gagnon, Dennis. HIKE The Santa Barbara Backcountry. 3rd ed. Santa Cruz, Ca: Western Tanager Press, 1991.

5. Lewellyn, Harry. Glovebox Guide to Unpaved Southern California. 1st ed. Santa Ana, Ca: Glovebox Publications, 1987.

6. McAuley, Milt. Guide to the Backbone Trail. 1st ed. Canoga Park, Ca: Canyon Publishing Company, 1990.

7. McAuley, Milt. Hiking in Topanga State Park. 3rd ed. Canoga Park, Ca: Canyon Publishing Company, 1991.

8. McAuley, Milt. Hiking Trails of Point Mugu State Park. 1st ed. Van Nuys, Ca: Canyon Publishing Company, 1982.

9. McAuley, Milt. Hiking Trails of the Santa Monica Mountains. 4th ed. Canoga Park, Ca: Canyon Publishing Company, 1987.

10. McAuley, Milt. Wildflower Walks of the Santa Monica Mountains Vol. 1. 1st ed. Canoga Park, Ca: Canyon Publishing Company, 1988.

11. McKinney, John. COAST WALKS 100 Adventures Along the California Coast. 2nd ed. Santa Barbara, Ca: Olympus Press, 1990.

12. Tway, Linda E., Ph.D. TIDEPOOLS Southern California. 1st ed. Santa Barbara, Ca: Capra Press, 1991.

Mountain Biking References:

1. Berlund, Carol. Mountain Biking the Central Coast Vol 1. 1st ed. Ed. Ed Zolkoski. San Luis Obispo, Ca. EZ NatureBooks, 1990.

2. Hasenauer, Jim. Mountain Biking the Coastal Range, Guide 7 the Santa Monica Mountains. 1st ed. Ed Reanne Douglass and Don Douglass. Bishop, Ca: Fine Edge Productions, 1989.

3. Immler, Robert. Mountain Bicycling Around Los Angeles. 1st. ed. Berkeley, Ca: Wilderness Press, 1990.

4. McTigue, Mickey. Mountain Biking the Coastal Range, Guide 4 Ventura County & the Sespe. 3rd ed. Ed. Sue Irwin. Bishop, Ca: Fine Edge Productions, 1993.

Children and Nature References:

1. Cornell, Joseph. Listening to Nature. 1st ed. Nevada City, Ca: Dawn Productions, 1987.

2. Cornell, Joseph. Sharing the Joy of Nature. 1st ed. Nevada City, Ca: Dawn Productions, 1989.

3. Cornell, Joseph Bharat. Sharing Nature with Children. 1st ed. Ed. George Beinhorn. Nevada City, Ca: Ananda Publications, 1979.

4. Docents of Nursery Nature Walks. Trails, Tails & Tidepools in Pails. 1st ed. Pacific Palisades, Ca: Nursery Nature Walks, 1992.

Native American References:

1. De Angulo, Jaime. Indian Tales. 18th ed. New York, NY: The Noonday Press, 1991.

2. Duvall, Jill D. THE CHUMASH A New True Book. 1st ed. Chicago, IL: Children's Press, 1994.

3. Margolin, Malcolm. The Way We Lived: California Indian Stories, Songs & Reminiscences. 2nd ed. Berkeley, Ca: Heydey Books, 1993.

4. Maxwell, Thomas J., Ph.D. The Temescals of Arroyo Conejo. 1st ed. Thousand Oaks, Ca: California Lutheran College, 1982.

5. Miller, Bruce. The Gabrielino. 1st ed. Los Osos, Ca: Sand River Press, 1991.

6. Mourning Dove. Coyote Stories. 2nd ed. Ed Heister Dean Guie. Lincoln, NA: University of Nebraska Press, 1990.

7. Sanger, Kay. When the Animals Were People. 1st ed. Banning, Ca: Malki Museum Press, 1983.

8. Santa Barbara Museum of Natural History, Docent Project. California's Chumash Indians. 1st ed. Santa Barbara, Ca: John Daniel, publisher, 1986.

9. Santa Barbara Museum of Natural History, Docent Project. The Chumash People: Materials for Teachers & Students. 2nd ed. Santa Barbara, Ca: Santa Barbara Museum of Natural History, 1989.
10. Trazer, Clifford E. California's Indians and the Gold Rush. 1st ed. Sacramento, Ca: Sierra Oaks Publishing Company, 1989.

Fauna References:

1. Brown, Philip R. Seashore! A Guide and Activity Book. 1st ed. Ventura, Ca: City of San Buenaventura Department of Parks & Recreation, 1988.
2. Clarke, Herbert. An Introduction to Southern California Birds. 1st ed. Missoula, Mt: Mountain Press Publishing, 1989.
3. Ehorn Nancy, Channel Islands National Park. 1st ed. Ventura, Ca: Friends of Channel Islands National Park. 1991.
4. Fitch, John E. and Robert J. Lavenberg. Tidepool and Nearshore Fishes of California. 1st ed. Berkeley, Ca: University Press, 1975.
5. Holt, Harold. A Birder's Guide to Southern California. 1st ed. Ed. Paul J. Baicich. Colorado Springs, Co: American Birding Association, 1990.
6. Lowery, Jim. Tracking Workbook: with Common Animal Tracks of Southern California. San Pedro, Ca: Earth Skills, 1990.
7. Steinhart, Peter. California's Wild Heritage: Threatened and Endangered Animals in the Golden State. 1st ed. Sacramento, Ca: California Department of Fish and Game, 1990.
8. Udvardy, Miklos D. The Audubon Society Field Guide to North American Birds: Western Region. 13th ed. New York, NY: Alfred a. Knopf, 1987.

Wildflower and Plant References:

1. Bakker, Elna. An Island Called California. 2nd ed. Berkeley, Ca: University of California Press, 1984.
2. Balls, Edward K. Early Uses of California Plants. 1st ed.

251

Berkeley, Ca: University of California Press, 1962.

3. Belzer, Thomas J. Roadside Plants of Southern California. 1st ed. Missoula, Mt: Mountain Press Publishing Company, 1986.

4. Clarke, Charlotte Bringle. Edible and Useful Plants of California. 1st ed. Ed. Arthur C. Smith. Berkeley, Ca: University of California Press, 1980.

5. Dale, Nancy. Flowering Plants of the Santa Monica Mountains, Coastal & Chaparral Regions of Southern California. 1st ed. Ed. Jo Kitz and Linda Hardie-Scott. Santa Barbara, Ca: Capra Press, 1986.

6. Heizer, Robert F. and Albert B. Elasser. The Natural World of the California Indians. 1st ed. Ed. Arthur C. Smith. Berkeley, Ca: University of California Press, 1980.

7. James, Wilma Roberts. Know your Poisonous Plants. 2nd ed. Happy Camp, Ca: Naturegraph Publishers, 1993.

8. Keator, Glenn. Native Shrubs of California. 1st ed. San Francisco, Ca: Chronicle Books, 1994.

9. Lenz, Lee W. and John Dourley. California Native Trees & Shrubs. 1st ed. Claremont, Ca: Rancho Santa Ana Botanic Garden, 1981.

10. McAuley, Milt. Wildflowers of the Santa Monica Mountains. 1st ed. Canoga Park, Ca:Canyon Publishing Company, 1985.

11. Munz, Philip A. California Spring Wildflowers. 1st ed. Berkeley, Ca: University of California Press, 1961.

12. Niehaus, Theodore F. Sierra Wildflowers Mt. Lassen to Kern County. 1st ed. Berkeley, Ca: University of California Press, 1974.

13. Ornduff, Robert. Introduction to California Plant Life. 1 st ed. Berkeley, Ca: University of California Press, 1974.